PLANT SECONDARY METABOLITES

Volume 2

Stimulation, Extraction, and Utilization

PLANT SECONDARY METABOLITES

Volume 2

Stimulation, Extraction, and Utilization

Edited by

Mohammed Wasim Siddiqui, PhD
Vasudha Bansal, PhD
Kamlesh Prasad, PhD

AAP | APPLE ACADEMIC PRESS

Apple Academic Press Inc.
3333 Mistwell Crescent
Oakville, ON L6L 0A2
Canada

Apple Academic Press Inc.
9 Spinnaker Way
Waretown, NJ 08758
USA

First issued in paperback 2021

Exclusive worldwide distribution by CRC Press, a member of Taylor & Francis Group

ISBN 13: 978-1-77463-108-9 (pbk)
ISBN 13: 978-1-77188-354-2 (hbk)

Library and Archives Canada Cataloguing in Publication

Plant secondary metabolites.

Includes bibliographical references and indexes.
Contents: Volume 2. Stimulation, extraction, and utilization / edited by Mohammed Wasim Siddiqui, PhD, Vasudha Bansal, PhD, Kamlesh Prasad, PhD.
Issued in print and electronic formats.
ISBN 978-1-77188-354-2 (v. 2 : hardcover).--ISBN 978-1-77188-355-9 (v. 2 : pdf)
1. Plant metabolites. 2. Plants, Edible--Metabolism. 3. Medicinal plants--Metabolism. 4. Metabolism, Secondary. I. Siddiqui, Mohammed Wasim, author, editor II. Bansal, Vasudha, author, editor III. Prasad, Kamlesh, author, editor

QK881.P63 2016 572'.42 C2016-904969-8 C2016-904970-1

Library of Congress Cataloging-in-Publication Data

Names: Siddiqui, Mohammed Wasim, editor. | Prasad, Kamlesh, editor.
Title: Plant secondary metabolites / editors, Mohammed Wasim Siddiqui, Kamlesh Prasad.
Other titles: Plant secondary metabolites (Siddiqui)
Description: New Jersey : Apple Academic Press, Inc., [2017-] | Includes bibliographical references and index.
Identifiers: LCCN 2016030295 (print) | LCCN 2016031056 (ebook) | ISBN 9781771883528 (hardcover : alk. paper) | ISBN 9781771883535 (ebook) | ISBN 9781771883535 ()
Subjects: LCSH: Plant metabolites. | Plants, Edible--Metabolism. | Medicinal plants--Metabolism. | Metabolism, Secondary. | MESH: Plants, Edible--metabolism | Plant Extracts--chemistry | Phytochemicals | Plants, Medicinal Classification: LCC QK881 .P5526 2017 (print) | LCC QK881 (ebook) | NLM QK 98.5.A1 | DDC 572/.42--dc23
LC record available at https://lccn.loc.gov/2016030295

Apple Academic Press also publishes its books in a variety of electronic formats. Some content that appears in print may not be available in electronic format. For information about Apple Academic Press products, visit our website at **www.appleacademicpress.com** and the CRC Press website at **www.crcpress.com**

Plant Secondary Metabolites:
Volume 1: Biological and Therapeutic Significance
Editors: Mohammed Wasim Siddiqui, PhD, and Kamlesh Prasad, PhD

Plant Secondary Metabolites:
Volume 2: Stimulation, Extraction, and Utilization
Editors: Mohammed Wasim Siddiqui, PhD, Vasudha Bansal, PhD, and Kamlesh Prasad, PhD

Plant Secondary Metabolites:
Volume 3: Their Roles in Stress Ecophysiology
Editors: Mohammed Wasim Siddiqui, PhD, and Vasudha Bansal, PhD

ABOUT THE EDITORS

Mohammed Wasim Siddiqui, PhD

Dr. Mohammed Wasim Siddiqui is an Assistant Professor and Scientist in the Department of Food Science and Postharvest Technology, Bihar Agricultural University, Sabour, India, and the author or co-author of more than 33 peer-reviewed journal articles, 24 book chapters, and 18 conference papers. He has six edited and one authored books to his credit, published by Elsevier, USA; Springer, USA; CRC Press, USA; and Apple Academic Press, USA. Dr. Siddiqui has established an international peer-reviewed *Journal of Postharvest Technology*. He has been honored to be the Editor-in-Chief of two book series, Postharvest Biology and Technology and Innovations in Horticultural Science, both published by Apple Academic Press, New Jersey, USA, where he is a Senior Acquisitions Editor for Horticultural Science. He is also as an editorial board member of several journals.

Recently, Dr. Siddiqui received the Young Achiever Award 2014 for outstanding research work by the Society for Advancement of Human and Nature (SADHNA), Nauni, Himachal Pradesh, India, where he is also an Honorary Board Member. He has been an active member of organizing committees of several national and international seminars, conferences, and summits.

Dr. Siddiqui acquired BSc (Agriculture) degree from Jawaharlal Nehru Krishi Vishwa Vidyalaya, Jabalpur, India. He received MSc (Horticulture) and PhD (Horticulture) degrees from Bidhan Chandra Krishi Viswavidyalaya, Mohanpur, Nadia, India, with specialization in the Postharvest Technology. He was awarded a Maulana Azad National Fellowship Award from the University Grants Commission, New Delhi, India. He is a member of the Core Research Group at the Bihar Agricultural University (BAU), providing appropriate direction and assisting with sensiting priority of the research. He has received several grants from various funding agencies to carry out his research projects that are associated with postharvest technology and processing aspects of horticultural crops. Dr. Siddiqui is dynamically involved in teaching (graduate and doctorate students) and research, and he has proved himself as an active scientist in the area of Postharvest Technology.

Vasudha Bansal, PhD

Vasudha Bansal, is working as a Postdoctoral Researcher in the Department of Environmental Engineering at Hanyang University in Seoul, South Korea. She has worked as a study coordinator in clinical trials for diabetic and osteoporotic patients in the Endocrinology Department in the Postgraduate Institute of Medical Education and Research (PGIMER) in Chandigarh, India. She has been awarded a young scientist talent scholarship from the Ministry of Education in Brazil (2014), a Bio-Nutra Junior award for best oral presentation from the National Institute of Food Technology and Entrepreneurship Management, Kundli, Haryana, India (2013), the Mrs. Gupta Physics award and the Mrs. Handa Zoology award for her BSc. She is an editorial member and peer reviewer of the *Journal of Food Research,* a peer reviewer of *Food Composition and Analysis* and the *Journal of Food Bioprocess and Technology*. She is a member of the Indian Science Congress Association, the Nutrition Society of India, and the Indian Dietetic Association. She acquired her PhD (food science) from the AcSIR-CSIO, Chandigarh (India) and her BSc and MSc in food and nutrition from Panjab University, Chandigarh, India.

Kamlesh Prasad, PhD

Prof. Kamlesh Prasad is a food technologist with a postgraduate degree in Food Technology from the Central Food Technological Research Institute, Mysore, India. He received a doctorate in Food Technology and Process and Food Engineering from GB Pant University of Agriculture and Technology, Pantnagar, India. His specialization is in fruits and vegetable processing. His research interest includes nondestructive testing, image analysis, nanotechnology, engineering properties of foods, and various related aspects of food processing.

Presently, he is a Professor in the Department of Food Engineering and Technology, Sant Longowal Institute of Engineering and Technology, Longowal, India. As an author or co-author, he has published more than 100 research papers in national and international journals of repute. He has five authored books and two edited books, and has contributed several chapters in books, published by leading presses of India and abroad. He has conducted a short-term course on "soft computing in process and product optimization" and chaired several conferences.

He is an active board member of the Agricultural and Food Management Institute (AFMI), Mysore, India. He is also the reviewer of many international and national journals and has edited many books and proceedings. He is associated with many reputed professional societies such as ASABE, MSI, ISNT, AFST (I), NSI, PAS, USI, ICC, and DTS (I). He is recognized as one of 2000 Outstanding Intellectuals of the 21st Century and one of the Top 100 Educators in 2010 by the International Biographical Centre, Cambridge, England. Also, his biography is published by Asia/Pacific International Biographical Centre, New Delhi and Who's Who in Science and Engineering by Marquis Who's Who, USA. As a co-coordinator, he has engaged himself in bringing the World Bank project under technical education quality improvement program (TEQIP) project to the institute for technical reform in education. He is involved in the process of social reform and in providing justice to deprived segments of society.

DEDICATION

This Book

Is

Affectionately Dedicated

to Our Beloved Parents

CONTENTS

LIST OF CONTRIBUTORS

Sachin R. Adsare
Food Engineering and Technology Department, Institute of Chemical Technology, Mumbai 400019, Maharashtra, India.

Roberta Ansorena
Consejo Nacional de Investigaciones Científicas y Técnicas, Argentina; Grupo de Investigación en Ingeniería en Alimentos, Facultad de Ingeniería, Universidad Nacional de Mar del Plata. J. B. Justo 4302, Mar del Plata, Buenos Aires, Argentina.

J. Fernando Ayala-Zavala
Centro de Investigación en Alimentación y Desarrollo, AC, Carretera a la Victoria km. 0.6. Apartado Postal 1735, Hermosillo 83000, Sonora, México.

Vasudha Bansal
Department of Civil & Environmental Engineering, Hanyang University, 222 Wangsimni-Ro, Seoul 133791, South Korea.

Amit K. Das
Department of Grain Science and Technology, CSIR—Central Food Technological Research Institute, Mysore 570020, Karnataka, India. E-mail: amitkdas12@gmail.com.

Madhuchhanda Das
DoS in Botany, Manasagangothri, University of Mysore, Mysore 570006, Karnataka, India.

P. Ganesan
Department of Molecular Nutrition, CSIR—Central Food Technological Research Institute, Mysore 570020, Karnataka, India. E-mail: ganesanp@cftri.res.in; ganesan381980@yahoo.com.

C. Ghanshyam
Academy of Scientific and Innovative Research, CSIR—Central Scientific Instruments Organisation, Sector 30, Chandigarh 160030, India.

Gustavo A. Gonzalez-Aguilar
Centro de Investigación en Alimentación y Desarrollo, AC, Carretera a la Victoria km. 0.6. Apartado Postal 1735, Hermosillo 83000, Sonora, México.

Humberto Gonzalez-Rios
Centro de Investigación en Alimentación y Desarrollo, AC, Carretera a la Victoria km. 0.6. Apartado Postal 1735, Hermosillo 83000, Sonora, México.

Raees-Ul Haq
Department of Food Engineering and Technology, SLIET, Longowal 148106, Punjab, India.

Pankaj S. Kulthe
Urmin Group of Companies, Ahmedabad 380054, Gujarat, India.

Sudhir Kumar
Department of Plant Breeding and Genetics, Bihar Agricultural University, Sabour, Bhagalpur 813210, Bihar, India.

Mona Kumari
Department of Plant Breeding and Genetics, Bihar Agricultural University, Sabour, Bhagalpur 813210, Bihar, India.

Julio Cesar Lopez-Romero
Centro de Investigación en Alimentación y Desarrollo, AC, Carretera a la Victoria km. 0.6. Apartado Postal 1735, Hermosillo 83000, Sonora, México.

Vera-López Obdulia
Departamento de Bioquímica-Alimentos, Facultad de Ciencias Químicas, Benemérita Universidad Autónoma de Puebla, 14 Sur y Av. San Claudio, Ciudad Universitaria, Col. San Manuel, 72420 Puebla, Puebla, México.

Carlos Enrique Ochoa-Velasco
Departamento de Bioquímica-Alimentos, Facultad de Ciencias Químicas, Benemérita Universidad Autónoma de Puebla, 14 Sur y Av. San Claudio, Ciudad Universitaria, Col. San Manuel, 72420 Puebla, Puebla, México. E-mail: carlosenriqueov@hotmail.com.

Hernández-Carranza Paola
Colegio de Ingeniería en Alimentos, Facultad de Ingeniería Química, Benemérita Universidad Autónoma de Puebla, 14 Sur y Av. San Claudio, Ciudad Universitaria, Col. San Manuel, 72420 Puebla, Puebla, México.

K. Yogendra Prasad
Department of Molecular Nutrition, CSIR—Central Food Technological Research Institute, Mysore 570020, Karnataka, India.

Kamlesh Prasad
Department of Food Engineering and Technology, SLIET, Longowal 148106, Punjab, India. E-mail address: dr_k_prasad@rediffmail.com.

Gayathri R.
SRM Research Institute, SRM University, Kattankulathur 603203, Tamilnadu, India.

Niranjana R.
SRM Research Institute, SRM University, Kattankulathur 603203, Tamilnadu, India.

Ávila-Sosa Raúl
Departamento de Bioquímica-Alimentos, Facultad de Ciencias Químicas, Benemérita Universidad Autónoma de Puebla, 14 Sur y Av. San Claudio, Ciudad Universitaria, Col. San Manuel, 72420 Puebla, Puebla, México.

Navarro-Cruz Addi Rhode
Departamento de Bioquímica-Alimentos, Facultad de Ciencias Químicas, Benemérita Universidad Autónoma de Puebla, 14 Sur y Av. San Claudio, Ciudad Universitaria, Col. San Manuel, 72420 Puebla, Puebla, México.

Chandan Roy
Department of Plant Breeding and Genetics, Bihar Agricultural University, Sabour, Bhagalpur 813210, Bihar, India.

Mohammed Wasim Siddiqui
Department of Food Science and Postharvest Technology, Bihar Agricultural University, Sabour, Bhagalpur 813210, Bihar, India.

Ravi Shankar Singh
Department of Plant Breeding and Genetics, Bihar Agricultural University, Sabour, Bhagalpur 813210, Bihar, India.

M. L. Singla
Academy of Scientific and Innovative Research, CSIR—Central Scientific Instruments Organisation, Sector 30, Chandigarh 160030, India.

Nimish Mol Stephen
SRM Research Institute, SRM University, Kattankulathur 603203, Tamilnadu, India; Department of Fish Processing Technology, Institute of Fisheries Technology, Tamilnadu Fisheries University, Ponneri 601204, Tamilnadu, India.

Baskaran V.
Department of Biochemistry and Nutrition, CSIR—Central Food Technological Research Institute, Mysore 570020, Karnataka, India.

Aneeta Yadav
Department of Plant Breeding and Genetics, Bihar Agricultural University, Sabour, Bhagalpur 813210, Bihar, India.

LIST OF ABBREVIATIONS

5-CQA	5-caffeoylquinic acid
ACTs	artemisinin-based combination therapies
ADI	acceptable daily intake
ANS	anthocyanidin synthase
Asa-POD	ascorbate peroxidase
ASE	accelerated solvent extraction
CAT	catalase
CC	column chromatography
CHS	chalcone synthase
COP1	constitutive photomorphogenic 1
CVD	cardiovascular diseases
DFR	dihydroflavonol 4-reductase
DHAR	dehydroascorbate reductase
DMAPP	dimethylallyl diphosphate
DSHEA	Dietary Supplement Health and Education Act
EAE	enzyme-assisted extraction
EAFUS	Everything Added to Food in the US
Eos	essential oils
ESE	enhanced solvent extraction
F3H	flavanone-3-hydroxylae
FAO	Food and Agriculture Organization
FID	flame-ionizing detector
GERD	gastroesophageal reflux disease
GGPP	geranylgeranyl pyrophosphate
GI tract	gastrointestinal tract
G-POD	guaiacol peroxidase
GR	glutathione reductase
GRAS	generally recognized as safe
GSH	glutathione
GSH-POD	glutathione peroxidase
GSL	glucosinolase
HDL-C	high-density lipoprotein cholesterol
HHP	high hydrostatic pressure

HHPE	high-hydrostatic pressure extraction
HPHT	high pressure high temperature
HPLC	high-performance liquid chromatography
HPP	high-pressure processing
HSPE	high-pressure solvent extraction
IPP	isopentenyl diphosphate
IU	international unit
LA	linoleic acid
MAE	microwave-assisted extraction
MDA	malondialdehyde
MDAR	monodehydroascorbate reductase
MEP	methyl-d-erythritol 4-phosphate
MS	mass spectrometry
MUFA	monounsaturated fatty acids
NOEL	no observed adverse effect level
O	olfactory
OBC	oat bran concentrate
OSC	oxidosqualene cyclase
PAL	phenylalanine ammonia-lyase
PDA	photodiode array
PEF	pulsed electric field
PFE	pressurized fluid extraction
PLEs	pressurized liquid extractions
PUFA	polyunsaturated fatty acids
ROS	reactive oxygen species
SC-CO2	supercritical carbon dioxide
SCFE	supercritical fluid extraction
SCFs	supercritical fluids
SFA	saturated fatty acids
SFE	supercritical fluid extraction
SOD	superoxide dismutase
TIAs	terpenoid–indole alkaloids
TLC	thin layer chromatography
TPC	total phenolic content
UAE	ultrasound-assisted extraction
UHPLC	ultra-high-performance liquid chromatography
US	ultrasound
UV	ultraviolet

UV-A	ultraviolet-A light
UV-B	ultraviolet-B light
UV-C	ultraviolet-C light
UVR8	ultraviolet resistances locus 8
VOCs	volatile organic compounds
WUPs	water-unextractable

PREFACE

The medicinal values of natural plant products such as secondary metabolites are well known. The secondary metabolites have been explored from antiquity due to their proven beneficial physiological roles in improving human health. These active phytochemicals have been proven crucial in curing various ailments. The presence of such pharmacologically active molecules has thus been used in replacing the artificial chemicals/medicines with the natural ones. Primitive people learned through trial and error about the causes and effects of various plant products in their various forms, such as aqueous, alcoholic, or dried extracts, and the gained knowledge has passed through generations. The large number of available phytochemicals in nature, having varying structural dissimilarities and properties, are of immense interest and attraction for their prevalence.

The plants contain several active compounds in their crude form with limited curative role, which are refined in various ways during course of time, leading ultimately to being used as a pure single active ingredient. Phenolic-rich foods are widely known for their cardioprotective, neuroprotective, and chemopreventive actions. Naturally available secondary metabolites as drugs take effective action against various ailments in comparatively lesser time than their synthetic counterparts. Hence, the plants are rich in natural secondary metabolites that could be used as dietary supplements, nutraceuticals, and cosmoceuticals. The herbal remedies could be the preferable options, available cheaply and abundantly, for their exploitation. Presently, limited information is available on the effective exploration of plant secondary metabolites.

Understanding the necessity to fill this knowledge gap, the present book, *Plant Secondary Metabolites: Stimulation, Extraction, and Utilization*, discusses the perspectives and technologies involved in the explorations of various bioactive plant metabolites.

The chapters included in this book provide an in-depth discussion on secondary plant metabolites. Chapter 1 deals with evolutionary perspective, *in vitro* production, and technological advances of secondary metabolites in plants. Several methods have been developed for the extraction of biologically active substances from the various plants. In this curriculum,

Chapter 2 discusses the advanced techniques in the extraction of phenolics from cereals, pulses, fruits, and vegetables. The types, sources, and biosynthesis of carotenoids have thoroughly been covered in Chapter 3. A comprehensive review of secondary metabolites of carrot along with its prospective health benefits has been given in Chapter 4. The plant-originated active compounds have great industrial potentialities. Chapter 5 precisely discusses the diverse utilization of secondary metabolites from plant sources. In addition to source and extraction methods, there are several other factors that might affect the fate of these compounds during processing. Chapter 6 presents a critical discussion on the effects of different food processing techniques on secondary metabolites.

The editors are confident enough that this book will be highly beneficial to guide academicians and researchers for exploring the possibilities of natural secondary metabolites in plants. The editors welcome and invite readers' feedback and suggestions on the material presented in the book for the development of future editions of this book.

ACKNOWLEDGMENTS

It was almost impossible to express the deepest sense of veneration to all without whose precious exhortation this book project could not be completed. At the onset of the acknowledgment, we ascribe all glory to the gracious "Almighty God" from whom all blessings come. We would like to thank for His blessing to write this book.

With a profound and unfading sense of gratitude, we sincerely thank the Bihar Agricultural University (BAU), India, CSIR—Central Scientific Instruments Organisation (CSIO), India, and Sant Longowal Institute of Engineering & Technology (SLIET), India, for providing us the opportunity and facilities to execute such an exciting project, and for supporting us toward our research and other intellectual activities around the globe. We convey special thanks to our colleagues and other research team members for their support and encouragement for helping us in every footstep to accomplish this venture. We would like to thank Mr. Ashish Kumar, Ms. Sandy Jones Sickels, and Mr. Rakesh Kumar of Apple Academic Press for their continuous support to complete the project.

Our vocabulary will remain insufficient in expressing our indebtedness to our beloved parents and family members for their infinitive love, cordial affection, incessant inspiration, and silent prayer to "God" for our well-being and confidence.

CHAPTER 1

SECONDARY METABOLITES: EVOLUTIONARY PERSPECTIVE, *IN VITRO* PRODUCTION, AND TECHNOLOGICAL ADVANCES

CHANDAN ROY, RAVI SHANKAR SINGH, ANEETA YADAV, SUDHIR KUMAR, and MONA KUMARI

Department of Plant Breeding and Genetics, Bihar Agricultural University, Sabour, Bhagalpur 813210, Bihar, India

CONTENTS

ABSTRACT

Secondary metabolites showed a range of biological activity such as antioxidant, antiproliferative, immunosuppressant, anti-infective, cholesterol lowering and growth promoting also termed as bioactive molecules. Higher organisms in the process of evolution have faced a diverse kind of biotic and abiotic stresses. Various genetic consequences have resulted in the evolution of specific kinds of secondary metabolites in specific organisms. Currently, the majority of pharmaceutically essential bioactive molecules are isolated from wild or cultivated plants as their chemical synthesis is uneconomical. Many of these medicinally important chemicals in limited quantity are produced in the natural population, making the drugs expensive to harvest. Advancement of *in vitro* production system for some of the important drugs has shown the challenges for profitable production of such chemicals.

1.1 INTRODUCTION

Secondary metabolites showed a range of biological activity such as antioxidant, antiproliferative, immunosuppressant, anti-infective, cholesterol lowering and growth promoting also termed as bioactive molecules. Higher organisms in the process of evolution have faced a diverse kind of biotic and abiotic stresses. Various genetic consequences have resulted in the evolution of specific kinds of secondary metabolites in specific organisms.

Plants do interact with the environment using such molecules either to resist pests or to promote pollination. During evolutionary process, organisms gained special types of genes and that too for higher fitness and resistance mechanism within it against various environmental stresses. Such molecules have a wide range of commercial and pharmaceutical utility, for example, shikonins as dyes/therapeutic, vanillin, and capsaicin as flavors, nicotine as stimulants, taxol, artemisinin, and atropine and quinine as therapeutic agents.

Some wild plants difficult to grow are becoming endangered due to exploitation. Climate change is adversely affecting plants and, due to rise in temperatures, some cold-adapted alpine species are migrating toward higher altitude. In such a scenario, *in vitro* production system with modulated biosynthetic pathway could be a good choice.

Besides clinical importance, secondary metabolites have been identified that are used for agricultural insect pest management. For example, phytoalexines are being synthesized by plants upon wounding and used in preventing the invasion of fungal and bacterial growth into the plants cells. These could be useful to induce hypersensitive response inside the plants for preventing pathogen attack.

Organic compounds produced in a species which do not directly associate with growth and development of the species are generally known as secondary metabolites (Table 1.1). Pheromone, the well-known secondary metabolites released by insect while mating, is well studied for the social behavior of insects. It has been reported that most of the species from higher organism like plants to the lower taxa like fungus and bacteria are producing secondary metabolites. Secondary metabolites in plants are species-specific and originate in the process of evolution. Very few plant species are associated with production of secondary metabolites. Use of secondary metabolites for several aesthetic and therapeutic purposes is in practice for many centuries. However, the scientific study for structural and functional characterization of secondary metabolites in plants was started in the middle of the 19th century when scientists started to determine the structure of secondary metabolites mostly monotermpenes. At that time, researchers relied upon crystallization techniques and degradation of products to determine the structure of metabolites. With the technological advances and use of chromatographic techniques, today's secondary metabolites production has gain the industrial importance.

The basic features distinguishing the secondary metabolites from primary metabolites of a species are structural differences; precursor molecules or biosynthetic origin of molecules. Based on the precursor molecule, secondary metabolites can be categorized into three major classes, that is, terpenoids, alkaloids, and phenolic compounds. The biochemical pathways related with the production of all the three groups of metabolites are well characterized in plants. Several well-characterized metabolites are produced *in vitro* conditions. Recently, microbial-released secondary metabolites are identified for disease establishment in plant. Toxins produced as a result of infection are a kind of secondary metabolite secretions by the fungus upon infection (Nelson et al., 1993; Yu & Keller, 2005). Genetic mechanism regulating the secondary metabolite production in microbes and plants has been unrevealed. A detailed study has been performed for penicillin biosynthetic pathway. Similarly, biosynthetic pathway of aflatoxin in *Aspergillus* spp. found that a cluster of genes are regulating the production of mycotoxin in the fungus (Minto & Townsend, 1997). In plants, all terpenoids are derived from the five-carbon precursor molecule isopentenyl diphosphate (IPP). Most of the known alkaloids with one or more nitrogen atoms are synthesized primarily from amino acids. Shikimic acid pathway or the malonate/acetate pathway plays the route for formation of phenolic compounds.

TABLE 1.1 Secondary Metabolites Produced by Plant Species.

Compound	Species	Activity
Capsaicin	*Capsicum frutescens*	Chilli
Vanillin	*Vanilla* spp.	Vanilla
Quinine	*Cinchona officinalis*	Antimalarial
Digoxin, reserpine	*Digitalis lanata*	Cardiac tonic
Diosgenin	*Dioscorea deltoidea*	Antifertility
Morphene	*Papaver somniferum*	Analgesic
Thebaine	*Papaver bracteatum*	Source of codeine
Scopolamine	*Datura stramonium*	Antihypertension
Atropine	*Atropa belladonna*	Muscle relaxant
Codeine	*Papaver* spp.	Analgesic
Shikonine	*Lithospermum erythrorhizon*	Dye, pharmaceutical
Anthraquinones	*Morinda citrifolia*	Dye, laxative
Rosmarinic acid	*Coleus blumei*	Spice, antioxidant
Jasmine	*Jasminum* spp.	Perfume
Stevioside	*Stevia rebaudiana*	Sweetener
Crocin	*Crocus sativus*	Saffron
Pyrethrins	*Chrysanthemum cinerariaefolium*	Insecticidal
Lutein	*Tagetes erecta*	Insecticidal
Nicotine	*Nicotiana rustica*	Insecticidal
	Nicotiana tabacum	Insecticidal
Azadirachtin	*Azadirachta indica*	Antineoplastic
Taxol	*Taxus brevifolia*	Antineoplastic
Fagaronine	*Fagara zanthoxyloides*	Antineoplastic

1.2 EVOLUTION OF SECONDARY METABOLITES IN PLANTS

As we have discussed, the function of secondary metabolites in plants does not associate with any activities related with its normal growth and development. If not, the question arises how the genes encoding secondary metabolites evolve in the organism or what kind of selection pressure has occurred in the process of evolution of organism. Primary metabolites like sugars, fatty acid, amino acid, and nucleic acids are the basic building blocks of an organism. However, an organism experiences various types of environmental pressure related with biotic and abiotic factors. With interactions of several entities of environment, plants developed specialized genes other than the conserved regions of genome. In the evolutionary progress, such genes became important to impart interactions with the environment. Though these molecules are not essential for the growth and development of an organism but it has been observed that plants lacking appropriate secondary metabolites in the cell have drastically reduced fitness. About 140 million years ago, since angiosperms started their evolution with the presence of microbes and other herbivores, both the plants and microbes evolved natural products. But the evolution of secondary metabolites is highly specific to the organism and even specific to the tissue of expression. In plants, it has been found that production of secondary metabolites is specific to developmental stages. Organs important for survival and fitness are expressing more amounts of secondary metabolites. Secondary metabolites may be present in an active state or may be activated into plants upon wounding (like phytoalexines are activated after wounding in plants). Even in the same organ different secondary metabolites are produced for different activities. Like, in plants flower produces secondary metabolites as chemo-attractants to attract pollinators (like fragrant monoterpenes, colored anthocyanine, or carotenoids in flower); whereas, nector and fruit pulp are often containing phenolic compounds, essential oils, tannins, saponins, and others, as secondary metabolites which help in prevention of attacks from bacteria and molds. Besides, protection of few secondary metabolites like alkaloids, lactins protease inhibitors, and others helps in transpiration of toxic nitrogenous compounds into the plants. Such a large structural and functional diversity of secondary metabolites in organisms reflects strong evolutionary mechanism followed by directional selection existing during the process of evolution.

1.3 BASIS OF GENE EVOLUTION

Evolutionary force directs the origin and existence of living species in nature. Evolution of any species and its relationship with its lineages is directly related with the genetic makeup of the organisms. Two major forces carried out during evolutionary process are creation of genetic variability and selection of suitable genotype by the nature. Several events have been identified that play a major role in evolutionary process for creation of genetic variability. Duplication, deletion, mutation, and polyploidy of genes or genome play the main role to change the genetic makeup of the species. Among these, partial or whole genome duplication in the ancestral species was found to be the main player for species evolution (Kellis et al., 2004; Roy & Deo, 2014). Ohno (1972) provided conclusive evidence that gene duplication is the major force for evolution of species. Following duplication, several events take place in the duplicated region of chromosome that decided the fate of duplicated gene in the genome. It may be either retained in the genome as such or may get lost or may remain with modified or new function. Genes related with the developmental stages are found duplicated in the genome as products of such genes are required in larger quantities. The product of few genes important for all types of tissues throughout the growing period are called housekeeping genes. Few genes are tissue and developmental stage specific. Most of the secondary metabolites are produced in specific types of tissues and growing conditions. Evolution of genes responsible for production of secondary metabolite takes place in two directions; one is the gene producing secondary metabolites and another is genes related for their genetic regulation. Several examples have been found with different functions of gene carrying common nucleotide sequences in the lineages. The way of gene family evolution where the members of families remain similar in sequence and functions is known as concerted evolution. Now question arises how these genes remain in the genome after duplication throughout the period of evolution. This can be explained by gene conversion and purifying selection. Under frequent gene conversion, the two paralogous genes will have similar sequence and function; and this resulted in concerted evolution of gene.

1.4 CONVERGENT VERSUS DIVERGENT EVOLUTION

Though most of the metabolites have been evolved by the plant species but the evolutionary process has lead to the development of species with a number of specific metabolites. These metabolites are the result of evolution of specific precursor or specific enzyme in the pathway of production. Sometimes species evolved a kind of product/trait playing similar functions that lack in their ancestors; this is called convergent evolution. In the early evolutionary stages, different kinds of plant species have faced an environment with common biotic and abiotic factors. That is how they all have evolved a kind of specialized metabolites for defense mechanism. Caffeine has evolved to defend the predators by two distantly related species, that is, coffee and tea. One classical example is evolution of anthocyanine pigments in plant. Many plants species produce anthocyanin as pigments in flowers and fruits. All of them are of different colors—red, pink, blue, orange, or magenta—observed in flowers. Plants belonging to the order Caryophyllales (except the plants belonging to the family Caryophylaceae) do produce betacyanin instead of anthocyanin that mimics the orange-colored pigment of anthocyanin.

Mechanisms of divergent evolution for the secondary metabolites biosynthetic pathways are rather simple than the convergent pathways. Accumulation of different pathways in the lineages from a common origin is indicative of divergent evolution. Large number of metabolic diversifications in plants has suggested that tandem duplication followed by mutation in the duplicated genes has created functionally diverse enzymes for metabolic pathways (Pichersky & Lewinsohn, 2011). Neo-functionalization (creation of new genes after duplication) immediate after duplication observed a strong mechanism followed for metabolic pathway evolution (Ober, 2005).Triterpenes in plants are synthesized by oxidosqualene cyclase (OSC) gene family. Lower plants like chlamydomonus and mosses contain only a single OSC gene whereas higher plants contain 16 duplicates in the genome. Phylogenetic analysis suggested in higher plants OSC gene family has occurred through tandem duplication followed by positive selection and diversifying evolution (Zheyong et al., 2011).

1.5 *IN VITRO* PRODUCTION OF SECONDARY METABOLITES

Plant cells are most prolific factories for synthesis of small molecules (200,000–1000,000 molecules). Some of them directly associated with growth and development of plants are termed as primary metabolites, in contrast to the secondary metabolites, which play an indirect role. Majority of the secondary metabolites are produced from three basic biosynthetic pathways: (1) shikimate, (2) isoprenoid, and (3) polyketide. Secondary metabolites that show a biological activity such as antioxidant, antiproliferative, immunosuppressant, anti-infective, cholesterol lowering and growth promoting are termed as bioactive molecules.

These molecules have a broad range of commercial and pharmaceutical utility; for example, shikonins as dyes/therapeutic, vanillin and capsaicin as flavors, rose and lavender oils as fragrances, caffeine, and nicotine as stimulants, morphine, and cocaine as hallucinogens, nicotine, and piperine as insecticides, coniine, and strychnine as poisons, taxol, artemisinin and atropine, and quinine as therapeutic agents. Currently, most pharmaceutically essential bioactive molecules are isolated directly from plants because their chemical synthesis is not economically feasible. The natural metabolites are the origin of as many as 60% of successful drugs. Most of the drugs available in market today are simple modifications or copies of the natural metabolites. Few of the most important anticancer, antifungal, and antibacterial drugs are directly used as natural products. Furthermore, these drugs are produced in very less amounts in the natural population, making the drugs expensive to harvest. Plant-derived drugs have huge market value US$30 billion in the USA. Of the 252 basic and essential drugs (WHO), 11% are exclusively derived from plants. Cultivation of some wild plants is difficult and these plants are becoming endangered due to exploitation of these metabolites, for example, a species of *Texus brevifolia* (taxol) and *Arnebia euchroma* (shikonins).

The relevance of the *in vitro* production systems lies in the fact that these are now being considered (1) as an alternative to field cultivation of plants for producing valuable secondary metabolites, (2) in conserving some wild plants difficult to cultivate, (3) to safeguard against the adverse effects of climate change on plants, (4) useful in production of natural products that are too complex to be chemically synthesized, and (5) in enhancing the yield of metabolites that are otherwise produced insufficiently in their native host.

In vitro plant cell cultures of several important medicinal plants are reported for the mass multiplication of the target compound. Important factor for *in vitro* production is the identification of suitable high-yielding cell culture. *In vitro* production is done from callus culture, cell culture, and cell suspension culture from plant leaves or flowers. However, the selection of explants also depends on the kind of metabolites we are targeting as most of them are synthesized by specialized tissues in the plants.

Several factors are responsible for accumulation of metabolites in the callus or in the cell suspension culture. The major determining factors are kind of growth regulators, carbon source, nitrogen source, and their concentration used in the culture media. *In vitro* production of several metabolites has been standardized, and has found commercial importance. Besides, many medicinally and agriculturally important metabolites remain untouched for their mass multiplication. The feasibility of *in vitro* production for important metabolites is necessary for their application in the disease and pest management in agricultural crops.

1.6 NEED FOR *IN VITRO* CULTURE

1.6.1 PLANT CELL CULTURE

Plant cells are amenable to *in vitro* culture for regeneration of parts/whole plants and production of all of the compounds found in the plants. Totipotency is the inherent property in a cell by which a cell can divide, differentiate, and develop into specialized tissue/organ/a plant. Plant cell cultures are not affected by climatic conditions and ecological specificity unlike agricultural system of production. Cell cultures system could be used round the year for various secondary metabolites production like colors, flavors, and sweeteners (Dornenburg & Knorr, 1996; Singh et al., 2010). The modulation of biosynthetic pathways for secondary metabolites in plants has become feasible nowadays by genetic engineering, but it requires detailed knowledge of the biosynthetic pathways and strategies to overcome the bottlenecks of the rate-limiting steps, reducing flux through competitive pathways, reducing catabolism and over expression of regulatory genes (Verpoorte et al., 2000). The compounds of diverse nature and origin like aromatics, alkaloids, steroids, coumarins, and terpenoids could be bio-transformed using plant cells, organ cultures, and enzymes into an array of compounds by various reaction types involving oxidations, reductions, methylations, hydroxylations, acetylations, isomerizations, glycosylations, and esterfications (Giri et al., 2001). Many of diverse secondary metabolites such as phenylpropanoids (anthocyanins, coumarins, flavonoids), alkaloids (acridines, betalaines, tropane, indoles), terpenoids (carotenes, monoterpenes, sesquiterpenes, diterpenes), quinones (naphthoquinones, anthroquinones), and so on, were isolated from plant cell culture (Rao & Ravishankar, 2002). Some of the notable examples for production of secondary metabolite using plant cell culture are given below.

1.6.2 SHIKONIN: A CLASSICAL EXAMPLE OF IN VITRO PRODUCTION

Shikonin is the *world's first successful industrialization of material production using plant cell cultures.* This is a natural product derived from the rare herbaceous plants like *Lithospermum erythrorhizon* and *Arnebia euchroma.* Shikonin is difficult to synthesize chemically, and there had been no way before that time to obtain the chemical except by extraction from the harvested plant. In the 1980s, however, Mitsui Petrochemical Industries Ltd., Japan, succeeded in the large-scale production of cell cultures from the plant, allowing shikonin to be produced in culture systems.

1.6.3 ARTEMISININ: A POWERFUL ANTIMALARIAL DRUG

An isoprenoid group of natural compound, artemisinin is a powerful antimalarial drug extracted from *Artemisia annua.* This has been very effective and useful in combating malaria worldwide. In 2010, the WHO estimated about 219 million cases of malaria and 660,000 deaths (about 90% of all malaria deaths occur in Africa) (Newman, 2012). Most effective and faster-acting artemisinin-based drugs as currently produced directly from plant-based products are too expensive for large-scale use. Artemisinin is an isoprenoid natural product extracted from *A. annua*. The WHO has recommended artemisinin-based combination therapies (ACTs) as the first-line treatment for malaria caused by *P. falciparum,* the most risky of the Plasmodium parasites that infect humans. By 2011, 79 countries and territories had adopted ACTs as first-line treatment against *P. falciparum* malaria. Chloroquine is used as effective drug and acts agaisnt

Plasmodium vivax malaria as *P. vivax* is resistant to chloroquine (source: WHO, World Malaria Report 2012). Cell culture-based *in vitro* production of artemisinin from *A. annua* was reported by Baldi and Dixit (2008).

1.6.4 VINBLASTINE/VINCRISTINE: ANTICANCER DRUGS

Catharanthus roseus (L.) G. Don (*Madagascar periwinkle*) is a major source of these anticancer drugs. *C. roseus* is known for its production of terpenoid–indole alkaloids (TIAs), many of which are pharmaceutically important. Vinblastine, vincristine, and 3′,4′-anhydrovinblastine have antineoplastic activity and are used in the treatment of various types of cancers. Economic data show that the requirement of vinblastine and vincristine that are derived by the oxidative coupling of vindoline and catharanthine is about 5–10 kg annually and industrial cost is US$ 5 million/kg (thus estimated market value is US$ 25–50 million annually). TIAs are produced in little amount in *C. roseus*, which make them expensive.

1.6.5 TAXOL®: ANTICANCER DRUG

World's first billion-dollar anticancer drug was from the bark of Pacific yew tree *Taxus brevifolia*, used in the treatment of refractory ovarian and metastatic breast cancer. Each Pacific Yew tree could supply about 2 kg of bark. Twelve kilograms of bark is required to produce one half gram of taxol. The Yew tree dies after stripping the bark (two to four fully grown trees needed for sufficient quantity of dosage for one patient). In 2011, the International Union for Conservation of Nature designated a species of Asian Yew Tree used to harvest paclitaxel as an endangered (http://www.paclitaxel.org/). Unfortunately, the structural complexity of taxol molecule precluded chemical synthesis for commercial use and was extremely expensive.

1.6.6 LYCOPENE

Carotenoids, natural metabolites of isoprenoids, are the most commonly found pigments of various color like orange, yellow, and red. These pigments serve as food colorants, feed supplements, and also for nutritional, phytoceuticals, and cosmetic purposes. These carotenoids also possess other characteristics like antioxidant properties and playing roles in preventing the occurrence of chronic diseases (Rao & Rao, 2007).

1.6.7 HAIRY ROOT CULTURES

This is another tool for the *in vitro* production of secondary metabolites. Hairy roots are differentiated cultures of transformed roots which are generated by the infection of *Agrobacterium rhizogenes* into wounds of higher plants. Many expensive secondary metabolites are synthesized in roots under natural conditions and often synthesis is linked to root differentiation. Secondary metabolites found in aerial part could be produced in hairy root cultures. These hairy roots are highly stable and productive. Due to climate change, when many plants are facing threat, the *in vitro* production systems including plant hairy root culture are comparatively least affected. Plant hairy root culture is a promising area of biotechnology and amenable to metabolic engineering for modulation of production of important secondary metabolites.

1.7 LIMITATION OF *IN VITRO* PRODUCTION USING PLANT CELL CULTURE

In vitro production of secondary metabolites is hampered by the insufficient knowledge on the biosynthetic pathways that leads to erratic production. For rational engineering of metabolic pathways, a thorough knowledge of the total biosynthetic pathways and their regulatory mechanisms controlling the onset and the flux of the pathways in needed. Till date, only a few pathways (e.g., terpenoid, flavonoids, indole, and isoquinoline alkaloids) into plant system are well understood as a result of several years' classical biochemical research.

1.8 METABOLIC ENGINEERING

Metabolic engineering can be intended for modification of biochemical reactions involved in metabolic processes with the help of genetic engineering techniques for the economic production of certain molecules within the cells. Nowadays, it has become a trend to apply the metabolic engineering to alleviate the constraint in the biochemical pathway to enhance *in vitro* production of bioactive molecules in plants as well as in microbes. Despite structural complexity and many other challenges, some notable landmarks in metabolic engineering of bioactive molecules have been achieved: (1) anticancer compounds vincristine and vinblastine production by engineering the terpenoid and indole pathways of *C. roseus* in hairy roots culture, and (2) engineering the biosynthesis of artemisinin in *E. coli*. Rational metabolic engineering in plants and microbes requires a thorough knowledge of the biosynthetic pathways and their regulation. Metabolic engineering has opened the new avenue for successful *in vitro* production of valuable bioactive molecules conserving the natural plant population and meeting their commercial and pharmaceutical demand.

1.8.1 WAYS TO METABOLICALLY ENGINEER AN ORGANISM

Several approaches are followed in metabolic engineering of an organism, such as (1) enhancement of the expression or activity of rate-limiting enzyme(s) of biosynthetic pathway, (2) introduction of a hetrologous pathway, (3) discovery of transcription factors which regulate the entire pathway, (4) prevention of feedback/feed forward inhibition of key enzyme, (5) balancing the heterologous pathway, and (6) use of suitable promoter for required level of gene expression.

1.8.2 METABOLIC ENGINEERING IN PLANTS

Attempts were made for changing the TIA Pathway through metabolic engineering in *Catharanthus roseus* (Zhou et al., 2009). Cell and metabolic engineering in *C. roseus* have been focused on increasing flux through modification of the TIA pathway by various means like optimization of medium composition, precursor feeding and elicitation, construction of novel culture systems, and introduction of genes responsible for specific metabolic enzymes into the *C. roseus* genome. Genes encoding rate-limiting enzymes and some key transcription factors were cloned to improve TIA production by overexpressing them in transgenic *C. roseus* cultures.

1.8.3 METABOLIC ENGINEERING IN MICROBES

Genetic engineering of biosynthetic pathways for the production of complex chemicals and pharmaceutical compounds in microbes is economically feasible option for chemical synthesis. There are several advantages with microbes in comparison to plants such as high yields of metabolites, possibility to use simple carbon sources (glucose/glycerol) and their faster growth, for speedy production (bioreactors). But often challenges come while expressing plant genes in microbes and for that codons have been optimized as per microbial system.

1.8.4 METABOLIC ENGINEERING FOR IN VITRO PRODUCTION OF ARTEMISININ

At the Keasling laboratory (Team leader, Jay D Keasling, Professor, Department of Chemical & Biomolecular Engineering, University of California, Berkeley), the metabolic engineering of both *E. coli* and *Saccharomyces cerevisiae* was carried out to produce a precursor to artemisinin, artemisinic acid, which can be easily converted into artemisinin. In this effort of synthetic biology, a modified mevalonate pathway was used, and the yeast cells were engineered to express the enzyme amorphadiene synthase and a cytochrome P450 monooxygenase (CYP71AV1), both from *Artemissia annua*. A three-step oxidation of amorpha-4,11-diene gives the resulting artemisinic acid. *In vitro* synthesis of artemisinic acid into microbial system will eventually reduce the cost of ACTs significantly below their current price and stabilize the supply of artemisinin while controlling access (Keasling, 2012).

1.8.5 METABOLIC ENGINEERING OF TAXOL PREXURSOR IN E. COLI

Metabolic engineering of several precursor molecules of taxol has been addressed in *E. coli* by P. K. Ajikumar and colleagues, who have reported taxadiene production of up to 300 mg/L. This was achieved after systematically modifying expression levels of one of the bacteria's own pathways (MEP), and also using a series of heterologous enzymes in taxol biosynthesis. To enhance the flux through the upstream MEP pathway, Ajikumar et al. (2011) modified known enzymatic bottlenecks (dxs, idi, ispD, and ispF) for over expression by an operon (dxs-idi-ispDF). To overflow the flux from the universal isoprenoid precursors like IPP and DMAPP for taxol biosynthesis, they constructed a synthetic operon of downstream genes GGPP synthase (G) and taxadiene synthase (T). Both pathways were placed under the control of inducible promoters in order to control their relative gene expression. The cyclic olefin taxadiene undergoes multiple rounds of stereospecific oxidations, acylations, and benzoylation to form the late intermediate Baccatin III and side chain assembly to, ultimately, form taxol.

1.8.6 RECENT TECHNOLOGY BOOSTING METABOLIC ENGINEERING

Metabolic engineering requires the information generated from the use of various tools such as Next Generation Sequencer (transcriptomics), Microarray (for global gene expression analysis), 2D gel electrophoresis and MALDI-TOF (proteomics), LC–MS (metabolomics). These tools help the discovery of genes/proteins and related metabolites, understanding the regulatory mechanism and devising strategies for metabolic engineering.

1.8.7 LIMITATIONS OF METABOLIC ENGINEERING

Many times while transforming large pathways into alternate hosts and manipulating their expression level, the native regulation of carbon flux through the pathway may be lost leading to imbalances in the pathways. The over expression of a gene may cause depletion of precursors, resources necessary for growth and production, and induce a stress response from excessive heterologous protein. In case of a multi gene heterologous pathway expression, challenges occur as the activity of a single enzyme may be out of balance with that of the other enzymes in the pathway, leading to unbalanced carbon flux and the accumulation of an intermediate.

Plant cell culture methods could be considered an attractive choice and alternative to agricultural methods for producing valuable secondary metabolites for color, flavor, fragrance, phytoceuticals, and nutraceuticals. But their commercial success is limited due to genetic instability, lack of understanding on their synthesis and regulation in plants, and heterologous hosts such as *E. coli* and *S. cerevisiae*. In this direction, understanding genetic networks involving the genes of biosynthetic pathways, transcription factors, cytochrome P450s, transporters, and localization studies on enzymes and substrates will give lots of insight. The economic feasibility of synthesis of simple to more complex chemicals and pharmaceuticals in plants and microbes raised future hope for meeting the demand for valuable molecules.

1.9 SIGNIFICANCE OF *IN VITRO* PRODUCTION OF SECONDARY METABOLITES

Very little attention was paid on secondary metabolites production in different organism and the enhancement of their production. At the end of the last century, with the advancement of technologies for detection of molecules and their metabolic pathways, *in vitro* production of secondary metabolites increased tremendously into the laboratory conditions.

In the past, secondary metabolites were directly harvested from the plant grown under natural habitat. This process is associated with several limitations:

1. Quantities of produce
2. Seasonal and climatic factors
3. Quality of produce
4. Microbial contamination
5. Tissue specificity of produce
6. Extraction and purification

Metabolic pathways are different for different types of metabolites and efforts have been taken to enhance the production under *in vitro* system. Recent advances in molecular biology, enzymology, and fermentation technology of plant cell culture provide a viable system for the production of secondary metabolites. Advancement in the cell culture techniques in laboratory conditions kept the promises for the efficient production of targeted metabolites. To maximize production system in cell culture, we need to focus on the biosynthetic activities of cell cultures, suitable explants, standardization of optimum culture condition, extraction and purification of the produce. The following advantages are associated with the production under controlled conditions:

1. Rapid and efficient production
2. Higher quantity and quality of produce
3. Structural modification of produce through metabolic engineering
4. Reduces the interfering compounds
5. Extraction is relatively easy
6. Reduces microbial contamination
7. Mass production with reduced cost
8. In-depth study of metabolic pathways

Transgenic research has made enhancing the target production and the modification of produce relatively easy. Among which transgenic hairy root culture is extensively used for the secondary metabolites production.

1.10 AGRICULTURALLY IMPORTANT SECONDARY METABOLITES AND THEIR FUTURE SCOPE

Plant produces a high diversity of secondary metabolites having prominent function in the protection against predators and microbial pathogens as they are toxic in nature and repellent to herbivores and microbes. A few of them have significant role in defense against abiotic stress (e.g., UV-B exposure) and important for the communication of plants with other organisms (Schafer & Wink, 2009). Plants have developed surveillance systems so that they can recognize particular insect pests and further respond with specific defense mechanisms. Chewing insects secrete elicitors in their saliva which help the plant to distinguish between general wounding and insect feeding.

Feeding on one particular part of the plant can induce systemic production of these chemicals in undamaged plant tissues, and after release, these chemicals act as signals to neighboring plants to begin producing similar compounds. Production of these chemicals exerts a high metabolic cost on the host plant, so many of these compounds are not produced in large quantities until after insects have begun to feed. There are many valuable secondary metabolites produced by the plants already in practice for pest management.

Furthermore, many of the secondary metabolites have been identified for diseases and pest management used by the plants themselves. Plants secrete volatile organic compounds (VOCs) like monoterpenoids, sesquiterpenoids, and homoterpenoids in response to the feeding by insects, which may repel harmful insects or attract beneficial predators that prey on the destructive pests.

For example, wheat seedlings infested with aphids release VOCs that repel other aphids. Lima beans and apple trees secrete chemicals that attract predatory mites when damaged by spider mites, and cotton plants produce volatile substances that act as attractant for predatory wasps when damaged by moth larvae.

Plant secondary metabolites can be divided into three chemically distinct groups, namely, terpenes, phenolics, N-and S-containing compounds. A few important metabolites belonging to these groups are discussed in this chapter.

1.10.1 TERPENOIDS (TERPENES)

Terpenoids represent the largest class of secondary metabolites and occur in all plants. The simplest terpenoid, the hydrocarbon isoprene (C5H8), is a volatile gas produced by leaves in large quantity during photosynthesis that may protect cell membranes from damage due to high temperature or light. Terpenoids are classified by the number of isoprene units used to construct them.

1.10.1.1 MONOTERPENOIDS AND SESQUITERPENOIDS

These are the primary components of essential oils which are highly volatile compounds and contribute to the fragrance (essence) of plants that produce them. Essential oils often act as insect toxins and many of them protect against fungal or bacterial attack. Mint plants (*Mentha* spp.) produce large amount of the monoterpenoids menthol and menthone which are stored in glandular trichomes on the epidermis. Pyrethrins are monoterpenoid esters produced by chrysanthemum and act as insect neurotoxins (Isman, 2006). Many commercially available insecticides are actually synthetic analogues of pyrethrins, called pyrethroids, which include the insecticides permethrin and cypermethrin.

Pine tree resin contains large amount of the monoterpenoids alpha- and beta-pinene, which are potent insect repellents. The characteristic sharp odor of the organic solvent turpentine is due to these compounds. Many spices, seasonings, condiments, and perfumes are made using essential oils that act as insect toxins in plants but are relatively harmless to humans. Examples include peppermint and spearmint (*Mentha* spp.), basil (*Ocimum* spp.), oregano (*Origanum* spp.), rosemary (*Rosmarinus* spp.), sage (*Salvia* spp.), black pepper (*Piper* spp.), cinnamon (*Cinnamomum* spp.), and bay leaf (*Laurus* spp.).

1.10.1.2 DITERPENOIDS

It includes gossypol, a terpenoid produced by cotton (*Gossypium hirsutum*) having strong antifungal and antibacterial activities.

1.10.1.3 TRITERPENOIDS

The molecular structures of triterpenoids are similar in plant and animal sterols and steroid hormones. Phytoecdysones are mimics of insect molting hormones. Production of these compounds by plants such as spinach (*Spinacia oleracea*) disrupts larval development and increases insect mortality. A class of triterpenoids called limonoids results in the fresh scent of lemon and orange peels. Azadirachtin, a limonoid isolated from neem trees (*Azadirachta indica*), acts as insect repellant by concentrations as low as a few parts per million (Defago et al., 2006; Kumar et al., 2003). Citronella, an essential oil isolated from lemon grass (*Cymbopogon citratus*), also contains high limonoid levels and is a popular insect repellent in the United States due to its low toxicity in humans and biodegradable properties.

1.10.2 PHENOLICS

Phenolics is another large class of secondary metabolites produced primarily via the shikimic acid and malonic acid pathways in plants to defend themselves against pathogen attack. It includes a wide variety of defense-related compounds including flavonoids, anthocyanins, phytoalexins, tannins, lignin, and furanocoumarins. Tannins are water-soluble flavonoid polymers produced by plants and stored in vacuoles. They are toxic to insects as they bind to salivary proteins and digestive enzymes (trypsin and chymotrypsin) resulting in protein inactivation. Insect herbivores that ingest high amounts of tannins fail to gain weight and eventually die.

1.10.2.1 PHYTOALEXINS

Most plant families produce organic phytoalexins of diverse chemistry, for example, sesquiterpenoids from Solanaceae, isoflavonoids from Leguminosae. In response to pathogen infection or other forms of stress, phytoalexins are produced that help in limiting the spread of the disease by accumulating around the site of infection. This mechanism appears to be a common mechanism of resistance to pathogenic microbes in a wide range of plants (Van Etten et al., 1994; Grayer & Harborne, 1994; Bailey & Mansfield, 1982; Darvill and Albersheim, 1984).

1.10.2.2 LIGNINS

They are found primarily in the secondary cell walls of plants and are highly branched heterogeneous polymers, although primary walls can also become lignified. Because it is insoluble, rigid, and virtually indigestible, it provides an excellent physical barrier against pathogen attack.

1.10.2.3 FURANOCOUMARINS

These phenolic compounds are produced by a wide variety of plants in response to pathogen or herbivore attack. They are activated by ultraviolet light and can be highly toxic to certain vertebrate and invertebrate herbivores as they integrate into DNA, which 33 results in rapid cell death.

1.10.3 SULPHUR-CONTAINING SECONDARY METABOLITES

They include glucosinolase (GSL), glutathione (GSH), phytoalexins, thionins, defensins, and allinin which have been linked directly or indirectly with the defense of plants against microbial pathogens.

1.10.3.1 GSH

It is one of the major forms of organic S in the soluble fraction of plants. It plays important role as a mobile pool of reduced S in the regulation of plant growth and development, and also acts as a cellular antioxidant in stress responses. Specialized cells like trichomes possess high activities of enzymes for synthesis of GSH and other phytochelatins necessary to detoxify heavy metals.

To mitigate oxidative stress, GSH functions as a reducing agent for other antioxidants such as ascorbic acid as well as an integral weapon in the defense against ROS generated by O_3 or as a reaction to biotic and abiotic stress.

1.10.3.2 GSL

These are a group of low molecular mass N-and S-containing plant glucosides that are produced by higher plants to contribute resistance against the unfavorable effects of predators, competitors, and parasites. Broken down products of GSL are released as volatile defensive substances exhibiting toxic or repellent effects (Mithen, 1992; Wallsgrove et al., 1999; DeVos & Jander, 2009), for example, mustard oil glucosides in Cruciferae and allylcys sulfoxides in *Allium* (Leustek, 2002).

1.10.3.3 DEFENSINS, THIONINS, AND LECTINS

All these are S-rich nonstorage plant proteins synthesized and accumulated after the attack of microbes and other related situations (Loon et al., 1994). All of which seize the growth of a broad range of fungi. Additionally, defensins genes are partly pathogen-inducible and others that are involved in resistance can be expressed constitutively. These components seem to be involved in the natural defense system of plants as they can be highly toxic to microorganisms, insects, and mammals. Accumulation of thionins in the cell wall of infected spikes of resistant wheat cultivars by *Fusarium culmorum* indicates that it is involved in defense responses to infections (Kang & Buchenauer, 2003).

Some plant species produce lectins which act as defensive proteins that bind to carbohydrate or carbohydrate-containing proteins. After being ingested by herbivores, lectins bind to epithelial cell lining of the digestive tracts and start interfering with nutrient absorption (Peumans & Van Damme, 1995).

1.10.4 NITROGEN-CONTAINING SECONDARY METABOLITES

This group includes alkaloids, cyanogenic glucosides, and nonprotein amino acids. Most of them are biosynthesized from common amino acids. All of them have significant role in the anti-herbivore defense and toxicity to humans.

1.10.4.1 ALKALOIDS

A large class of bitter-tasting nitrogenous compounds that are found in many vascular plants includes caffeine, cocaine, morphine, and nicotine. Caffeine, is an alkaloid found in coffee (*Coffee arabica*), tea (*Camellia sinensis*), and cocoa (*Theobroma cacao*). It is toxic to both insects and fungi. High levels of caffeine produced by coffee seedlings are liable to inhibit the germination of other seeds in the vicinity of the growing plant (allelopathy). Allelopathy allows one plant species to "defend" itself against other plants competing for growing space and nutrient resources.

1.10.4.2 NICOTINE

It is produced in the roots of tobacco plants (*Nicotiana tabacum*) and transported to leaves for storage in vacuoles. It is produced when herbivores graze on the leaves and break open the vacuoles. Capsaicin and related capsaicinoids produced by the genus *Capsicum* are the active components of chili peppers which are used for their characteristic burning sensation in hot, spicy foods.

1.10.4.3 CYANOGENIC GLUCOSIDES

A group of N-containing protective compounds other than alkaloids which release the poison HCN and usually occur in members of families, namely, Graminae, Rosaceae, and Leguminosae. They are not toxic as such but are readily broken down to produce off volatile poisonous substances like HCN and H_2S after crushing of the plants. Amygdalin is a common cyanogenic glucoside found in the seeds of almonds, apricot, cherries, and peaches, while dhurrin is, found in *Sorghum bicolor*. Upon damaging as during herbivore feeding, the cell contents of different tissues mix and form HCN, a toxin of cellular respiration, by binding to the Fe-containing heme group of cytochrome oxidase and other respiratory enzymes.

Similarly, the presence of cyanogenic glucosides in cassava make it suitable for long-time storage without being attacked by pests (Pearce et al., 1991). Lima bean (*Phaseolus lunatus* L.) is a model plant for studies of inducible indirect anti-herbivore defenses including the production of VOCs (Ballhorn et al., 2009). Plants have also evolved multiple defense mechanisms against microbial pathogen attacks and various types of environmental stress.

1.10.5 FUTURE PROSPECTS

The identification of the mechanisms causing systemic acquired resistance will be an important milestone for sustainable agricultural production, as the use of fungicides could then be minimized or eliminated. Therefore, additional research in area of natural pesticides development is needed in current scenario. In the long term, perhaps it will be possible to generate gene cassettes for complete pathways, which could then be utilized for the production of important defensive secondary metabolites in bioreactors or for metabolic engineering of crop plants. This will improve their capacity to cope up with herbivores and microbial pathogens as well as other environmental stresses.

Today, advanced tools are the need to dissect the exact correlation between N and S fertilization and crop-resistance management. In a number of previous research articles and review papers, it has been shown that the N-and S-containing secondary metabolites are influenced by optimum supply of N and S and their good nutrition can enhance the capability of a plant to cope with biotic and abiotic stress.

KEYWORDS

- **climate change**
- **secondary metabolites**
- **toxic nitrogenous compounds**
- **gene conversion**
- **oxidosqualene cyclase**

REFERENCES

Ajikumar, P. K.; Xiao, W. H.; Tyo, K. E.; Wang, Y.; Simeon, F.; Leonard, E.; Mucha, O.; Phon, T. H.; Pfeifer, B.; Stephanopoulos, G. Isoprenoid Pathway Optimization for Taxol Precursor Overproduction in *Escherichia coli*. *Science* **2010,** *330*(6000), 70–74.

Bailey, J. A.; Mansfield, J. W. *Phytoalexins*. Wiley: New York, 1982.

Baldi, A.; Dixit, V. K. Enhanced Artemisinin Production by Cell Cultures of Artemisia Annua. *Curr. Trends Biotechnol. Pharmacy* **2008,** *2*(2), 341–348.

Ballhorn, D. J.; Kautz, S.; Heil, M.; Hegeman, A. D. Cyanogenesis of Wild Lima Bean (*Phaseolus lunatus* L.) Is an Efficient Direct Defence in Nature. *Plant Signal. Behav.* **2009,** *4*(8), 735–745.

Darvill, A. G.; Albersheim, P. Phytoalexins and their Alicitors—A Defence against Microbial Infection in Plants. *Ann. Rev. Plant Physiol.* **1984,** *35*, 243–275.

Defago, M.; Valladares, G.; Banchio, E.; Carpinella, C.; Palacios, S. Insecticide and Antifeedant Activity of Different Plant Parts of *Melia azedarach* on *Xanthogaleruca luteola*. *Fitoterapia* **2006,** *77*, 500–505.

DeVos, M.; Jander G. Myzuspersicae (Green peach aphid) Salivary Components Induce Defence Responses in *Arabidopsis thaliana*. *Plant Cell Environ.* **2009,** *32*(11), 1548–1560.

Dornenburg, H.; Knorr, D. Generation of Colors and Flavors in Plant Cell and Tissue Cultures. *Crit. Rev. Plant Sci.* **1996,** *15*(2), 141–168.

Giri, A.; Dhingra, V.; Giri, C. C.; Singh, A.; Ward, O. P.; Narasu, M. L. Biotransformations Using Plant Cells, Organ Cultures and Enzyme Systems: Current Trends and Future Prospects. *Biotechnol. Adv.* **2001,** *19*(3), 175–199.

Grayer, R. J.; Harborne, J. B. A Survey of Antifungal Compounds from Higher Plants, 1982–1993. *Phytochemistry* **1994,** *37*, 19–42.

Isman, M. B. Botanical Insecticides, Deterrents, and Repellents in Modern Agricultural and an Increasingly Regulated World, *Annu. Rev. Entomol.* **2006,** *51*, 45–56.

Kang, Z.; Buchenauer, H. Immonocytochemical Localization of Cell Wall-Bound Thionins and Hydroxyproline Rich Glycoproteins in *Fusarium culmorum*-Infected Wheat Spikes. *J. Phytopathol.* **2003,** *151*(3), 120–129.

Keasling, J. D. Synthetic Biology and the Development of Tools for Metabolic Engineering. *Metab. Eng.* **2012,** *14*, 189–195.

Kellis, M.; Birren, B. W.; Lander, E. S. Proof and Evolutionary Analysis of Ancient Genome Duplication in the Yeast *Saccharomyces cerevisiae*. *Nature* **2004,** *428*, 617–624.

Kumar, A. R. V.; Jayadevi, H. C.; Ashoka, H. J.; Chandrashekara, K. Azadirachtin use Efficiency in Commercial Neem Formulations. *Curr. Sci.* **2003,** *84*(11), 1459–1464.

Leustek, T. Sulfate Metabolism. In *The Arabidopsis Book*; Somerville, C. R., Meyerowitz, E. M., Eds.; American Society of Plant Biologists: Rockville, MD, 2002. DOI:10.1199/tab.0009.

Loon, L. Mechanisms of Resistance to Plant Diseases, Kluwer Academic Publishers: Netherlands. 1994, pp 325–370.

Minto, R. E.; Townsend, C. A. Enzymology and Molecular Biology of Aflatoxin Biosynthesis. *Chem. Rev.* **1997,** *97*, 2537–2556.

Mithen, R. Leaf Glucosinolate Profiles and their Relationship to Pest and Disease Resistance in Oilseed Rape. *Euphytica* **1992,** *63*, 71–83.

Nelson, P. E.; Desjardins, A. E.; Plattner, R. D. Fumonisins, Mycotoxins Produced by *Fusarium* species: Biology, Chemistry, and Significance. *Annu. Rev. Phytopathol.* **1993,** *31*, 233–252.

Newman, R. D. *World Malar. Rep.* **2012,** *2011*, 1–14.

Ober, D. Seeing Double: Gene Duplication and Diversification in Plant Secondary Metabolism. *Trends Plant Sci.* **2005,** *10*, 444–449.

Ohno, S. So Much "Mjunk" DNA in Our Genome. *Brookhaven Symp. Biol.* **1972,** *23*, 366–370.

Pearce, G.; Strydom, D.; Johnson, S.; Ryan, C. A. A Polypeptide from Tomato Leaves Induces Wound Inducible Protienase Inhibitor Proteins. *Science* **1991,** *253*, 895–898.

Peumans, W. J.; Van Damme, E. J. M. Lectins as Plant Defence Proteins. *Plant Physiol.* **1995,** *109*, 342–347.

Pichersky, E.; Lewinsohn, E. Convergent Evolution in Plant Specialized Metabolism. *Annu. Rev. Plant Biol.* **2011,** *62*, 549–566.

Rao, A. V.; Rao, L. G. Carotenoids and Human Health. *Pharmacol. Res.* **2007,** *55*, 207–216.

Rao, S. R.; Ravishanka, G. A. Plant Cell Cultures: Chemical Factories of Secondary Metabolites. *Biotechnol. Adv.* **2002,** *20*, 101–153.

Roy C.; Deo I. Gene Duplication: A Major Force in Evolution and Bio-diversity. *Int. J. Biodivers. Conserv.* **2014,** *6*(1), 41–49.

Schafer, H.; Wink, M. Medicinally Important Secondary Metabolites in Recombinant Microorganisms or Plants: Progress in Alkaloid Biosynthesis. *Biotechnol. J.* **2009,** *4*(12), 1684–1703.

Singh, R. S.; Gara, R. K.; Bhardwaj, P. K.; Kaachra, A.; Malik, S.; Kumar, R.; Sharma, M.; Ahuja, P. S.; Kumar, S. Expression of *3-hydroxy-3-methylglutaryl-CoA* Reductase, *p-hydroxybenzoate-m-geranyltransferase* and Genes of Phenyl Propanoid Pathway Exhibits Positive Correlation with Shikonins Content in Arnebia [*Arnebia euchroma* (Royle) Johnston]. *BMC Mol. Biol.* **2010,** *11*, 88.

Van Etten, H.; Temporini, E.; Wasmann, C. Phytoalexin (and Phytoanticipin) Tolerance as a Virulence Trait: Why Is It Not Required by All Pathogens? *Physiol. Mol. Plant Pathol.* **1994,** *59*, 83–93.

Verpoorte, R.; van der Heijden, R.; Memelink, J. Engineering the Plant Cell Factory for Secondary Metabolite Production. *Transgen. Res.* **2000,** *9*(4–5), 323–343.

Wallsgrove, R.; Benett, R.; Kiddle, G.; Bartlet, E.; Ludwig-Mueller, J. Glucosinolate Biosynthesis and Pest Disease Interactions. In Proceedings of the 10th International Rapeseed Congress, Canberra, Australia, 1999.

Yu, J.-H.; Keller, N. Regulation of Secondary Metabolism in Filamentous Fungi. *Annu. Rev. Phytopathol.* **2005,** *43*, 437–58.

Zheyong X.; Lixin, D.; Dan L.; Jie G.; Song G.; Jo D.; Paul O. M.; Anne O.; Xiaoquan Q. Divergent Evolution of Oxidosqualene Cyclases in Plants. *New Phytol.* **2011,** *193*, 1022–1038.

Zhou, M. L.; Shao, J. R.; Tang, Y. X. Production and Metabolic Engineering of Terpenoid Indole Alkaloids in Cell-Cultures of the Medicinal Plant *Catharanthusroseus* (L.) G. Don (*Madagascar periwinkle*). *Biotechnolnol. Appl. Biochem.* **2009,** *52*, 313–323.

CHAPTER 2

ADVANCED TECHNIQUES IN EXTRACTION OF PHENOLICS FROM CEREALS, PULSES, FRUITS, AND VEGETABLES

AMIT K. DAS[1*], SACHIN R. ADSARE[2], MADHUCHHANDA DAS[3], PANKAJ S. KULTHE[4], and GANESAN P.[5*]

[1]*Department of Grain Science and Technology, CSIR—Central Food Technological Research Institute, Mysore 570020, Karnataka, India*
[2]*Food Engineering and Technology Department, Institute of Chemical Technology, Mumbai 400019, Maharashtra, India*
[3]*DoS in Botany, Manasagangothri, University of Mysore, Mysore 570006, Karnataka, India*
[4]*Urmin Group of Companies, Ahmedabad 380054, Gujarat, India*
[5]*Department of Molecular Nutrition, CSIR—Central Food Technological Research Institute, Mysore 570020, Karnataka, India*
**Corresponding author, E-mail: amitkdas12@gmail.com, ganesanp@cftri.res.in.*

CONTENTS

ABSTRACT

Phenolic compounds are naturally occurring antioxidants, usually found in fruits, vegetables, whole grain cereals, and pulses. Optimization of the extraction procedure is essential for an accurate assay of phenolic compounds from these food matrices. Analytical methodology of phenolics generally consists of an extraction with aqueous–organic solvents to obtain free phenolics, followed by a hydrolysis treatment, to obtain the bound ones. These traditional methods are carried out with different solvent system such as ethanol, acetone, methanol, and/or a combination with water. In these processes, the degradation can be triggered both by external and internal factors. Light, air, and temperature are the most important factors that facilitate degradation reactions. In addition to this, the extraction must be performed with the most adequate solvent and under ideally predetermined analytical conditions of temperature and pH. Likewise, solid-phase extraction is used for liquid samples. However, modern techniques such as ultrasound (US), microwave, pulsed electric field (PEF), enzyme, high hydrostatic pressure, pressurized liquid, and supercritical fluid-assisted extraction methods are replacing the conventional ones reducing the extensive use of solvents and accelerating the extraction process. This chapter discusses the literature on implications of these modern methods to extract the phenolics.

2.1 INTRODUCTION

Polyphenolic compounds classified as phenolic acids and its derivatives, tannins, and flavonoids are some of the most abundant and widely distributed groups of natural compounds in the plant kingdom. The flavonoids are sub-classified into anthocyanins, flavones, flavonols, and related substances. In recent years, phenolics in many edible plant products such as cereals, pulses, fruit, and vegetables have received increasing attention because of their influence on nutritional value and quality of foods, biochemical and physiological functions, and pharmacological implications (Table 2.1).

Epidemiological studies have consistently shown that diets rich in whole grains are associated with a decreased risk of a number of chronic diseases such as coronary heart disease (Vita, 2005; Mellen et al., 2008), type II diabetes (Liu et al., 2000; Tapola et al., 2005), and certain cancers

TABLE 2.1 Different Phenolic Compounds Found in Cereals, Pulses, Fruits, and Vegetables.

Sources	Phenolic compounds	Reference
Apple peel	Flavonols, quercetin-3-*O*-glycosides, phloridzin, cyanidin-3-*O*-galactoside, and chlorogenic acid	Khanizadeh et al. (2008)
Apple peel and flesh	Peel: anthocyanins and flavonols, flesh: chlorogenic acid, phloridzin and flavanols	Chen et al. (2012)
Apple pomace	Chlorogenic acid, quercetin-3-glucoside/quercetin-3-glacaside, quercetin-3-xylo-side, phloridzin , quercetin-3-arabinoside, and quercetin-3-rhamnoside	Cao et al. (2009)
Strawberry	Anthocyanins, flavan-3-ols, flavonols, ellagitannins, ellagic acid glycosides, and cinnamic acid conjugates	Aaby et al.(2012)
Mandrin	Hydroxycinnamic acid, flavonoids, flavones, flavanones and flavonols, hydroxy-benzoic acid, and chlorogenic acid	Zhang et al. (2014)
Berries	Anthocyanins, flavanols, hydroxycinammic acids, and hydroxybenzoic acids	Määttä-Riihinen et al. (2004)
European berries	Trans-resveratrol, cinnamic acid, ferulic acid, *p*-coumaric acid, quercetin, and morin	Ehala et al. (2005)
Olives	Hydroxytyrosol, luteolin 7-*O*-glucoside, oleuropein, rutin, apigenin 7-*O*-glucoside, and luteolin	Vinha et al. (2005); Seabra et al. (2010)
Cherries	Anthocyanins and hydroxycinnamic acids	Manach et al. (2004)
Black grapes	Anthocyanine (cyanidin-3-*O*-glucosides, malvanidin-3-*O*-glucosides and delph-inidin 3-*O*-glucosides) and flavonols especially resveratrol	Belitz and Grosch (1999); Yang et al. (2009)
Citrus peel (*Citrus unshiu* Marc)	Hydroxycinnamates and polymethoxylated flavones, cinnamic acids (caffeic, *p*-coumaric, ferulic, sinapic acid), and benzoic acids (protocatechuic, *p*-hydroxyben-zoic, vanillic acid)	Manthey and Grohmann (2001); Ma et al. (2009)
Red raspberries	Anthocyanins (cyanidin-3-sophoroside, cyanidin-3-(2G-glucosylrutinoside), cyanidin-3-sambubioside, cyanidin-3-rutinoside, cyanidin-3-xylosylrutinoside, cyanidin-3-(2G-glucosylrutinoside), and cyanidin-3-rutinoside)	Chen et al. (2007)

TABLE 2.1 *(Continued)*

Sources	Phenolic compounds	Reference
Black, green tea	Flava-3-ols and flavonols, four tea catechinepimers, namely, gallocatechingallate, catechingallate, gallocatechin, and catechin, caffeine and theobromine, flavonol glycosides	Sajilata et al.(2008)
Citrus fruits	Flavanones, flavonols, and phenolic acid	Beecher (2003); Manach et al. (2004)
Plums, prunes, apples, pears, kiwi	Hydroxycinnamic acids and catechins	Belitz and Grosch (1999); Manach et al. (2004)
Chestnut	Phenolic acids and flavonoids	Barreira et al. (2008)
Orange juice	Flavanols	Manach et al. (2004)
Artemisia leaves	Hydroxybenzoic acids, hydroxycinnamic acids, flavonols, and catechins. Ferulic and caffeic conjugates acid, gallic acid, and catechin	Carvalho et al. (2011)
Aubergin	Anthocyanins and hydroxycinnamic acids	Manach et al. (2004)
Tomato wastes	Caffeic, chlorogenic, *p*-coumaric, ferulic, and rosmarinic acid, quercetin, and rutin	Cetkovic et al. (2012)
Chicory, artichoke	Hydroxycinnamic acids	Manach et al. (2004)
Parsley	Flavones	Beecher (2003); Manach et al. (2004)
Rhubarb	Anthocyanins	Manach et al. (2004)
Sweet potato leaves	Flavonols	Chu et al. (2000)
Yellow onion	Flavonols	Manach et al. (2004)
Whole oats	Catechin	Zieliński and Kozłowska (2000)

TABLE 2.1 *(Continued)*

Sources	Phenolic compounds	Reference
Wheat flour, wheat bran	Hydroxybenzoic, caffeic, cinammic, ferulic and protocatechuic acids	Pellegrini et al. (2006). Arranz and Calixto (2010)
Rice bran	Derivatives of ferulic acid	Wang et al. (2015)
Finger millet	Polyphenols	Chethan and Malleshi (2007)
Flaxseed	*p*-Coumaric acid, *O*-coumaric acid, frulic, *p*-hydroxybenzoic acid, vanillic acid	Naczk and Shahidi (2004); Hao and Beta (2012)
Peanuts	Daidzein, genistein, biochanin A, and trans-resveratrol	Chukwumah et al. (2009)
sorghum	ferulic, *p*-coumaric acids	Chiremba et al. (2012),
Maize	Ferulic, *p*-coumaric, vanillic, syringic, *p*-hydroxybenzoic, caffeic, acids and quercetin, kaempferol, and cyaniding 3-*O*-glucoside	Das and Singh (2015)

(Dashwood, 2007; Watson, 2008; Haas et al., 2009). The beneficial effect of whole grains and cereal products has often been attributed to their fiber content (Jacobs et al., 2000). Foods rich in photochemical such as whole grain cereals, legumes, fruits, and vegetables tend to be rich sources of many other health-promoting components such as vitamins, minerals, trace elements, and antioxidants (Sidhu & Kabir, 2007) which may work synergistically to optimize human health (Fardet, 2010). Antioxidants such as vitamins E and C are thought to protect the body from damaging free radicals and may have a role to play in disease prevention. Traditionally, phenolics have also been considered as potent antioxidants. However, this view is currently being revised and increasingly it is emerging that phenolics may have far more important effects *in vivo* such as enhancing endothelial function (Caton et al., 2010), cellular signaling, and anti-inflammatory properties (Williams et al., 2004; Sies et al., 2005; Ramos, 2008).

Since the huge variations among bioactive compounds and large number of plant resources are available, the existing processes are needed to be upgraded for their suitability for extraction, isolation, purification, and identification of these compounds. An appropriate extraction process consists of separation, identification, and characterization of bioactive compounds. Different extraction techniques ought to be used in diverse conditions for understanding the extraction selectivity from various natural sources. Numbers of techniques including traditional ones are being used since ancient time and these are potent to extract bioactive compounds.

The chapter discusses the techniques, which have been developed with common objectives such as extraction of targeted bioactive compounds from complex plant material, enhancing selectivity of analytical methods, increasing the concentration of targeted compounds, transferring the bioactive compounds into a more suitable form for detection, separation, and to provide a strong and reproducible method.

2.2 EXTRACTION SYSTEMS FOR PHENOLIC COMPOUNDS

The ultimate goal of sample preparation is to eliminate or reduce potential matrix interferences. In the development of extraction techniques for phenolic compounds, the extraction must be performed with the most adequate solvent and under ideally predetermined analytical conditions of temperature and pH. The extraction procedure is sequential and systematically carried out using an aqueous organic solvent to extract phenolic compounds. In addition, it is important to consider the polyphenolic structure because these compounds may have multiple hydroxyl groups that can be conjugated to sugars, acids, or alkyl groups. Thus, due to varying polarities of phenolic compounds, it is difficult to develop a single method for optimum extraction of all phenolic compounds. Hence, the optimization of the extraction procedure is essential for an accurate assay of phenolic compounds from different food matrices. Cost-effectiveness and reduction in sampling time during the extraction is also important because of its stability, shelf life, and economics (Table 2.2).

2.3 CONVENTIONAL EXTRACTION TECHNIQUES

Various classical extraction techniques have been used for the extraction of phenolic compounds from various plant sources. Most of these techniques are based on the choice of solvent coupled with the application of heat and/or agitation. Selection of solvent for the extraction of phenolic compounds depends upon the solubility in solvent. Large number of organic, inorganic, polar, and nonpolar solvents alone or in combinations have been used for the extraction of varied range of phenolic compounds. The existing classical techniques which have been used for extraction of phenolics includes the following:

1. Soxhlet extraction
2. Maceration
3. Hydrodistillation

In 1879, German chemist Franz Ritter Von Soxhlet proposed Soxhlet extractor for the extraction of lipid, but later it became the widely used apparatus for extraction of various bioactive compounds from different natural sources. Soxhlet extraction is standard and model technique for evaluating the performance of new extraction methods. Generally, in a conventional Soxhlet system, dry plant material is placed in a thimble-holder and filled with suitable condensed fresh solvent from a distillation flask. When the liquid reaches the overflow level, the solution of the thimble-holder is aspirated by a siphon, which unloads it back into the distillation flask, carrying extracted solutes into the bulk liquid. In the solvent flask, distillation process separates the solute from solvent,

TABLE 2.2 Phenolic Compounds Extracted Using Different Solvents.

Source	Solvent	Compound	Reference
Wheat bran	70% methanol or ethanol, absolute ethanol and 50% acetone	Phenolic antioxidants	Zhou and Yu (2004)
Barley	Acidified methanol (HCl/methanol/water, 1:80:10, v/v/v)	Total phenolic content	Lahouar et al. (2014)
Milled oat groats	Methanol or isopropanol	Total phenolic compounds	Auerbach and Gray (1999)
Barley	80% methanol or ethanol	Total phenolic compounds	Bonoli et al. (2004b), Madhujith et al. (2006)
Whole oats	80% methanol and water	Catechin	Zieliński and Kozłowska, (2000)
Wheat flour	Acid hydrolysis	Hydroxybenzoic, caffeic, cinammic, ferulic and protocatechuic acids	Hartzfeld et al. (2002), Adom and Liu (2002), Pellegrini et al. (2006) Arranz and Calixto (2010)
Wheat bran	Methanol/H_2SO_4 90:10 (v/v) at 85°C for 20 h		
	Alkali hydrolysis (2 M NaOH for 4 h shaking under nitrogen gas)		
Rice bran	Alkaline hydrolysis, extraction with 5 times ethyl acetate	*Para*-hydroxy methyl benzoate glucoside, cycloeucalenol *cis*-ferulate, cycloeucalenol *trans*-ferulate, *trans*-ferulic acid, *trans*-ferulic acid methyl ester, *cis*-ferulic acid, *cis*-ferulic acid methyl ester, methyl caffeate, vanillicaldehyde, and *para*-hydroxybenzaldehyde	Wang et al. (2015)
Whole grain wheat and its bran	0–100% (v/v; water/ethanol, methanol or acetone)	Total phenolic	Liyana-Pathirana and Shahidi (2005)
Finger millet	1% HCl–methanol	Polyphenols	Chethan and Malleshi (2007)

TABLE 2.2 *(Continued)*

Source	Solvent	Compound	Reference
Wheat	Solvent composition (water, methanol, 70% methanol, ethanol, 70% ethanol, acetone and 70% acetone), extraction temperature (30–60°C), extraction time (15–90 min) and solid-to-solvent ratio (1:2.5 to 1:20, w/v)	Extraction of PCS was improved by solid-state fermentation (SSF) of wheat By *Rhizopus oryzae* RCK2012 which helped to release the bound compounds from matrix	Dey and Kuhad (2014)
Soybeans, flaxseed and olives	Water, methanol, dilute NaOH. and dilute HCl	Phenolic	Alu'datt et al. (2013)
Olives		Secoiridoids, iridoids, demethyloleuropein, oleuropein, tyrosol, hydroxytyrosol and gallic, protocatechuic, *p*-hydroxybenzoic, vanillic, caffeic, syringic, *p*-coumaric, *O*-coumaric, ferulic, and cinnamic acids	Soler-Rivas et al. (2000), Dağdelen et al. (2013)
Soybean flour		Ferulic, syringic, *O*-coumaric, *p*-coumaric, ferulic and vanillic acids, and daidzein, glycitein and genistein isoflavones	Johnsson et al. (2000), Lee et al. (2004)
Flaxseed		Phenylpropanoids such as *p*-coumaric acid, *O*-coumaric acid, ferulic, *p*-hydroxybenzoic acid, vanillic acid, sinapic acid, secoisolariciresinol, matairesinol, enterolactone and enterodiol	Johnsson et al. (2000), Naczk and Shahidi (2004), Hao and Beta (2012),

where solute is left in the flask and fresh solvent passes back into the plant material holder. The process runs repeatedly to achieve complete extraction. Selection of suitable extracting solvent for the extraction of targeted phenolic compound using the Soxhlet extraction method is an important step because yield of extracts and extract compositions changes with different solvents (Zarnowski & Suzuki, 2004). Various solvents including water, hexane, isopropanol, ethanol, methanol, hydrocarbons, and so on have been used for extraction of various phenolic compounds.

Soxhlet extraction is a well-established technique with advantages of wide industrial applications, better reproducibility and efficiency, and less extract manipulation in contrast to other novel extraction methods such as US-assisted, microwave-assisted, supercritical fluid, or pressurized solvent extractions. However, Soxhlet extraction is an old technique which is time-and solvent-consuming.

Bonoli et al. (2004a) extracted free phenolic compounds from barley flour by simple acetone-based solid–liquid extraction method which led to higher extraction yields of flavan-3-ols and proanthocyanidins, while use of alcohol-based methods (aqueous ethanol or methanol) produced a higher recovery index for all the phenolic classes (catechins and hydrolysable tannins). The prolonged alkaline hydrolysis digestion time results into extraction of higher amounts of hydroxycinnamic acids and flavonols. However, pressurized liquid extractions (PLEs) did not produce a satisfactory recovery of the free phenolic compounds in barley.

Generally, the acidic hydrolysis treatment is used to hydrolyze polysaccharides which disrupts the cell wall structure and allows the release of phenolics bound to cell wall constituents, mainly polysaccharides and protein. In the extraction of phenolics, an alkali or weak acidic hydrolysis treatment of cereals may allow a complete hydrolysis of phenolics esterified or bound to soluble carbohydrates, protein, and other constituents to release free phenolics. However, these treatments may obtain a low and partial release of insoluble or non-extractable phenolics (Adom & Liu, 2002).

Arranz and Calixto (2010) observed that the usage of enhanced sulfuric acid hydrolysis produced a high yield of phenolics that were trapped within cores or bound to cell wall constituents (dietary fiber, proteins) and found that hydrolysable polyphenol (hydroxybenzoic, caffeic, cinammic, ferulic, and protocatechuic acids) content increased up to nine fold for wheat bran and two fold in the case of wheat flour compared with content in methanol/acetone extracts. Das et al. (2014) also observed two- to threefold higher yield of bound phenolic content from maize samples than previous reports on this grain using enhanced acidic hydrolysis. However, they have replaced the use of sulfuric acid, as was used by Arranz and Calixto (2010) with hydrochloric acid citing the reason that the former produces a toxic ester dimethyl sulfate through the reflux method of extraction with methanol.

Wang et al. (2015) reported that the bound phenolic compounds in rice bran were released while extracted with ethyl acetate based on alkaline digestion. Their investigation led to the isolation of a new compound, *para*-hydroxy methyl benzoate glucoside, along with other nine known compounds, cycloeucalenol *cis*-ferulate, cycloeucalenol *trans*-ferulate, trans-ferulic acid, *trans*-ferulic acid methyl ester, *cis*-ferulic acid, *cis*-ferulic acid methyl ester, methyl caffeate, vanillic aldehyde, and *para*-hydroxybenzaldehyde.

Solid substrate fermentation using microorganisms such as *Saccharomyces cerevisieae*, and others, could be a promising technology to enhance the production and extraction of phenolic compounds for the design of different functional foods and for the specific use as nutraceuticals (Dey & Kuhad, 2014). Chethan and Malleshi (2007) reported that the polyphenol contents of finger millet are concentrated in the seed coat and the acidic methanol solvent is an effective solvent for the extraction of millet phenolics. The phenolics of the millet are heat-stable but pH-sensitive and are largely unstable under alkaline conditions.

Rao and Muralikrishna (2004) isolated nonstarch polysaccharide–phenolic acid complexes from native and germinated cereals, including rice, maize, wheat, and millet-ragi, designated as water-extractable and water-unextractable (WUPs) nonstarch polysaccharides having yield of 0.60–3.56% and 7.49–37.8%, respectively. Ferulic and coumaric acids were the main bound phenolic acids predominantly bound (90%) to WUPs. The content of ferulic acid is several-fold higher than that of coumaric acid and their contents decreased upon germination.

Increased extraction efficiency was found for phenolic extraction from wheat flour and bran using successive acidified methanol/water (50:50, v/v, pH 2) and acetone/water (70:30, v/v) than 70:30 (v/v) of either ethanol:water or methanol:water (Pérez-Jiménez & Sauro-Calixto, 2005).

In comparison to extraction of phenolic antioxidants from wheat bran, 70% methanol or ethanol, absolute ethanol, and 50% acetone, it was found that 50% acetone was most effective (Zhou & Yu, 2004). Whole oats in 80% methanol extracted substantially higher total phenolic compounds than water extracts, resulting in an increase in the antioxidant capacity (Zieliński & Kozłowska, 2000), while both 80% methanol or ethanol was found to be efficient at extracting phenolic compounds from barley (Bonoli et al., 2004b; Madhujith et al., 2006). Methanol extraction of milled oat groats yielded higher total phenolics than isopropanol (Auerbach & Gray, 1999).

Preparation of plant extract using maceration is one of popular, inexpensive techniques and has been used since long. For small-scale extraction, maceration generally consists of several steps. This technique is widely used for extraction of volatile or thermal instable products such as color pigment, tannin, aroma compounds, and so on (Canals et al., 2005; Del Llaudy et al., 2008). The process of maceration consists of mainly three steps: first step consists of grinding of plant materials into small particle to increase the surface area for proper mixing with solvent, addition of solvent in material is second step, while in third step, liquid is strained off by pressing of solid residue to recover solution which is containing compound of interest. The obtained strained and pressed-out liquids are mixed together and filtered for separation of impurities. If this process is done at higher temperatures it is called digestion, which is similar to daily life example of tea preparation. The main disadvantage of this process is that it is time-consuming and requires solvent in a batch stirred tank.

Hydrodistillation is a traditional method for extraction of bioactive compounds and essential oils from plants. This is the simplest and usually the cheapest process of distillation. Hydrodistillation seems to work best for powders and very tough materials like roots, wood, or nuts; so easy to use for extraction of bioactive compounds such as phenolics from plant matrices. The main advantages of this method are that less steam is used, shorter processing time, and a higher yield. Organic solvents are not involved and it can be performed before dehydration of plant materials. There are three types of hydrodistillation, namely, water distillation, water and steam distillation, and direct steam distillation (Vankar, 2004). In distillation, the plant material is heated either by placing it in water, which is brought to the boiling temperature, or by passing steam through it. The heat and steam cause the cell structure of the plant material to burst and break down, thus releasing the bioactive compound and essential oils. Indirect cooling by water condenses the vapor mixture of water and oil. Condensed mixture flows from condenser to a separator, where oil and bioactive compounds separate automatically from the water (Silva et al.,

2005). Hydrodiffusion, hydrolysis, and decomposition by heat are three main physicochemical processes in hydrodistillation process. Extraction using high temperature affects recovery of volatile components; this drawback limits its use for extraction of thermolabile compounds.

Extraction efficiency of any conventional method mainly depends on the choice of solvents to be used. The polarity of the targeted compound is the most important factor for choosing the solvent. Structural affinity between solvent and solute, mass transfer, use of cosolvent, environmental safety, human toxicity, and cost of solvent are also important in selection of solvent for the extraction of bioactive compounds. Some examples of phenolic compounds extracted using different solvents are given in Table 2.1.

2.4 NONCONVENTIONAL OR MODERN EXTRACTION TECHNIQUES

Disadvantages of existing extraction technologies, like increased consumption of energy, more consumption of harmful chemicals, longer extraction time, low-extraction selectivity, and thermal decomposition of thermolabile compounds, have forced the food and chemical industries to find new extraction techniques which typically use less solvent and energy. To overcome these limitations, new techniques are being introduced, which are considered as nonconventional extraction techniques such as microwave extraction, supercritical fluid extraction, US extraction, high-hydrostatic pressure extraction (HHPE), accelerated solvent extraction, enzyme-assisted extraction (EAE), and subcritical water extraction, and so on (Azmir et al., 2013). Extraction and separation under extreme or non-classical conditions is currently a dynamic developing area in applied research and industry.

Number of novel extraction techniques have been developed for the isolation of phenolics from plants to shorten the extraction time, reduce the solvent consumption, and enhance the extraction yield and the quality.

2.4.1 ULTRASOUND-ASSISTED EXTRACTION

Ultrasonic radiation is a special type of sound wave beyond upper limit of human detection, which has frequencies higher than 20 kHz (20 kHz–100

MHz). The main benefit of ultrasound-assisted extraction (UAE) can be observed in solid plant sample because US energy facilitates leaching of organic and inorganic compounds from plant matrix which facilitates the extraction of organic and inorganic compounds from solid matrices using liquid solvents (Herrera & Luque de Castro, 2005). US waves pass through a medium by creating compression and shearing forces giving rise to a phenomenon known as cavitation, which means production, growth, and collapse of bubbles. A large amount of energy can produce from the conversion of kinetic energy of motion into heating the contents of the bubble (Suslick & Doktycz, 1990). Only liquid and liquid containing solid materials have cavitation effect, so plant material containing liquid shows efficient extraction of phenolic compound. Ultrasonic baths or closed extractors fitted with an ultrasonic horn transducer are two general designs used for US-assisted extraction of phenolics. Probable mechanism is that US induces a greater penetration of solvent into cellular materials and improves mass transfer. It also disrupts biological cell wall and facilitates the release of contents. Therefore, the enhancement of extraction with ultrasonic power involves two major factors such as efficient cell disruption and effective diffusion across the cell wall (Mason et al., 1996). US can increase extraction kinetics and even improve the quality of extracts. Moisture content and particle size of sample along with temperature, pressure, frequency, and time of sonication are very important factors for obtaining efficient and effective extraction (Table 2.3).

Moreover, temperature, pressure, frequency, and time of sonication are the major factors for the action of US. UAE has also been incorporated along with various classical techniques as they are reported to enhance the efficiency of a conventional system. In a solvent extraction unit, an US device is placed in an appropriate position to enhance the extraction efficiency (Vinatoru, 2001).

The advantages of UAE include reduction in extraction time, energy, and use of solvent. US energy for extraction also facilitates more effective mixing, faster energy transfer, reduced thermal gradients and extraction temperature, selective extraction, reduced equipment size, faster response to process extraction control, quick start-up, increased production, and eliminates process steps (Chemat et al., 2008).

The UAE of mustard seed varieties produced higher recoveries of total phenolic content (TPC) in comparison with the conventional solid–liquid extraction (Szydłowska-Czerniak et al., 2014). Wang et al. (2008)

TABLE 2.3 Operating Conditions and Salient Results for the Ultrasound-Assisted Extraction of Phenolic Compounds.

Source / product	Compound	Conditions	Salient results	Reference
Wheat bran	Phenolic compound	Sonic power 100 W, sonication time of 5 or 10 min and ultrasonic intensity of ~8 W/cm^2 (0.5%, 2% or 5% NaOH)	Extraction of phenolics-rich heteroxylans from wheat bran was increased using ultrasound power	Hromádková et al. (2008)
Mustard seeds	Total phenolic content	Ultrasound power-to-sonication time ratio, 4.5, 4.8, and 4.3 W min^{-1}	Extraction using ultrasonication with 50% methanol, extract of mustard seed showed increase in phenolic content	Szydłowska-Czerniak et al. (2014)
Peanuts	Daidzein, genistein, biochanin A and trans-resveratrol	Sonication at 80 kHz	The results obtained showed that sample-to-solvent ratio, frequency, and duration of sonication had significant effect on the extraction efficiency of the phytochemicals from peanut	Chukwumah et al. (2009)
Peanut skins	Phenolic antioxidants	Microwave power, 10%, 50%, 90% nominal; irradiation time, 30, 90, 150 s; and sample mass, 1.5, 2.5, 3.5 g; and peanut skins were extracted with of 30% ethanol in water	The maximum total phenolic content (TPC) under the optimized conditions (90% microwave power, 30 s irradiation time and 1.5 g skins) was 143.6 mg gallic acid equivalent (GAE)/g skins	Ballard et al. (2010)
Rice grains	15 different phenolic compounds	Temperature, 125–175°C; microwave power, 500–1000 W; time, 5–15 min; solvent, 10–90% EtOAc in MeOH; and solvent-to-sample ratio, 10:1 to 20:1	MAE has been optimized for the extraction of phenolic compounds from rice grains and optimized parameters showed significant yield of phenolic compound	Setyaningsih et al. (2015)

TABLE 2.3 *(Continued)*

Source / product	Compound	Conditions	Salient results	Reference
Red raspberries	Anthocyanins	Ultrasonic horn (22 kHz, 650 W),1.5 M HCl–95% ethanol	Scanning electron microscopy (SEM) results showed strong disruption of fruit tissue structure under ultrasonic acoustic cavitation, this condition facilitated the efficient and rapid to extract ACYS from red raspberry using UAE	Chen et al. (2007)
Cabernet Franc (CF) Grapes	Phenolics (anthocyanins and tannins)	US1 (24 kHz, 5 min, 121 kJ/kg), US2 (24 kHz, 10 min, 242 kJ/kg), US3 (24 kHz, 15 min, 363 kJ/kg)	Pretreatment using ultrasonication increased the phenolics extraction yield by 7% compared with control	Darra et al. (2013)
Cocos nucifera	Phenolic compounds	Ultrasonic cleaning bath (25 kHz, 150 W), water/ethanol 50/50% (v/v) pH = 6.5 (HCl)	Extraction of phenolic compounds (22.44 mg/g) by ultrasound using a 50% (v/v) ethanol:water solution was maximized	Rodrigues et al. (2008)
Citrus peel	Caffeic, *p*-coumaric, ferulic, sinapic acid, and benzoic acids (protocatechuic,*p*-hydroxy-benzoic, vanillic acid)	Ultrasonic cleaning bath 60 kHz, 80% methanol	Increase in yields of phenolic compounds with increase in both ultrasonic time and temperature was observed	Ma et al. (2009)
Peanuts (*Arachis hypogaea*)	Isoflavones and transresveratrol	Ultrasonic cleaning bath, dual or triple frequency combinations: 25 and 40 kHz; 25 and 80 kHz; 40 And 80 kHz and 25, 40, and 80 kHz, 80% ethanol	Study investigated the potential of PSC as a biomarker for polyphenol content and antioxidant capacity	Chukwumah et al. (2009)

TABLE 2.3 *(Continued)*

Source / product	Compound	Conditions	Salient results	Reference
Soy beverages	Isoflavones	Solvent, methanol and ethanol; sample:solvent ratio, 5:1 to 0.2:1; temperature, 10–60°C; and extraction time, 5–30 min	Method for isoflavone extraction from soy beverages blended with fruit juices developed in a simple and reproducible way	Rostagno et al. (2007b)
Soybean	Isoflavones	Ultrasound probe system of 200 W and 24 kHz, temperatures (10 and 60°C), solvent systems, ethanol, methanol, and acetonitrile	Efficiency of the extraction of soy isoflavones was improved by ultrasound	Rostagno et al. (2003)
Wheat bran	Phenolic compound	Ethanol concentration, 64%; extraction temperature, 60°C; and extraction time, 25 min	At optimum extraction conditions about 12 mg gallic acid equivalents/g of experimental total phenolic content was extracted from wheat bran	Wang et al. (2008)
Tea	Polyphenol	Sonic power 40 kHz, water	Increased yield at 65°C, compared with 85°C X	Xia et al. (2006)

extracted the phenolic compounds from wheat bran by UAE technology using response surface methodology (RSM) considering parameters solvent concentration, extraction temperature, and extraction time with the application of US for maximum extraction of phenolic compounds. It was found that 64% ethanol concentration, 60°C extraction temperature, and 25 min extraction time are the optimum extraction conditions with the 12 mg gallic acid equivalents/g of experimental TPC.

Extraction efficiency of four isoflavone derivatives named as daidzin, glycitin, genistin, and malonylgenistin from freeze–dried ground soybeans were compared for mix-stirring extraction and UAE using different solvents and extraction temperatures. It was found that the efficiency of the extraction of soy isoflavones was improved by US but types of solvent have significant effect on extraction (Rostagno et al., 2003).

Chen et al. (2007) carried out the extraction of anthocyanin from red raspberry by UAE process and the effects of operating conditions, such as the ratio of solvents to materials, ultrasonic power, and extraction time, on the extraction yield were studied using RSM. Approximately, 78.13% of the total red pigments could be obtained by UAE using optimized parameters.

Rodrigues et al. (2008) observed maximum extraction of phenolic compounds (22.44 mg/g) by US using a 50% (v/v) ethanol:water solution for 15 min of sonication at US intensity of 4870 W/m^2 was used.

2.4.2 MICROWAVE-ASSISTED EXTRACTION

The microwave-assisted extraction (MAE) appears to be one of the novel methods for extracting soluble products into a fluid from a wide range of materials using microwave energy (Paré et al., 1994). Microwaves are made up of two oscillating fields that are perpendicular such as electric fields and magnetic fields and so called electromagnetic fields and are in the frequency range from 300 MHz to 300 GHz. In MAE system, electromagnetic energy is rapidly converted to heat and different chemical substances have different capacity to absorb microwaves, which makes MAE an efficient method for extractions of selectively targeted compounds from complex food matrices (Barbero et al., 2006; Hemwimon et al., 2007; Liazid et al., 2011). The extraction mechanism of MAE is supposed to involve three sequential steps: first, separation of solutes from active sites of sample matrix under increased temperature and pressure; second, diffusion of solvent across sample matrix; and third, release of solutes from sample matrix to solvent (Alupului, 2012; Table 2.4). The efficiency of the MAE process depends on extraction time, extraction temperature, solid–liquid ratio, and the type and composition of solvent used (Pizarro et al., 2006; Rostagno et al., 2007a; Song et al., 2011). Rapid heating for the extraction of bioactive substances from food matrices, reduced thermal gradients, reduced equipment size, and better recovery than conventional extraction processes are some of the advantages of MAE system (Cravottoa et al., 2008).

Mandal et al. (2007) explained the phenomenon of increase in TPC using MAE, after exposure of plant cells to microwave heating. Plant material absorbs microwave energy and subsequently converts into heat and the moisture begins to evaporate. This process of water vaporization generates pressure within the cell wall that eventually leads to cell rupture, thus facilitating the leaching out of active constituents into the surrounding solvent and improving extraction yield.

Extractions using solvent methanol, acetone, and hexane for both conventional and microwave-assisted methods were studied for total phenolic and tocopherol contents and free radical-scavenging capability of wheat bran (Oufnac et al., 2007). In comparison to conventional method, microwave-assisted solvent extraction using methanol significantly increases the total phenolic compound content to 467.5 and 489.5 μg of catechin equivalent, total tocopherol content to 18.7 and 19.5 μg, and free radical-scavenging capability to 0.064 and 0.072 μmol of trolox equivalent/g of wheat bran at extraction temperatures of 100 and 120°C, respectively.

An MAE method has been applied for the first time to the extraction of the phenolic compounds from rice grains. The effects of microwave power, temperature, extraction time, and solvent treatment were investigated. The extraction variables were optimized by the response surface methodology. It was shown that at 185°C extraction temperature, 1000 W microwave power, 20 min of extraction time, and 10:1 solvent-to-sample ratio resulted in higher yield of 15 phenolic compounds. The results demonstrated that MAE was more effective in terms of both yield and time consumption.

MAE to any number of materials can significantly reduce extraction time compared to conventional extraction methods. The MAE method is capable of extracting nearly 30% more phenolic compounds from peanut skins in 1/12th of the time required for SLE (Kerem et al., 2005; Martino et al., 2006).

TABLE 2.4 Operating Conditions and Salient Results for the Microwave-Assisted Extraction of Phenolic Compounds.

Source /product	Compound	Conditions	Salient results	Reference
Maize, wheat, barley and oilseed rape stems	Ferulic acid	Microwave digestion (750 W for 90 s) with 4 M NaOH	Microwave digestion was shown to be more effective than dioxane–HCl at liberating β-ether bound phenolic acids	Provan et al. (1994)
Soybean	Isoflavones	Solvents (methanol (MeOH) and ethanol (EtOH), 30–70% in water and water), temperatures (50–150°C), time (5–30 min), sample size (1.0–0.1 g) and optimized extraction conditions were: 0.5 g of sample, 50°C, 20 min and 50% ethanol as extracting solvent	Using optimized condition approximately 75% of total isoflavones were extracted in 10 min	Rostagno et al. (2007a)
Fava beans	Phenolic content	Microwave power 500 W; time 15, 30, 45, 60, and 75 s	The phenolic content of germinated flava seeds was increased by 700%	Randhir and Shetty (2004)
Peanut skins	Phenolic antioxidants	Microwave power, 10%, 50%, 90% nominal; irradiation time, 30, 90, 150 s; and sample mass, 1.5, 2.5, 3.5 g; and peanut skins were extracted with of 30% ethanol in water	The maximum total phenolic content (TPC) under the optimized conditions (90% microwave power, 30 s irradiation time and 1.5 g skins) was 143.6 mg gallic acid equivalent (GAE)/g skins	Ballard et al. (2010)
Cereal brans (wheat, rice and oat)	Total phenolic content	Solvent: methanol, acetone, and hexane; temperature, 50–70°C; microwave energy 2450 mHz for 3.5 min	Microwave-assisted solvent extraction significantly increased the total phenolic content in solvent	Dar and Sharma (2011)
Rice grains	Phenolic compounds	Temperature 125–175°C, microwave power 500–1000 W, time 5–15 min, solvent 10–90% EtOAc in MeOH, and solvent-to-sample ratio 10:1 to 20:1	MAE has been employed and optimized for the extraction of phenolic compounds from rice grains and optimized parameters showed significant yield of phenolic compound	Setyaningsih et al. (2015)

TABLE 2.4 *(Continued)*

Source /product	Compound	Conditions	Salient results	Reference
Wheat bran	Total phenolic compound	Solvents, methanol, acetone, and hexane; temperature, 100–120°C; microwave power, 500–1000 W	The total phenolic content was increased compared to conventional solvent extraction method	Oufnac et al. (2007)
Rye and wheat brans	Total phenolic content	Extraction at 10.3 MPa pressure and 80°C temperature, solvent:hexane, acetone, and methanol:water (80:20%)	Methanol:water extract showed highest antioxidant potential and particle size of ground bran have significant effect on yield	Povilaitis et al. (2015)
Barley	Total phenolic content	Roasting (280°C, 20 s)	8% loss in phenol content	Sharma and Gujral (2010)
		Microwave cooking (900 W, 120s)	49.6% loss in phenol content	
Sorghum and maize bran	Ferulic and coumaric acids	MAE (45 s, and 1400 W) in 2 M NaOH, temperature, 150–190°C	Temperature around 170°C is sufficient to release bound phenolics in presence of MAE	Chiremba et al. (2012)
Flaxseed	Lignans and phenolic acids	Microwave power 300 W, 70% methanol supplemented with 0.1 or 1 M NaOH	MAE was more effective in terms of both yield and time consumption compared to conventional method	Beejmohun et al. (2007)

Phenolic antioxidants were extracted using microwave-assisted system from peanut skins by Ballard et al. (2010). They investigated the effects of microwave power, irradiation time, and sample mass by RSM on TPC, ORAC level, and resveratrol content of peanut skin extracts using 30% ethanol in water. Optimized conditions were 90% microwave power, 30 s irradiation time, and 1.5 g skins with TPC, 143.6 mg GAE/g skins and showed highest ORAC and resveratrol values.

During extraction, particle size of the plant matrices plays important role. Smaller the size, larger will be the surface area to get exposed to the solvent resulting in increased extraction yield. In a study conducted by Povilaitis et al. (2015), rye and wheat brans were ground to different particle size fractions and extracted at 10.3 MPa pressure and 80°C temperature by consecutive application of hexane, acetone, and methanol:water (80:20%). The highest extract yield was obtained from rye bran using methanol–water. In most cases, particle size had a significant effect, as smaller particle size gives higher extraction yield.

Moreira et al. (2012) reported a novel application of MAE of phenolics from brewer's spent grains and extraction yield of ferulic acid was investigated through response surface methodology. At optimal conditions (15 min extraction time, 100°C extraction temperature, 20 mL of solvent, and maximum stirring speed), the yield of ferulic acid was 1.31 ± 0.04% (w/w), which was five fold higher than that obtained with conventional solid–liquid extraction techniques.

Randhir and Shetty (2004) reported that microwave treatment can significantly stimulate the phenolic antioxidant activity and Parkinson's relevant L-DOPA content of fava beans sprouts. Their study reveals that the phytopharmaceutical value was improved during germination by a microwave treatment of the seeds and the phenolic content of the germinated sprouts was increased by 700% and L-DOPA content by 59% compared to control.

Provan et al. (1994) used microwave digestion (750 W for 90 s) with 4 M NaOH to release esterified and etherified hydroxycinnamic acids from cell wall of maize, wheat, barley, and oilseed rapeseed stems. Digestion result provided a measure of β-ether linked units and it was shown that microwave digestion was more effective than dioxane–HCl at liberating β-ether bound phenolic acids.

Phenolic compound extraction was carried out from oat bran concentrate (OBC) by Stevenson et al. (2008) using integrated treatment of supercritical carbon dioxide (SC-CO_2), then microwave-irradiation (MI) at 50, 100, or 150°C for 10 min in water, 50% or 100% ethanol. Defatted OBC in 50% ethanol and MI at 150°C extracted higher phenolic content than any other combination.

Chiremba et al. (2012) performed the extraction of bound phenolic acids from sorghum and maize bran and endosperm with MAE (45 s and 1400 W) in 2M sodium hydroxide with the aim of releasing ferulic and coumaric acids at 190°C and found that this temperature is sufficient to break ether bonds which are labile at 170°C.

Similarly, extraction of lignans and phenolic acids using alkaline hydrolysis by MAE was compared with the conventional treatment. Lower extraction times and better yields were observed since alkaline hydrolysis of ferulic acid is accomplished in 6 h, whereas it is 15 min for MAEs. Furthermore, in dual treatment two fold increase in the yield (2.1–4.4 mg/g) of ferulic acid in the extract was observed (Beejmohun et al., 2007).

2.4.3 PULSED ELECTRIC FIELD (PEF) EXTRACTION

Application of high-voltage pulsed electric fields (PEFs) to plant tissues increases porosity of the plant cells, enhancing extraction of bioactive compounds from plant matrices (Table 2.5).

During PEF treatment, suspension of plant matrices is placed in a treatment chamber confined between electrodes. An electric potential passes through the membrane of that cell which separates molecules according to their charge in the cell membrane. Increasing electric potential beyond threshold value of approximately 1 V leads to repulsion between the charged molecules that creates pores in weak areas of the membrane and causes drastic increase of permeability (Vorobiev & Lebovka, 2008).

Typical systems for the treatment of pumpable fluids consist of a PEF generation unit, which is, in turn, composed of a high-voltage generator, a pulse generator, a treatment chamber, a suitable product handling system, and a set of monitoring and controlling devices (Soliva-Fortuny et al., 2009).

PEF process can operate in either continuous or in batch mode depending on the design of treatment chamber (Puértolas et al., 2010). The effectiveness of PEF treatment depends on various process parameters, including field strength, specific energy input, pulse number, treatment temperature, and properties of the sample to be treated (Heinz et al., 2003).

TABLE 2.5 Operating Conditions and Salient Results for the Pulsed Electric Field (PEF)-Assisted Extraction of Phenolic Compounds.

Source/product	Compound	Conditions	Salient results	Reference
Wine grapes	Polyphenolic content	0.5 to 2.4 kV/cm and 50 pulses	Low-intensity PEF increase the total polyphenolic content in fresh pressed grape juice by 13–28%	Balasa et al. (2006)
Merlot grapes	Polyphenols and anthocyanins	PEF pretreatment (500–700 V/cm) of the grapes with a short treatment duration (40–100 ms)	Application of a pulsed electric field PEF treatment resulted in permeabilization of merlot skin leading to increased extraction of polyphenols and anthocyanins	Delsart et al. (2012)
Tempranillo grapes	Anthocyanins	The PEF treatment consisted of 50 pulses at a frequency of 1 Hz and electric field strengths of 5 and 10 kV/cm	PEF remarkably enhanced the color intensity, anthocyanins, and index of total polyphenols with respect to the untreated control along the maceration carried out during vinification	López et al. (2008)
Grape (by-products- skins, stems, and seeds)	Anthocyanin	PEF (3 kV cm^{-1})	The application of PEF increased the antioxidant activity of the extracts because of increase in anthocyanin extraction using PEF by four fold than the control extraction	Corrales et al. (2008)
Soybeans	Isoflavonoids	1.3 kV/cm, 50 pulses, 1.857 kJ/kg 1.3 kV/cm, 20 pulses, 0.743 kJ/kg	Amount of isoflavonoid diadzein increased by 20% in comparison to the referent sample. A rise of 21% of isoflavonoid genistein in comparison to the reference was reached	Guderjan et al. (2005)
Rapeseed	Tocopherols polyphenols	5 kV/cm, 60 pulses 7 kV/cm, 120 pulses	Increased content of polyphenols and tocopherols, which are important antioxidants in rapeseed, most likely caused increased antioxidant capacity, measured in PEF-treated samples	Guderjan et al. (2007)

TABLE 2.5 *(Continued)*

Source/product	Compound	Conditions	Salient results	Reference
Apple	Polyphenols	1, 3 5 kV/cm, 30 pulses untreated control juice and juice after pectolytic mash treatment	Polyphenolic contents of PEF-treated samples with 1 and 3 kV/cm were insignificant, and slightly lower contents were partially observed after treatment of 5 kV/cm	Schilling et al. (2007)
Grapes	Juice and anthocyanins	3 kV/cm, 50 pulses	Total anthocyanin content was almost 3 times higher than of untreated grapes	Tedjo et al. (2002)
Tea leaves	Polyphenols	0.1 to 1.1 kV/cm, a pulse duration PD from 0.0001 to 0.1 s, number of pulses N from 10 to 50 and pause between pulses from 0.5 to 5 s.	Experimental results show that longer pulses are more effective and the PEF treatment accelerates kinetics of the extraction of polyphenolic compounds from fresh tea leaves	Zderic et al. (2013)
Cabernet Franc (CF) grapes.	Phenolics (anthocyanins and tannins)	PEF1 0.8 kV/cm, 100 ms, 42 kJ/kg, and PEF2 5 kV/cm, 1 ms, 53 kJ/kg	PEFs were the most effective pretreatments and increased the phenolics extraction yield (51 and 62% by PEF1 and PEF2, respectively) compared to moderate thermal (20%), ultrasound (7%)	Darra et al. (2013)
Blueberry	Total phenolic content	PEF treatments carried out at field strengths of 1, 3, and 5 kV/cm and an energy input of 10 kJ/kg	The juice and press cake extracts of berries obtained after PEF pretreatment had a significantly higher total phenolic content (+43%), total anthocyanin content (+60%), and antioxidant activity (+31%) up to 1 kV/cm PEF	Bobinaitė et al. (2014)
Flaxseed hulls	Polyphenols	PEF treatment applied at 20 kV/cm for 10 ms, rehydrated product for 40 min before PEF, and addition of solvent such as 20% of ethanol and 0.3 mol/L hydroxide sodium	PEF treatment allowed the extraction of up to 80% of polyphenols. Use of solvent (mostly alkaline hydrolysis) was more effective than the acidic hydrolysis	Boussetta et al. (2014)

Balasa et al. (2006) investigated secondary metabolite production together with extractability of plant pigments (anthocyanin) from wine grapes (*Vitis vinifera*) induced by low intensity PEFs. Increase of total polyphenolic content (13–28%) in fresh pressed grape juice obtained from PEF-treated grapes (0.5–2.4 kV/cm and 50 pulses) was observed. It was noticed that higher the electric field strength, higher is the degree of cell membrane permeabilization and subsequent release of intracellular compounds. Pretreatment of PEFs before mechanical pressing or solvent extraction of rapeseed and its impact on functional food ingredients were studied by Guderjan et al. (2007). Application of PEF treatments (60 pulses of 5 kV/cm, and 120 pulses of 7 kV/cm; pulse duration 30 ms) increases the oil yield by pressing as well as with solvent extraction and the TPC of PEF-treated samples after pressing was three times higher than that of nontreated rapeseed.

In another study, Tedjo et al. (2002) reported PEF-induced cell permeabilization and release of intracellular pigments (anthocyanins) from wine grapes. It was found that after applying electric field strength (3 kV/cm and 50 pulses), 65% of the total membrane area was permeabilized which increased the total anthocyanin content by three fold compared to untreated sample.

Guderjan et al. (2005) showed that the recovery of isoflavonoids (genistein and daidzein) from soybeans increased by 20–21% when PEF was used as pretreatment process. Corrales et al. (2008) investigated the extraction of bioactive compound such as anthocyanins from grape by-product using various techniques and their result showed that PEF gives better extraction of anthocyanin monoglucosides. Application of PEF treatment resulted into increase in permeabilization of Merlot skin which increased the extraction of phenolics and anthocyanins (Delsart et al., 2012). The use of a PEF treatment on grape skin before maceration step can reduce the duration of maceration and improve the stability of bioactives (anthocyanin and phenolics) during vinification (López et al., 2008).

Boussetta et al. (2014) extracted high level of phenolics from flaxseed hulls using PEFs treatment and studied the effect of the different operating parameters on the extraction of phenolics including the PEF treatment duration, the PEF electric-field strength, the solvent composition (ethanol, acid, or base content), and the rehydration duration of the product. Extraction of phenolics increased up to 80% when PEF treatment applied at 20 kV/cm for 10 ms. Rehydration of the product before PEF and addition of ethanol, citric acid, and sodium hydroxide have increased the extraction of phenolics.

Darra et al. (2013) investigated the effect of pretreatment such as moderate thermal (MT), US, and PEF with various conditions on extraction yield of phenolics from Cabernet Franc grapes. The results showed that in all the study, pretreatments improved phenolics extraction (anthocyanins and tannins content); however, the moderate (0.8 kV/cm) and high (5 kV/cm) PEFs were the most effective pretreatments and increased the phenolics extraction yield by 51% and 62%, respectively, while the MT and US3 pretreatments increased the phenolics extraction yield by 20% and 7%, respectively.

Araujo et al. (2000) studied the use of supercritical carbon dioxide for extraction and concentration of tocopherols from soybean sludge at temperature of 80°C and pressure of 76 bars to remove interferers during pre extraction, and 50°C at 197 bars to extract tocopherols. The results showed that soybean sludge initially containing about 9.2% of tocopherols could be enriched to about 40.6% under optimized conditions in 37 min.

2.4.4 ENZYME-ASSISTED EXTRACTION (EAE)

Enzyme-assisted extraction (EAE) of bioactive compounds from plant matrices is a potential alternative to conventional solvent-based extraction methods. These methods are gaining more attention because of the need for ecofriendly extraction technologies. Some phytochemicals in the plant matrices are dispersed in cell cytoplasm and some compounds are retained in the polysaccharide–lignin network by hydrogen or hydrophobic bonding, which are not accessible with a solvent in a routine extraction process (Table 2.6). Enzymatic pretreatment has been considered as a novel and an effective way to release bounded compounds and increases overall yield by breaking the cell wall and hydrolyzing the structural polysaccharides and lipid bodies (Rosenthal et al., 1996; Singh et al., 1999). Application of enzymes in extraction improves the effect of solvent pretreatment by increasing the yield of extractable compounds and reducing the amount of solvent needed for extraction. Various factors including enzyme composition and concentration, particle size of plant materials, solid to water ratio and hydrolysis time are recognized as key factors for extraction (Niranjan & Hanmoungjai, 2004).

TABLE 2.6 Enzyme Used and Salient Results of Enzyme-Assisted Extraction of Phenolic Compounds.

Source	Enzyme	Compound	Sailent results	Reference
Kinnow peel	A-L-Rhamnosidase	Naringin (prunin and rhamnose)	An enzyme that catalyzes the cleavage of terminal rhamnosyl groups from naringin to yield prunin and rhamnose	Puri et al. (2012)
5 Citrus peels (Yen Ben lemon, Meyer lemon, grapefruit, mandarin and orange)	Cellulase® MX, Cellulase® CL, and Kleerase® AFP	Total phenolic contents	Cell wall-degrading enzymes broke down the integrity of the cell walls of the citrus peels and eased the extraction. Celluzyme MX showed highest extraction (65.5%)	Li et al. (2006)
Grape pomace	Pectinolytic and cellulolytic enzymes	Anthocyanin and other phenolic compounds	Pre extraction of the pomace with hot water followed by treatment with cell wall-degrading enzymes increased the yields of phenolic compounds	Kammerer et al. (2005)
Blueberries	Econase CE, Biopectinase, Biopectinase, Pectinex BE-3L	Anthocyanins and polyphenolics	The enzyme-aided processing affected the flavonol extractability, improved yield of flavonol was found in juices while it was decreased in press residues	Koponen et al. (2008)
Black currant and bilberry	Commercial juice-processing enzymes	Flavonols	Enzymes were helpful in extraction of anthocyanin and total phenolics from skins of fruit	Lee and Wrolstad (2004)
Apple peel	Cellulase	Phenolics	The overall phenolic release was significantly increased in a synergistic manner with combined pretreatment, instead of individual pretreatment (such as boiling, acid, and pectinase treatment)	Kim et al. (2005)
Black currant	Grindamyl pectinase, Macer8 FJ, Macer8 R, and Pectinex BE, Novozym 89 protease	Antioxidative phenols (anthocyanins)	Enzyme pretreatment decreased particle sizes in pomace from 500–1000 μm to <125 μm resulting into increase in the phenol yields by 1.6–5 times	Landbo and Meyer (2001)

TABLE 2.6 *(Continued)*

Source	Enzyme	Compound	Sailent results	Reference
Grape pomace	Grindamyl pectinase	Phenolic compounds	Reduction in the particle size of grape pomace to 125–250 μm increased the enzymatic polysaccharide hydrolysis and enhanced the recovery of phenols	Meyer et al. (1998)
Grapes	Celluclast 1.5 L, Pectinex Ultra, and Novoferm		The highest antioxidant activities registered were 86.8%, 82.9%, and 90% at 12 h for Celluclast 1.5 L, Pectinex Ultra and Novoferm, respectively, it was due to the release of *O*-coumaric acid by enzymatic treatment	Gómez-García et al. (2012)
Grape pomace	Mixture of pectinolytic and cellulolytic enzyme (ratio 2:1)	Phenolic acids, nonanthocyanin flavonoids, and anthocyanins	Enzymatic treatment resulted in significantly improved extraction yields 91.9% for phenolic acids, 92.4%, and 63.6% for nonanthocyanin flavonoids and anthocyanins, respectively	Maier et al. (2008)
Tea beverages	Pepsin	Catechins	Recovery of total catechins was highest for pepsin-treated tea samples (89–102%) compared to methanol deproteination (78–87%) and acid precipitation (20–74%)	Ferruzzi and Green (2005)

Recent studies on EAE have shown faster extraction, higher recovery, reduced solvent usage, and lower energy consumption when compared to nonenzymatic methods. Various enzymes such as pectinases, cellulases, and hemicellulases are widely used in alcoholic beverage and juice-processing industries for clarification to degrade cell walls and improve juice extractability. During processing, the disruption of the cell wall matrix also releases components such as phenolic compounds into the juice, thus improving product quality. EAE methods have been shown to achieve high extraction yields for compounds including polysaccharides, oils, natural pigments, flavors, and medicinal compounds (Passos et al., 2009; Barzana et al., 2002; Sowbhagya & Chitra, 2010).

Enzyme application on agroindustrial residues with the aim to increase phenolic compounds has been studied. Meyer et al. (1998) showed an increase of phenolic compounds extracted from grape pomace by means of enzyme incorporation. They found a correlation between yield of total phenols and degree of plant cell wall breakdown by enzyme.

EAE of phenolics can be an alternative to the application of sulfite-assisted extraction process from grape pomace. Maier et al. (2008) observed that a mixture of pectinolytic and cellulolytic enzyme preparations (ratio 2:1) yielded the highest amounts of phenolic compounds and results showed aqueous pre extraction of the pomace followed by enzymatic treatment resulted in significantly improved extraction yields about 91.9%, 92.4%, and 63.6% for phenolic acids, nonanthocyanin flavonoids, and anthocyanins, respectively. Kammerer et al. (2005) developed a novel process for enzyme-based extraction of phenolics from grape pomace. Optimization of enzymatic hydrolysis of grape skins was carried out for parameters such as selection of pectinolytic and cellulolytic enzymes, enzyme–substrate ratio, and time–temperature regime of enzymatic treatment and at optimal conditions 5000 ppm of a pectinolytic and 2500 ppm of a cellulolytic enzyme preparation showed higher extraction of phenolics at 50°C.

Another important finding of that study was that the extraction of phenolic antioxidants improved significantly with higher enzyme concentration. Laroze et al. (2010) found that application of enzyme in hydro-alcoholic extraction of phenolic antioxidant from raspberry solid waste was increased compared to control (nonenzymatic). Gómez-García et al. (2012) used commercial enzymes such as Celluclast 1.5 L, Pectinex Ultra, and Novofermto to release phenolic compounds from grape wastes.

They found a good correlation between antioxidant activity and phenolics released. Ultra and Novoferm showed the highest antioxidant activities, that is, 90 ± 0.37% while for Celluclast 1.5 L and Pectinex, it was 86.8 ± 0.81 and 82.9 ± 0.31, respectively.

2.4.5 HIGH-HYDROSTATIC PRESSURE EXTRACTION (HHPE)

High-hydrostatic pressure extraction (HHPE) is currently considered as an attractive, innovative, energy-efficient, environment-friendly technique used for the high-pressure processing (HPP) of foods in the extraction of functional ingredients from natural products (Knorr, 1993; Toepfl et al., 2006; Huang et al., 2013; Table 2.7). The application of high hydrostatic pressure for processing food products consists of utilization of nonthermal super-high hydraulic pressure between 1000 and 8000 bars or even higher (Cretnik et al., 2005; Toepfl et al., 2006).

HHP as an extraction technique showed excellent advantages in the field of natural product extraction. Unlike the other two high-pressure extraction techniques, namely, supercritical fluid extraction, which is usually performed at pressures from 20 to 80 MPa and temperature range of 40–80°C (Cretnik et al., 2005), and PLE, which is performed at pressures from 5–10 MPa and at temperatures of 50–200°C, the HHP technique could be operated at room temperature without any heating treatment.

According to phase behavior theory, solubility of compound increases with increase in pressure, this condition favors an increase on the permeability of the cells of the food being extracted, also enhancing the values of mass transfer rates using this techniques (Cheftel, 1995; Knorr, 1999; Aertsen et al., 2009). The pressure applied increases plant cell permeability, leading to diffusion of cell component according to mass transfer and phase behavior theories (Shouqin et al., 2005; Khoddami et al., 2013). Requirement of expensive equipment is the main disadvantage of HHP which includes a solvent transporting pump, a pressure vessel and system controller, and a collection device for the extract (Smith, 2002). However, in the case of phenolic, which is a low-volume, high-cost product and is in great demand, extraction methods with high purity and processing efficiency are expected. The cost of equipment might not play a critical role in selection of these methods.

HHPE creates a large pressure difference between the cell membrane interior and exterior and allows solvent to penetrate into the cell causing

TABLE 2.7 Operating Conditions and Salient Results for High Hydrostatic Pressure (HHP)-Assisted Extraction of Phenolic Compounds.

Source	Compound	Condition	Sailent results	Reference
Maclura pomifera fruits	Phenolic compounds	HHP (100–800 MPa/10 min), solvent cocktail (dH_2O:ethanol:methanol:acetone:CH_2Cl_2), methanol	The highest amount of phenolic compounds (913.173 μg gallic acid equivalent (GAE/mL) was observed in HHPE at 500 MPa using solvent cocktail	Altuner et al. (2012)
Strawberry and blackberry purées	Anthocyanin (cyanidin-3-glycoside, pelargonidin-3-glucoside)	400, 500, 600 MPa/15 min/10–30 °C and thermal treatments (70 °C/2 min)	No significant changes in anthocyanins were observed between pressure-treated and unprocessed purées ($p > 0.05$)	Patras et al. (2009)
Green tea	Polyphenols	Solvents acetone, methanol, ethanol and water; pressure 100, 200, 300, 400, 500, 600 MPa; holding time 1, 4, 7, 10 min; ethanol concentration 0–100% mL/mL; and liquid/solid ratio 10:1 to 25:1 mL/g	At optimum condition the extraction yield of polyphenols was about 30 ± 1.3% for HHPE in 1 min, while same yield was observed for extraction at room temperature for 20 h, ultrasonic extraction for 90 min and heat reflux extraction for 45 min	Jun et al. (2009)
Cashew apples	Polyphenols	250 or 400 MPa for 3, 5 and 7 min	HPP at 250 or 400 MPa for 3, 5 and 7 min did not change pH, acidity, total soluble solids, ascorbic acid, or hydrolysable polyphenol contents. HPP can be used in the food industry for the generation of products with higher nutritional quality	Queiroz et al. (2010)
Strawberry	Anthocyanins	200, 400, 600, and 800 MPa for 15 min at a temperature controlled between 18 and 22°C	High-pressure treatment at 800 MPa led to the highest stability of the anthocyanins when strawberries were stored at a temperature of 4°C	Zabetakis et al. (2000)

TABLE 2.7 *(Continued)*

Source	Compound	Condition	Sailent results	Reference
Propolis	Flavanols	Ethanol concentration 35–95%, v/v, HHPE pressure 100–600 MPa, HHPE time 1–10 min, and solid/liquid ratio (1:5–1:45 g cm^{-3})	Extraction yield with HHPE for 1 min was higher than those using extraction at room temperature for 7 days and heat reflux extraction for 4 h, respectively. HHPE was more effective than the conventional extraction methods from the viewpoints of extraction time, the extraction efficiency, and the extraction yield of flavonoids	Shouqin et al. (2005)
Grape skins	Anthocyanins	Combined heat treatment at 70°C and high hydrostatic pressure (600 MPa)	The extraction of anthocyanine was found to increase three fold with HHP compared to control	Corrales et al. (2009)
Propolis	Polyphenols			Shouqin et al. (2005)
Litchi fruit (pericarp)	Flavonoids	HHPE at 200 and 400 MPa for 30 min and solvent—ethanol:HCl (85:15)	The application of HPE has demonstrated a potential for a higher extraction efficiency of flavonoids from LFP tissues with 40% yield of flavonoids compared to 1.83% of conventional extraction	Prasad et al. (2009)
Orange juice	Flavanone	400 MPa/40°C/1 min	HHP technologies were found more effective for extraction of flavonoids (15.46%) from orange juice during refrigerated storage	Plaza et al. (2011)

leakage of cell components (Fernandez-Garcia et al., 2001), which also cause deprotonation of charged groups and disruption of salt bridges and hydrophobic bonds, resulting in conformational changes and denaturation of proteins making the cellular membranes less and less selective, thereby rendering the compounds more accessible to extraction up to equilibrium (Corrales et al., 2009; Jun et al., 2009).

It had been successfully used for the extraction of flavonoid from propolis (Shouqin et al., 2005), anthocyanins from grape by-products (Corrales et al., 2008), and phenolic compounds from *Maclura pomifera* fruits (Altuner et al., 2012). Higher yields of phenolic compounds have been obtained using HHPE compared with conventional extraction methods.

Queiroz et al. (2010) evaluated the impact of HPP phenolic compound and ascorbic acid contents and antioxidant capacity of cashew apple juice at room temperature. Their study showed that HPP at 250 or 400 MPa for 3, 5, and 7 min did not change pH, acidity, total soluble solids, ascorbic acid, or hydrolysable polyphenol contents. However, juice pressurized for 3 and 5 min showed higher soluble polyphenol contents.

Patras et al. (2009) assessed the effect of high-pressure treatments and conventional thermal processing on antioxidant activity levels of key antioxidant groups (phenolics, ascorbic acid, and anthocyanins) and the color of strawberry and blackberry purées. Their study found that no significant changes in anthocyanins were observed between pressure-treated and unprocessed purées ($p > 0.05$) at HHP treatment (400, 500, 600 MPa/15 min/10–30°C) and thermal treatments (70°C/2 min), whereas conventional thermal treatments significantly reduced the levels ($p < 0.05$).

Altuner et al. (2012) observed that HHPE in solvent cocktail was the most effective method when compared to the other methods for extraction of phenolics from *Maclura pomifera* fruits at 500 MPa for 10 min. Prasad et al. (2009) reported that high-pressure treatment at 400–600 MPa obtained significantly more phenols, anthocyanins, and ascorbic acid from strawberry purées than thermal treatment. In blackberry purées, higher extraction of anthocyanins was observed in comparison to thermally treated purées. Shouqin et al. (2005) observed that the HHPE is suitable for fast extraction of flavonoids from propolis because of higher extraction yield, higher extraction selectivity, short extraction time, and being less labor intensive compared to conventional methods. Their results showed extraction yield using HHPE for 1 min was higher than that of those using extraction at room temperature for 7 days and heat reflux extraction for

4 h, respectively. Plaza et al. (2011) also found an increase in the total concentration of flavanone (15.46%) extracted from orange juice after applying such treatment (400 MPa/40°C/1 min).

2.4.6 PRESSURIZED LIQUID EXTRACTION (PLE)

PLE is widely considered as an advanced extraction technique based on the extraction of solid materials with solvents at high temperatures and high pressures, enough to maintain the solvents in the liquid state during the whole extraction procedure (Table 2.8). Richter et al. (1996) was the first to introduce the concept of PLE. This method is now known by several names, pressurized fluid extraction (PFE), accelerated solvent extraction (ASE), enhanced solvent extraction (ESE), and high-pressure solvent extraction (HSPE) (Nieto et al., 2013). Pressurized liquid extraction is termed as "green" technology for the extraction of bioactive compounds from foods and plant matrices.

The concept of PLE is the application of high pressure to maintain solvent liquid beyond its normal boiling point which facilitates the extraction process. PLE technique requires small amounts of solvents because of the combination of high pressure (>1000 psi) and temperature and provides faster extraction. Use of high extraction temperature in PLE is characterized with increase in the analyte solubility and mass transfer rate and decrease in the viscosity and surface tension of solvents, thus allowing better penetration into the matrix and then improving extraction rate (Richter et al., 1996, Ibanez et al., 2012). Advantages of PLE include improvement in the extraction yield, decrease in time and solvent consumption, and protection of sensitive compounds. It also facilitates the use of solvent mixtures and other extraction additives that would enhance the extraction efficiency. ASE method has potential for industrial extraction of phenolic compounds from plant matrices to be used in colorants, foods, and beverages with high levels of antioxidants for health benefits (Kaufmann & Christen, 2002; Ibanez et al., 2012; Barros et al., 2013).

PLE has been successfully applied to extract various phenolic compounds from different plant materials. PFE helps to increase the extraction rates of colorants (anthocyainins, carotenoids, betanines, etc.) and bioactive compounds (phenolics) from foods and food by-products and it also shortens extraction time and reduces solvent consumption and/

TABLE 2.8 Operating Conditions and Salient Results for Pressurized Liquid Extraction of Phenolic Compounds.

Source/ product	Compound	Conditions	Salient results	Reference
Jabuticaba skins	Anthocyanins	Pressure 5–10 MPa, temperature 313–393 K, and static extraction time 3–15 min	Compared to LPSE, PLE under optimized conditions yields 2.15-and 1.66-fold more anthocyanins and total phenolic compounds, respectively, with 40-fold lower cost	Santos et al. (2012)
Grapes	Trans-resveratrol	Water at 40°C and 40 atm of pressure (three cycles of 5 min), and then with methanol at 150°C and 40 atm (three cycles of 5 min)	The time required for the extraction stage is reduced to some 40 min compared with the traditional methods of maceration of solid samples	Pineiro et al. (2006)
Anatolia propolis	Gallocatechin (GCT), catechin, epicatechingallate, caffeic acid, chlorogenic acid, and myricetin and total phenolic contents	40°C, 1500 psi, ethanol:water:HCl; (70:25:5, v/v/v) containing 0.1% *tert*-butylhydroquinone (TBHQ)	PLE using optimized conditions results into recovery of polyphenols in the range of 97.2–99.7%	Erdogan et al. (2011)
Parsley (*Petroselinum crispum*)	Phenolic compound (apiin and malonyl-apiin)	Solvents: ethanol:water, 50:50, v/v and/or acetone:water, 50:50, v/v)	Various parameters including temperature, particle size, and solid-to-solvent ratio have significant effect on the extraction yield of the phenolic compounds	Luthria et al. (2008)

or lowers extraction temperatures (Boussetta et al., 2012; Puértolas et al., 2013). Using optimized conditions, isoflavones were extracted from soybeans (freeze-dried) without degradation by PLE (Rostagno et al., 2004).

The feasibility of PLE to extract phenolic compounds from plant material is clearly demonstrated in some studies. Barros et al. (2013) reported the application of ASE at temperatures above 100°C using water and ethanol/water significantly improved extraction of phenolic compounds from black sorghum bran compared to extractions using aqueous acetone and acidified methanol. Flavonoid extraction from spinach by PLE using a mixture of ethanol and water (70:30) solvent at 50–150°C was more effective than water solvent at 50–130°C (Howard & Pandjaitan, 2008). Luthria (2008) showed that temperature, pressure, particle size, flush volume, static time, and solid-to-solvent ratio parameters influenced the extraction of phenolic compounds from parsley (*Petroselinum crispum*) flakes by PLE. In similar studies, several extraction parameters of PLE such as temperature, pressure, solvent type, extraction time, and cell size were investigated for their effects on the extraction performances. Individual phenolic compounds such as gallocatechin, catechin, epicatechingallate, caffeic acid, chlorogenic acid, and myricetin and TPCs were recovered from various parts of *Anatolia propolis* using PLE at optimum condition (40°C, 1500 psi, ethanol:water:HCl (70:25:5, v/v/v) containing 0.1% *tert*-butylhydroquinone as solvent, three extraction cycles within 15 min, and a cell size of 11 mL) (Erdogan et al., 2011).

2.4.7 SUPERCRITICAL FLUID EXTRACTION

Supercritical fluid extraction (SFE) is a novel method with green approach behind its development for the extraction of solid materials using a supercritical fluid. SFE has been successfully used in environmental, pharmaceutical, and polymer applications and food analysis (Zougagh et al., 2004; Pourmortazavi & Hajimirsadeghi, 2007; Sahena et al., 2009). In supercritical state, the specific properties of gas and/or liquid disappear, so that supercritical fluid cannot be liquefied by modifying temperature and pressure. Also supercritical fluid possesses gas-like properties such as diffusion, viscosity, and surface tension, and liquid-like density and solvation power. These properties make it suitable for extracting compounds in a short time with higher yields (Shivonen et al., 1999).

In SFE, carbon dioxide is the most widely used solvent because it is nontoxic, nonflammable, cheap, easily eliminated after extraction, and endowed with a high solvating capacity for nonpolar molecules. Other solvents are freon, ammonia, and some organic solvents (Luque de Castro et al., 1994). In a typical SFE procedure, the supercritical fluid continuously enters the solid matrix where it dissolves the material of interest. The extraction can be achieved with a remarkably high selectivity by adjusting the solvating capacity of the supercritical fluid by changing the pressure or temperature. Major advantages of SFE include pre-concentration effects, cleanliness, safety, and simplicity (Luque de Castro & Jime´nez-Carmona (2000). CO_2-SFE is a nonconventional technique that can offer very good yields and is suitable for varied range of phenolics.

More expensive equipment with the difficulty of extracting polar molecules without adding modifiers to CO_2 are the drawbacks of SFE. The limitation of low polarity of carbon dioxide has been successfully overcome by the use of chemical modifier to enhance the polarity of CO_2 (Lang & Wai, 2001; Ghafoor et al., 2010). The properties of sample and targeted compounds and the previous experimental results are main basis for selection of the best modifier. In supercritical state, the specific properties of gas and/or liquid vanish. Usually, a small amount of modifier is considered as useful to significantly enhance the polarity of carbon dioxide. For example, 0.5 mL of dichloromethane can enhance the extraction which is same for 4 h hydrodistillation (Hawthorne et al., 1994). The properties of sample and targeted compounds and the previous experimental results are the main basis for selection of the best modifier. The successful extraction of phenolic compounds from plant materials relies upon several parameters including temperature, pressure, particle size, and moisture content of feed material, time of extraction, flow rate of CO_2, and solvent-to-feed-ratio (Ibanez et al., 2012).

The advantages of using supercritical fluids for the extraction of phenolics are as follows: supercritical fluid has a higher diffusion coefficient and lower viscosity and surface tension than a liquid solvent, higher penetration to sample matrix and favorable mass transfer, low extraction time, repeated reflux of supercritical fluid to the sample, higher selectivity of supercritical fluid, ideal method for thermolabile compound extraction, recycling and reuse of supercritical fluid is possible, small amount of sample is sufficient and is an environment-friendly method.

Rostagno et al. (2002) found the low extraction of isoflavones using SC-CO_2 from soybean flour. They evaluated the extraction of soybeans isoflavones (genistin, genistein, and daidzein) at different temperatures, pressures, and modifier percentages and compared with conventional extraction methods such Soxhlet and ultrasonification. The maximum amount of total isoflavonoids extracted by each method was 311.55 μg/g for ultrasonification, 212.86 μg/g for Soxhlet, and 86.28 μg/g for SC-CO_2.

2.5 CONCLUSION

Since the use of conventional methods for extracting the bioactive phenolics implies the consumption of excessive hazardous solvents and requires long extraction time, new methods such as those discussed in this chapter have promised the renewed approaches of bioactive extraction. However, most of these techniques need installation of specialized equipment which attracts set-up and maintenance costs resulting in increased cost of operation.

KEYWORDS

- **polyphenolic compounds**
- **flavonoids**
- **Soxhlet extraction**
- **maceration**
- **ultrasonic radiation**

REFERENCES

Aaby, K.; Mazur, S.; Nes, A.; Skrede, G. Phenolic Compounds in Strawberry (Fragaria x ananassa Duch.) Fruits: Composition in 27 Cultivars and Changes during Ripening. *Food Chem.* **2012,** *132*, 86–97.

Adom, K. K.; Liu, R. H. Antioxidant Activity of Grains. *J. Agric. Food Chem.* **2002,** *50*, 6182–6187.

Aertsen, A.; Meersman, F.; Hendrickx, M. E. G.; Vogel, R. F.; Michiels, C. W. Biotechnology under High Pressure: Applications and Implications. *Trends Biotechnol.* **2009,** *27*(7), 434–441.

Altuner, E.M.; Işlek, C.; Çeter, T.; Alpas, H. High Hydrostatic Pressure Extraction of Phenolic Compounds from *Maclura pomifera* Fruits. *Afr. J. Biotechnol.* **2012,** *11*, 930–937.

Alu'datt, M. H.; Rababah, T.; Ereifej, K.; Alli, I. Distribution, Antioxidant and Characterisation of Phenolic Compounds in Soybeans, Flaxseed and Olives. *Food Chem.* **2013,** *139*, 93–99.

Alupului, A. Microwave Extraction of Active Principles from Medicinal Plants. *U.P.B. Sci. Bull. Series B.* **2012,** *74*, 129–142.

Araujo, J. M. A.; Nicolino, A. P. N.; Blatt, C. Utilization of Supercritical Carbon Dioxide for Concentration of Tocopherols from Soybean Oil Deodorizer Distillate. *Pesq. Agropec. Bras* **2000,** *35*, 201–205.

Arranz, S.; Calixto, F. S. Analysis of Polyphenols in Cereals may be Improved Performing Acidic Hydrolysis: A Study in Wheat Flour and Wheat Bran and Cereals of the Diet. *J. Cereal Sci.* **2010,** *51*, 313–318.

Auerbach, R. H.; Gray, D. A. Oat Antioxidant Extraction and Measurement—Towards a Commercial Process. *J. Agric. Food Chem.* **1999,** *79*(3), 385–389.

Azmir, J.; Zaidul, I. S. M.; Rahman, M. M.; Sharif, K. M.; Mohamed, A.; Sahena, F.; Jahurul, M. H. A.; Ghafoor, K.; Norulaini, N. A. N.; Omar, A. K. M. Techniques for Extraction of Bioactive Compounds from Plant Materials: A Review. *J. Food Eng.* **2013,** *117*, 426–436.

Balasa, A.; Guderjan, M.; Janositz, A.; Volkert, M.; Knorr, D. Keine Qualitätsverluste-Schonende Aufbereitung und Verarbeitung von bioaktiven Inhaltsstoffen. *dei-dieErnährungsindustrie* **2006,** *1*, 10–13.

Ballard, T. S.; Mallikarjunan, P.; Zhou, K.; O'Keefe, S. Microwave-Assisted Extraction of Phenolic Antioxidant Compounds from Peanut Skins. *Food Chem.* **2010,** *120*, 1185–1192.

Barbero, G. F.; Palma, M.; Barroso, C. G. Determination of Capsaicinoids in Peppers by Microwave-Assisted Extraction-High Performance Liquid Chromatography with Fluorescence Detection. *Anal. Chim. Acta* **2006,** *5*, 227–233.

Barreira, J. C. M.; Ferreira I. C. F. R.; Oliveira, M. B. P. P.; Pereira, J. A. Antioxidant Activities of the Extracts from Chestnut Flower, Leaf, Skin and Fruit. *Food Chem.* **2008,** *107*, 1106–1113.

Barros, F.; Dykes, L.; Awika, J. M.; Rooney, L. W. Accelerated Solvent Extraction of Phenolic Compounds from Sorghum Brans. *J. Cereal Sci.* **2013,** *58*, 305–312.

Barzana, E.; Rubio, D.; Santamaria, R. I.; Garcia-Correa, O.; Garcia, F.; RidauraSanz, V. E.; López-Munguía, A. Enzyme-Mediated Solvent Extraction of Carotenoids from Marigold Flower (*Tagetes erecta*). *J. Agric. Food Chem.* **2002,** *50*, 4491–4496.

Beecher, G. R. Overview of Dietary Flavonoids: Nomenclature, Occurrence and Intake. *J. Nutr.* **2003,** *133*, 3248S–3254S.

Beejmohun, V.; Fliniaux, O.; Grand, É.; Lamblin, F.; Bensaddek, L.; Christen, P.; Kovvensky, J.; Fliniaux, M.; Mensnard, F. Microwave-Assisted Extraction of the Main Phenolic Compounds in Flaxseed. *Phytochem. Anal.* **2007,** *18*, 275–282.

Belitz, H. D.; Grosch, W. Phenolic Compounds. In *Food Chemistry*; Springer: Berlin, 1999; pp 764–775.

Bobinaitė, R.; Pataro, G.; Lamanauskas, N.; Šatkauskas, S.; Viškelis, P.; Ferrari, G. Application of Pulsed Electric Field in the Production of Juice and Extraction of Bioactive Compounds from Blueberry Fruits and Their By-products. *J. Food Sci. Technol.* **2014**. DOI:10.1007/s13197-014-1668-0.

Bonoli, M.; Verardo, V.; Marconi, E.; Caboni, M. F. Antioxidant Phenols in Barley (*Hordeum vulgare* L.) Flour: Comparative Spectrophotometric Study among Extraction Methods of Free and Bound Phenolic Compounds. *J. Agric. Food Chem.* **2004a,** *52*, 5195–5200.

Bonoli, M.; Verardo, V.; Marconi, E.; Caboni, M. F. Free and Bound Phenolic Compounds in Barley (*Hordeum vulgare* L.) Flours Evaluation of the Extraction Capability of Different Solvent Mixtures and Pressurized Liquid Methods by Micellar Electrokinetic Chromatography and Spectrophotometry. *J. Chromatogr. A* **2004b,** *1057*, 1–12.

Boussetta, N.; Soichi, E.; Lanoisellé, J. L.; Vorobiev, E. Valorization of Oilseed Residues: Extraction of Polyphenols from Flaxseed Hulls by Pulsed Electric Fields. *Ind. Crops Prod.* **2014,** *52*, 347–353.

Boussetta, N.; Vorobiev, E.; Le, L. H.; Cordin-Falcimaigne, A.; Lanoisselle, J. L. Application of Electrical Treatments in Alcoholic Solvent for Polyphenols Extraction from Grape Seeds. *LWT—Food Sci. Technol.* **2012,** *46*, 127–134.

Canals, R.; Llaudy, M. C.; Valls, J.; Canals, J. M.; Zamora, F. Influence of Ethanol Concentration on the Extraction of Color and Phenolic Compounds from the Skin and Seeds of Tempranillo Grapes at Different Stages of Ripening. *J. Agric. Food Chem.* **2005,** *53*, 4019–4025.

Cao, X.; Wang, C.; Pei, H.; Sun, B. Separation and Identification of Polyphenols in Apple Pomace by High-Speed Counter-Current Chromatography and High-Performance Liquid Chromatography Coupled with Mass Spectrometry. *J. Chromatogr. A.* **2009,** *1216*, 4268–4274.

Carvalho, I. S.; Cavaco, T.; Brodelius, M. Phenolic Composition and Antioxidant Capacity of Six Artemisia Species. *Ind. Crop Prod.* **2011,** *33*, 382–388.

Caton, P. W.; Nayuni, N. K.; Kieswich, J.; Khan, N. Q.; Yaqoob, M. M.; Corder, R. Metformin Suppresses Hepatic Gluconeogenesis through Induction of SIRT1 and GCN5. *J. Endocrinol.* **2010,** *205*, 97–106.

Cetkovic, G.; Savatovic, S.; Canadanovic-Brunet, J.; Djilas, S.; Vulic, J.; Mandic, A.; Cetojevic-Simin, D. Valorisation of Phenolic Composition, Antioxidant and Cell Growth Activities of Tomato Waste. *Food Chem.* **2012,** *133*, 938–945.

Cheftel, J. C. Review: High Pressure, Microbial Inactivation and Food Preservation. *Food Sci. Technol. Int.* **1995,** *1*, 75–90.

Chemat, F.; Tomao, V.; Virot, M. Ultrasound-Assisted Extraction in Food Analysis. In *Handbook of Food Analysis Instruments*; Otles, S., Ed.; CRC Press: Boca Raton, FL, 2008; pp 85–94.

Chen, C. S.; Zhang, D.; Wang, Y. Q.; Li, P. M.; Ma, F. W. Effects of Fruit Bagging on the Contents of Phenolic Compounds in the Peel and Flesh of "Golden Delicious", "Red Delicious", and "Royal Gala" Apples. *Sci. Hortic.* **2012,** *13*, 68–73.

Chen, F.; Sun, Y.; Zhao, G.; Liao, X.; Hu, X.; Wu, J.; Wang, Z. Optimization of Ultrasound-Assisted Extraction of Anthocyanins in Red Raspberries and Identification of Anthocyanins in Extract Using High-Performance Liquid Chromatography–Mass Spectrometry. *Ultrason. Sonochem.* **2007,** *14*, 767–778.

Chethan, S.; Malleshi N. G. Finger Millet Polyphenols: Optimization of Extraction and the Effect of pH on their Stability. *Food Chem.* **2007,** *105*, 862–870.

Chiremba, C.; Taylor, J. R. N.; Rooney, L. W.; Beta, T. Phenolic Acid Content of Sorghum and Maize Cultivars Varying in Hardness. *Food Chem.* **2012,** *134*, 81–88.

Chu, Y. H.; Chang, C. L.; Hsu, H. F. Flavonoid Content of Several Vegetables and Their Antioxidant Activity. *J. Sci. Food Agric.* **2000,** *80*, 561–566.

Chukwumah, Y. C.; Walker, L. T.; Verghese, M.; Ogutu, S. Effect of Frequency and Duration of Ultrasonication on the Extraction Efficiency of Selected Isoflavones and *Trans*-Resveratrol from Peanuts (*Arachis hypogaea*). *Ultrason. Sonochem.* **2009,** *16*, 293–299.

Corrales, M.; Garcıa, A. F.; Butz, P.; Tauscher, B. Extraction of Anthocyanins from Grape Skins Assisted by High Hydrostatic Pressure. *J. Food Eng.* **2009,** *90*, 415–421.

Corrales, M.; Toepfl, S.; Butz , P.; Knorr , D.; Tauscher, B. Extraction of Anthocyanins from Grape By-products Assisted by Ultrasonics, High Hydrostatic Pressure or Pulsed Electric Fields: A Comparison. *Innov. Food Sci. Emerg. Technol.* **2008,** *9*, 85–91.

Cravottoa, G.; Boffaa, L.; Mantegnaa, S.; Peregob, P.; Avogadrob, M.; Cintasc, P. Improved Extraction of Vegetable Oils under High-Intensity Ultrasound and/or Microwaves. *Ultrason. Sonochem.* **2008,** *15*, 898–902.

Cretnik, L.; Skerget, M.; Knez, Z. Separation of Parthenolide from feverfew: Performance of Conventional and High-Pressure Extraction Techniques. *Sep. Purif. Technol.* **2005,** *41*, 13–20.

Dağdelen, A.; Tumen, G.; Ozcan, M. M.; Dundar, E. Phenolics Profiles of Olive Fruits (*Olea europaea* L.) and Oils from Ayvalık, Domat and Gemlik Varieties at Different Ripening Stages. *Food Chem.* **2013,** *136*, 41–45.

Dar, B. N.; Sharma, S. Total Phenolic Content of Cereal Brans using Conventional and Microwave Assisted Extraction. *Am. J. Food Technol.* **2011,** *6*, 1045–1053.

Darra, N. E.; Grimi, N.; Maroun, R. G.; Louka, N.; Vorobiev, E. Pulsed Electric Field, Ultrasound, and Thermal Pretreatments for Better Phenolic Extraction During Red Fermentation. *Eur. Food Res. Technol.* **2013,** *236*, 47–56.

Das, A. K.; Singh, V. Antioxidative Free and Bound Phenolic Constituents in Pericarp, Germ and Endosperm of Indian Dent (*Zea mays* var. *indentata*) and Flint (*Zea mays* var. *indurata*) Maize. *J. Funct. Foods.* **2015,** *13*, 363–374.

Das, A. K.; Sreerama, Y. N.; Singh, V. Diversity in Phytochemical Composition and Antioxidant Capacity of Dent, Flint and Specialty Corns. *Cereal Chem.* **2014,** *91*, 639–645.

Dashwood, R. H. Frontiers in Polyphenols and Cancer Prevention. *J. Nutr.* **2007,** *137*, 267S–269S.

Del Llaudy, M. C.; Canals, R.; Canals, J. M.; Zamora, F. Influence of Ripening Stage and Maceration Length on the Contribution of Grape Skins, Seeds and Stems to Phenolic Composition and Astringency in Wine-Simulated Macerations. *Eur. Food Res. Technol.* **2008,** *226*, 337–344.

Delsart, C.; Ghidossi, R.; Poupot, C.; Cholet, C.; Grimi, N.; Vorobiev, E.; Milisic, V.; Peuchot, M. M. Enhanced Extraction of Phenolic Compounds from Merlot Grapes by Pulsed Electric Field. *Am. J. Enol. Vitic.* **2012,** *63*, 205–211.

Dey, T. B.; Kuhad R. C. Enhanced Production and Extraction of Phenolic Compounds from Wheat by Solid-State Fermentation with *Rhizopusoryzae* RCK2012. *Biotechnol. Rep.* **2014,** *4*, 120–127.

Ehala, S.; Vaher, M.; Kaljurand, M. Characterization of Phenolic Profiles of Northern European Berries by Capillary Electrophoresis and Determination of Their Antioxidant Activity. *J. Agric. Food Chem.* **2005,** *53*, 6484–6490.

Erdogan, S.; Ates, B.; Durmaz, G.; Yilmaz, I.; Seckin, T. Pressurized Liquid Extraction of Phenolic Compounds from *Anatolia propolis* and Their Radical Scavenging Capacities. *Food Chem. Toxicol.* **2011,** *49*, 1592–1597.

Fardet, A. New Hypotheses for the Health-Protective Mechanisms of Whole-Grain Cereals: What Is beyond Fibre? *Nutr. Res. Rev.* **2010,** *23*, 65–134.

Fernandez-Garcia, A.; Butz, P.; Tauscher, B. Effects of High Pressure Processing on Carotenoid Extractability, Antioxidant Activity, Glucose Diffusion and Water Binding of Tomato Puree (*Lycopersicum sculentum* Mill). *J. Food Sci.* **2001,** *66*, 1033–1038.

Ferruzzi, M. G.; Green, R. J. Analysis of Catechins from Milk–Tea Beverages by Enzyme Assisted Extraction Followed by High Performance Liquid Chromatography. *Food Chem.* **2005,** *99*, 484–491.

Ghafoor, K.; Park, L.; Choi, Y.-H. Optimization of Supercritical Fluid Extraction of Bioactive Compounds from Grape (*Vitis labrusca* B.) Peel by Using Response Surface Methodology. *Innov. Food. Sci. Emerg. Technol.* **2010,** *11*, 485–490.

Gómez-García, R.; Martínez-Ávila, G. C. G.; Aguilar, C. N. Enzyme-Assisted Extraction of Antioxidative Phenolics from Grape (*Vitis vinifera* L.) Residues. *3 Biotech* **2012,** *2*, 297–300.

Guderjan, M.; Elez-Martinez, P.; Knorr, D. Application of Pulsed Electric Fields at Oil Yield and Content of Functional Food Ingredients at the Production of Rapeseed Oil. *Innov. Food Sci. Emerg. Technol.* **2007,** *8*, 55–62.

Guderjan, M.; Toepfl, S.; Angersbach, A.; Knorr, D. Impact of Pulsed Electric Field Treatment on the Recovery and Quality of Plant Oils. *J. Food Eng.* **2005,** *67*, 281–287.

Haas, P.; Machado, M. J.; Anton, A. A.; Silva, A. S. S.; Francisco, A. Effectiveness of Whole Grain Consumption in the Prevention of Colorectal Cancer: Meta-Analysis of Cohort Studies. *Int. J. Food Sci. Nutr.* **2009,** *60*, 1–13.

Hao, M.; Beta, T. Qualitative and Quantitative Analysis of the Major Phenolic Compounds as Antioxidants in Barley and Flaxseed Hulls using HPLC/MS/MS. *J. Sci. Food Agric.* **2012,** *92*, 2062–2068.

Hartzfeld, P. W.; Forkner, R.; Hunter, M. D.; Hagerman, A. E. Determination of Hydrolyzable Tannins (gallotannins and ellagitannins) after Reaction with Potassium Iodate. *J. Agric. Food Chem.* **2002,** *50*, 1785–1790.

Hawthorne, S. B.; Yang, Y.; Miller, D. J. Extraction of Organic Pollutants from Environmental Solids with Sub- and Supercritical Water. *Anal. Chem.* **1994,** *66*, 2912–2920.

Heinz, V., Toepfl, S.; Knorr, D. Impact of Temperature on Lethality and Energy Efficiency of Apple Juice Pasteurization by Pulsed Electric Fields Treatment. *Innov. Food Sci. Emerg. Technol.* **2003,** *4*, 167–175.

Hemwimon, S.; Pavasant, P.; Shotipruk, A. Microwave-Assisted Extraction of Antioxidative Anthraquinones from Roots of *Morinda citrifolia. Sep. Purif. Technol.* **2007,** *54*, 44–50.

Herrera, M. C.; Luque de Castro, M. D. Ultrasound-Assisted Extraction of Phenolic Compounds from Strawberries Prior to Liquid Chromatographic Separation and Photodiode Array Ultraviolet Detection. *J. Chromatogr. A* **2005,** *1100*, 1–7.

Howard, L.; Panjaitan, N. Pressurized Liquid Extraction of Flavonoids from Spinach. *J. Food Sci.* **2008,** *73*, C151–C157.

Hromádková, Z.; Kostalová, Z.; Ebringerová, A. Comparison of Conventional and Ultrasound-Assisted Extraction of Phenolics-Rich Heteroxylans from Wheat Bran. *Ultrason. Sonochem.* **2008,** *15*, 1062–1068.

Huang, H. W.; Hsu, C. P.;Yang, B. B.; Wang, C. Y. Advances in the Extraction of Natural Ingredients by High Pressure Extraction Technology. *Trends Food Sci. Technol.* **2013,** *33*, 54–62.

Ibanez, E.; Herreoro, M.; Mendiola, J. A.; Castro-Puyana, M. Extraction and Characterization of Bioactive Compounds with Health Benefits from Marine Resources: Macro and Micro Algae, Cyanobacteria, and Invertebrates. In *Marine Bioactive Compounds: Sources Characterization and Applications*; Hayes, M., Ed.; Springer: U.S.A., 2012; pp 55–89.

Jacobs, D. R.; Pereira, M. A.; Meyer, K. A.; Kushi, L. H. Fiber from Whole Grains, but not Refined Grains, Is Inversely Associated with All-Cause Mortality in Older Women: The Iowa Women's Health Study. *J. Am. Coll. Nutr.* **2000,** *19*, 326S–330S.

Johnson, L. A. Recovery of Fats and Oils from Plant and Animal Sources. *Introduction to Fats and Oils.* AOCS Press: Champaign, 2000; pp 108–135.

Jun, X.; Deji, S.; Shou, Z.; Bingbing, L.; Ye, L.; Zhang, R. Characterization of Polyphenols from Green Tea Leaves Using a High Hydrostatic Pressure Extraction. *Int. J. Pharm.* **2009,** *382*, 139–143.

Kammerer, K. D.; Claus, A.; Schieber, A.; Carle R. A Novel Process for the Recovery of Polyphenols from Grape (*Vitis vinifera*) Pomace. *J. Food Sci.* **2005,** *70*, 157–163.

Kaufmann, B.; Christen, P. Recent Extraction Techniques for Natural Products: Microwave-Assisted Extraction and Pressurised Solvent Extraction. *Phytochem. Anal.* **2002,** *13*, 105–113.

Kerem, Z.; German-Shashoua, H.; Yarden, O. Microwave-Assisted Extraction of Bioactive Saponins from Chickpea (*Cicer arietinum* L.) *J. Sci. Food Agric.* **2005,** *85*, 406–412.

Khanizadeh, S.; Tsao, R.; Rekika, D.; Yang, R.; Charles, M. T.; Rupasinghe, V. H. P. Polyphenol Composition and Total Antioxidant Capacity of Selected Apple Genotypes for Processing. *J. Food Comp. Anal.* **2008,** *21*, 396–401.

Khoddami, A.; Wilkes, M. A.; Roberts, T. H. Techniques for Analysis of Plant Phenolic Compounds. *Molecules* **2013,** *18*, 2328–2375.

Kim, Y. J.; Kim, D. O.; Chun, O. K.; Shin, D. H.; Jung, H.; Lee, C. Y.; Wilson, D. B. Phenolic Extraction from Apple Peel by Cellulases from *Thermobifida fusca*. *J. Agric. Food Chem.* **2005,** *53*, 9560–9565.

Knorr, D. Effects of High Hydrostatic Pressure Processes on Food Safety and Quality. *Food Technol.* **1993,** *47*, 156–161.

Knorr, D. Novel Approaches in Food-Processing Technology: New Technologies for Preserving Foods and Modifying Function. *Curr. Opin. Biotechnol.* **1999,** *10*, 485–491.

Koponen, J. M.; Happonen, A. M.; Auriola, S.; Kontkanen, H.; Buchert, J.; Poutanen, K. S.; Törrönen, A. R. Characterization and Fate of Black Currant and Bilberry Flavonols in Enzyme-Aided Processing. *J. Agric. Food Chem.* **2008,** *56*, 3136–3144.

Lahouar, L.; El-Arem, A.; Ghrairi, F.; Chahdoura, H.; Salem, H. B.; Felah, M. E.; Achour, L. Phytochemical Content and Antioxidant Properties of Diverse Varieties of Whole Barley (*Hordeum vulgare* L.) Grown in Tunisia. *Food Chem.* **2014,** *145*, 578–583.

Landbo, A. K.; Meyer, A. S. Enzyme-Assisted Extraction of Antioxidative Phenols from Black Currant Juice Press Residues (*Ribes nigrum*). *J. Agric. Food Chem.* **2001,** *49*, 3169–3177.

Lang, Q.; Wai, C. M. Supercritical Fluid Extraction in Herbal and Natural Product Studies—A Practical Review. *Talanta* **2001,** *53*, 771–782.
Laroze, L.; Soto, C.; Zúñiga, M. E. Phenolic Antioxidants Extraction from Raspberry Wastes Assisted by Enzymes. *Electron. J. Biotechnol.* **2010,** *13*, 1–14.
Lee, J.; Wrolstad, R. E. Extraction of Anthocyanins and Polyphenolics from Blueberry Processing Waste. *J. Food Sci.* **2004,** *69*, 564–573.
Li, B. B.; Smith, B.; Hossain, M. M. Extraction of Phenolics from Citrus Peels: II. Enzyme-Assisted Extraction Method. *Sep. Purif. Technol.* **2006,** *48*, 189–196.
Liazid, A.; Guerrero, R. F.; Cantos, E.; Palma, M.; Barroso, C. G. Microwave Assisted Extraction of Anthocyanins from Grape Skins. *Food Chem.* **2011,** *125*, 1238–1243.
Liu, S.; Hu, F. B.; Colditz, G. A.; Willett, W. C. A Prospective Study of Whole-Grain Intake and Risk of Type 2 Diabetes Mellitus in US Women. *Am. J. Public Health.* **2000,** *90*, 1409–1415.
Liyana-Pathirana, C.; Shahidi, F. Optimization of Extraction of Phenolic Compounds from Wheat Using Response Surface Methodology. *Food Chem.* **2005,** *93*, 47–56.
López, N.; Puértolas, E.; Condón, S.; Álvarez, I.; Raso, J. Effects of Pulsed Electric Fields on the Extraction of Phenolic Compounds during the Fermentation of must of Tempranillo Grapes. *Innov. Food Sci. Emerg. Technol.* **2008,** *9*, 477–482.
Luque de Castro, M. D.; Jime´nez-Carmona, M. M. Where Is Supercritical Fluid Extraction Going? *Trend Anal. Chem.* **2000,** *19*, 223–228.
Luque de Castro, M. D.; Valcárcel, M.; Tena, M. T. *Analytical Supercritical Fluid Extraction*: Springer-Verlag: New York, 1994.
Luthria, D. L. Influence of Experimental Conditions on the Extraction of Phenolic Compounds from Parsley (*Petroselinum crispum*) Flakes Using a Pressurized Liquid Extractor. *Food Chem.* **2008,** *107*, 745–752.
Ma, Y. Q.; Chen, J. C.; Liu, D. H.; Ye, X. Q. Simultaneous Extraction of Phenolic Compounds of Citrus Peel Extracts: Effect of Ultrasound. *Ultrason. Sonochem.* **2009,** *16*, 57–62.
Määttä-Riihinen, K. R.; Kamal-Eldin, A.; Törrönen, A. R. Identification and Quantification of Phenolic Compounds in Berries of *Fragaria* and *Rubus* Species (Family Rosaceae). *J. Agric. Food Chem.* **2004,** *52*, 6178–6187.
Madhujith, T.; Izdorczyk, M.; Shahidi, F. Antioxidant Properties of Pearled Barley Fractions. *J. Agric. Food Chem.* **2006,** *54*, 3283–3289.
Maier, T.; Göppert, A.; Kammerer, D. R.; Schieber, A.; Carle, R. Optimization of a Process for Enzyme-Assisted Pigment Extraction from Grape (*Vitis vinifera* L.) Pomace. *Eur. Food Res. Technol.* **2008,** *227*, 267–275.
Manach, C.; Scalbert, A.; Morand, C.; Jimenez, L. Polyphenols: Food Sources and Bioavailability. *Am. J. Clin. Nutr.* **2004,** *79*, 727–747.
Mandal, V.; Mohan, Y.; Hemalatha, S. Microwave-Assisted Extraction—An Innovative and Promising Extraction Tool for Medicinal Plant Research. *Pharmacogn. Rev.* **2007,** *1*, 7–18.
Manthey, J. A.; Grohmann, K. Phenols in Citrus Peel Byproducts. Concentrations of Hydroxycinnamates and Polymethoxylated Flavones in Citrus Peel Molasse. *J. Agric. Food Chem.* **2001,** *49*, 3268–3273.
Martino, E.; Ramaiola, I.; Urbano, M.; Bracco, F.; Collina, S. Microwave Assisted Extraction of Coumarin and Related Compounds from *Melilotus officinalis* (L.) Pallas

as an Alternative to Soxhlet and Ultrasound-Assisted Extraction. *J. Chromatogr. A* **2006,** *1125*, 147–151.

Mason, T. J.; Paniwnyk, L.; Lorimer, J. P. The Uses of Ultrasound in Food Technology. *Ultrason. Sonochem.* **1996,** *3*, S253–S260.

Mellen, P. B. Walsh, T. F.; Herrington, D. M. Whole Grain Intake and Cardiovascular Disease: A Meta-Analysis. *Nutr. Metab. Cardiovasc. Dis.* **2008,** *18*, 283–290.

Meyer, A. S.; Jepsen, S. M.; Sorensen N. S. Enzymatic Release of Antioxidants for Human Low Density Lipoprotein from Grape Pomace. *J. Agric. Food Chem.* **1998,** *46*, 2439–2446.

Moreira M. M.; Morais S.; Barros A. A.; Delerue-Matos C.; Guido, L. F. A Novel Application of Microwave-Assisted Extraction of Polyphenols from Brewer's Spent Grain with HPLC-DAD-MS Analysis. *Anal. Bioanal. Chem.* **2012,** *403*, 1019–1029.

Naczk, M.; Shahidi. F. Extraction and Analysis of Phenolics in Food. *J. Chromatogr. A* **2004,** *1054*, 95–111.

Nieto, A.; Borrull, F.; Pocurull, E.; Marce, R. M. Pressurized Liquid Extraction: A Useful Technique to Extract Pharmaceuticals and Personal-Care Products from Sewage Sludge. *J. Supercrit. Fluids* **2013,** *81*, 245–253.

Niranjan, K., Hanmoungjai, P. Enzyme-aided aqueous extraction. In: *Nutritionally Enhanced Edible Oil Processing*; Dunford, N. T., Dunford, H. B., Eds.; AOCS Publishing: New York, 2004.

Oufnac, D. S.; Xu, Z.; Sun, T.; Sabliov, C.; Prinyawiwatkul, W.; Godber S. Extraction of Antioxidants from Wheat Bran Using Conventional Solvent and Microwave-Assisted Methods. *Cereal Chem.* **2007,** *84*, 125–129.

Paré, J. J. R.; Bélanger, J. M. R.; Stafford, S. S. Microwave-Assisted Process (MAP™): A New Tool for the Analytical Laboratory. *Trends Anal. Chem.* **1994,** *13*, 176–184.

Passos, C. P.; Yilmaz, S.; Silva, C. M.; Coimbra, M. A. Enhancement of Grape Seed Oil Extraction Using a Cell Wall Degrading Enzyme Cocktail. *Food Chem.* **2009,** *115*, 48–53.

Patras, N. P., Brunton, S. D.; Butler, F. Impact of High Pressure Processing on Total Antioxidant Activity, Phenolic, Ascorbic Acid, Anthocyanin Content and Colour of Strawberry and Blackberry Purees. *Innov. Food Sci. Emerg. Technol.* **2009,** *10*, 308–313.

Pellegrini, N.; Serafini, M.; Salvatore, S.; Del Rio, D.; Bianchi, M.; Brighenti, F. Total Antioxidant Capacity of Spices, Dried Fruits, Nuts, Pulses, Cereals and Sweets Consumed in Italy Assessed by Three Different *In Vitro* Assays. *Mol. Nutr. Food Res.* **2006,** *50*, 1030–1038.

Pérez-Jiménez, F.; Sauro-Calixto. Literature Data may Underestimate the Actual Antioxidant Capacity of Cereals. *J. Agric. Food Chem.* **2005,** *53*, 5036–5040.

Pineiro, Z.; Palma, M.; Barroso. C. G. Determination of Trans-Resveratrol in Grapes by Pressurised Liquid Extraction and Fast High-Performance Liquid Chromatography. *J. Chromatogr. A* **2006,** *1110*, 61–65.

Pizarro, C.; González-Sáiz, J. M.; Pérez-del-Notario, N. Multiple Response Optimisation Based on Desirability Functions of a Microwave-Assisted Extraction Method for the Simultaneous Determination of Chloroanisoles and Chlorophenols in Oak Barrel Sawdust. *J. Chromatogr. A* **2006,** *1132*, 8–14.

Plaza, L.; Sánchez-Moreno, C.; Ancos, B. D.; Elez-Martínez, P.; Martín-Belloso, O.; Cano, M. P. Carotenoid and Flavanone Content during Refrigerated Storage of Orange Juice

Processed by High-Pressure, Pulsed Electric Fields and Low Pasteurization. *LWT—Food Sci. Technol.* **2011,** *44*, 834–839.

Pourmortazavi, S. M.; Hajimirsadeghi, S. S. Supercritical Fluid Extraction in Plant Essential and Volatile Oil Analysis. *J. Chromatogr. A* **2007,** *1163*, 2–24.

Povilaitis, D.; Šulniūtė, V.; Venskutonis, P. R.; Kraujalienė, V. Antioxidant Properties of Wheat and Rye Bran Extracts Obtained by Pressurized Liquid Extraction with Different Solvents. *J. Cereal Sci.* **2015,** *62*, 117–123.

Prasad, K. N.; Yang, E.; Yi, C.; Zhao, M.; Jiang, Y. Effects of High Pressure Extraction on the Extraction Yield, Total Phenolic Content and Antioxidant Activity of Longan Fruit Pericarp. *Innov. Food Sci. Emerg. Technol.* **2009,** *10*, 155–159.

Provan, G. J.; Scobbie, L.; Chesson, A. Determination of Phenolic Acids in Plant Cell Walls by Microwave Digestion. *J. Sci. Food Agric.* **1994,** *64*, 63–65.

Puértolas, E.; Cregenzán, O.; Luengo, E.; Álvarez, I.; Raso, J. Pulsed electric-Field-Assisted Extraction of Anthocyanins from Purplefleshed Potato. *Food Chem.* **2013,** *136*, 1330–1336.

Puértolas, E.; López, N.; Condon, S.; Alvarez, I.; Raso, J. Potential Applications of PEF to Improve Red Wine Quality. *Trends Food Sci. Technol.* **2010,** *21*, 247–255.

Puri, M.; Sharma, D.; Barrow, C. J. Enzyme-Assisted Extraction of Bioactives from Plants. *Trends Biotechnol.* **2012,** *30*, 37–44.

Queiroz, C.; Moreira, F. F.; Lavinas, F. C.; Lopes, M. L. M.; Fialho, E.; Valente-Mesquita, V. L. Effect of High Hydrostatic Pressure on Phenolic Compounds, Ascorbic Acid and Antioxidant Activity in Cashew Apple Juice. *High Pressure Res.* **2010,** *30*, 507–513.

Ramos, S. Cancer Chemoprevention and Chemotherapy: Dietary Polyphenols and Signaling Pathways. *Mol. Nutr. Food Res.* **2008,** *52*, 507–526.

Randhir, R.; Shetty K. Microwave-Induced Stimulation of L-DOPA, Phenolics and Antioxidant Activity in Fava Bean (*Vicia faba*) for Parkinson's Diet. *Process Biochem.* **2004,** *39*, 1775–1784.

Rao, R. S. P.; Muralikrishna, G. Non-Starch Polysaccharide–Phenolic Acid Complexes from Native and Germinated Cereals and Millet. *Food Chem.* **2004,** *84*, 527–531.

Richter, B. E.; Jones, B. A.; Ezzell, J. L.; Porter, N. L.; Avdalovic, N.; Pohl, C. Accelerated Solvent Extraction: A Technology for Sample Preparation. *Anal. Chem.* **1996,** *68*(6), 1033–1039.

Rodrigues, S.; Pinto, G. A. S.; Fernandes, F. A. N. Optimization of Ultrasound Extraction of Phenolic Compounds from Coconut (*Cocos nucifera*) Shell Powder by Response Surface Methodology. *Ultrason. Sonochem.* **2008,** *15*, 95–100.

Rosenthal, A.; Pyle, D. L.; Niranjan, K. Aqueous and Enzymatic Processes for Edible Oil Extraction. *Enzyme Microb. Technol.* **1996,** *19*, 402–420.

Rostagno, M. A.; Araújo, J. M. A.; Sandi, D. Supercritical Fluid Extraction of Isoflavones from Soybean Flour. *Food Chem.* **2002,** *78*, 111–117.

Rostagno, M. A.; Palma, M.; Barroso, C. G. Microwave Assisted Extraction of Soy Isoflavones. *Anal. Chim. Acta* **2007a,** *588*, 274–282.

Rostagno, M. A.; Palma, M.; Barroso, C. G. Pressurized Liquid Extraction of Isoflavones from Soybeans. *Anal. Chim. Acta* **2004,** *522*, 169–177.

Rostagno, M. A.; Palma, M.; Barroso, C. G. Ultrasound-Assisted Extraction of Soy isoflavones. *J. Chromatogr. A* **2003,** *1012*, 119–128.

Rostagno, M. A.; Palma, M.; Barroso, C. G. Ultrasound-Assisted Extraction of Isoflavones from Soy Beverages Blended with Fruit Juices. *Anal. Chim. Acta* **2007,** *597,* 265–272.

Rostagno, M. A.; Palma, M.; Barroso, C. G. Ultrasound-Assisted Extraction of Isoflavones from Soy Beverages Blended with Fruit Juices. *Anal. Chim. Acta* **2007b,** *597,* 265–272.

Sahena, F.; Zaidul, I. S. M.; Jinap, S.; Karim, A. A.; Abbas, K. A.; Norulaini, N. A. N.; Omar A. K. M. Application of Supercritical CO_2 in Lipid Extraction—A Review. *J. Food Eng.* **2009,** *95,* 240–253.

Sajilata, M. G.; Bajaj P. R.; Singhal R. S. Tea Polyphenols as Nutraceuticals. *Compr. Rev. Food Sci. F* **2008,** *7,* 229–254.

Santos, D. T.; Veggi, P. C.; Meireles, M. A. A. Optimization and Economic Evaluation of Pressurized Liquid Extraction of Phenolic Compounds from Jabuticaba Skins. *J. Food Eng.* **2012,** *108*(3), 444–452.

Schilling, S.; Alber, T.; Toepfl, S.; Neidhart, S.; Knorr, D.; Schieber, A.; et al. Effects of Pulsed Electric Field Treatment of Apple Mash on Juice Yield and Quality Attributes of Apple Juices. *Innov. Food Sci. Emerg. Technol.* **2007,** *8,* 127–134.

Seabra, R. M.; Andra de, P. B.; Valentao, P.; Faria, M.; Paice, A. G.; Oliveira, M. B. P. P. Phenolic Profiles of Portuguese Olives: Cultivar and Geographics. In *Olives and Olive Oil in Health and Disease Prevention*; Preedy, V. R., Watson, R. R., Eds.; Academic Press: Oxford, UK, 2010; pp 177–186.

Setyaningsih, W.; Saputro, I. E.; Palma, M.; Barroso C. G. Optimisation and Validation of the Microwave-Assisted Extraction of Phenolic Compounds from Rice Grains. *Food Chem.* **2015,** *169,* 141–149.

Sharma, P.; Gujral, H. S. Antioxidant of Polyphenol Oxidase Activity of Germinated Barley Its Milling Fractions. *J. Food Chem.* **2010,** *120,* 673–678.

Shivonen, M.; Jarvenapaa; Hietaniem, V.; Houpalahti, R. Advances in Supercritical Carbon Dioxide Technologies. *Trends. Food. Sci. Tchnol.* **1999,** *10,* 217–222.

Shouqin, Z.; Jun, S.; Changzheng, W. High Hydrostatic Pressure Extraction of Flavonoids from Propolis. *J. Chem. Technol. Biotechnol.* **2005,** *80,* 50–54.

Sidhu, S. J.; Kabir, Y. Functional Foods from Cereal Grains. *Int. J. Food Prop.* **2007,** *10,* 231–244.

Sies, H.; Stahl, W.; Sevanian, A. Nutritional, Dietary and Post-Prandial Oxidative Stress. *J. Nutr.* **2005,** *135,* 969–972.

Silva, L. V.; Nelson, D. L.; Drummond, M. F. B.; Dufossé, L.; Glória, M. B. A. Comparison of Hydrodistillation Methods for the Deodorization of Turmeric. *Food Res. Int.* **2005,** *38*(8–9), 1087–1096.

Singh, R. K.; Sarker, B. C.; Kumbhar, B. K.; Agrawal, Y. C.; Kulshreshtha, M. K. Response Surface Analysis of Enzyme-Assisted Oil Extraction Factors for Sesame, Groundnut, and Sunflower Seeds. *J. Food Sci. Technol.* **1999,** *36,* 511–514.

Smith, R. M. Extractions with Superheated Water. *J. Chromatogr. A* **2002,** *975,* 31–46.

Soler-Rivas, C.; Espin, J. C.; Wichers, H. J. Oleuropein and Related Compounds. *J. Sci. Food Agric.* **2000,** *80,* 1013–1023.

Soliva-Fortuny R.; Balasa, A.; Knorr, D.; Martin-Belloso, O. Effects of Pulsed Electric Fields on Bioactive Compounds in Foods: A Review. *Trends Food Sci. Technol.* **2009,** *20,* 544–556.

Song, J.; Li, D.; Liu, C.; Zhang Y. Optimized Microwave-Assisted Extraction of Total Phenolics (TP) from *Ipomoea batatas* Leaves and Its Antioxidant Activity. *Innov. Food Sci. Emerg. Technol.* **2011,** *12*, 282–287.

Sowbhagya, H. B.; Chitra, V. N. Enzyme-Assisted Extraction of Flavorings and Colorants from Plant Materials. *Crit. Rev. Food Sci. Nutr.* **2010,** *50*, 146–161.

Stevenson, D. G.; Inglett, G. E.; Chen, D.; Biswas, A.; Eller, F. J.; Evangelista R. L. Phenolic Content and Antioxidant Capacity of Supercritical Carbon Dioxide-Treated and Air-Classified Oat Bran Concentrate Microwave-Irradiated in Water or Ethanol at Varying Temperatures. *Food Chem.* **2008,** *108*, 23–30.

Suslick, K. S.; Doktycz, S. J. The Effects of Ultrasound on Solids. In *Advances in Sonochemistry*; Mason; T. J., Ed.; JAI Press: New York, 1990; vol. 1; pp 197–230.

Szydłowska-Czerniak, A.; Tułodziecka, A.; Karlovitsb, G.; Szłyk, E. Optimisation of Ultrasound-Assisted Extraction of Natural Antioxidants from Mustard Seed Cultivars. *J. Sci. Food Agric.* **2014**. http://onlinelibrary.wiley.com/doi/10.1002/jsfa.6840/epdf

Tapola, N.; Karvonen, H.; Niskanen, L.; Mikola, M.; Sarkkinen, E. Glycemic Responses of Oat Bran Products in type 2 Diabetic Patients. *Nutr. Metab. Cardiovasc. Dis.* **2005,** *15*, 255–226.

Tedjo, W.; Eshtiaghi, M. N.; Knorr, D. Einsatz, nichtthermischerverfahrenzurzell-permeabilisierung von weintraubenunggewinnung von inhaltsstoffen. *FlussigesObst.* **2002,** *9*, 578–583.

Toepfl, S.; Mathys, A.; Heinz, V.; Knorr, D. Review: Potential of High Hydrostatic Pressure and Pulsed Electric Fields for Energy Efficient and Environmentally Friendly Food Processing. *Food Rev. Int.* **2006,** *22*, 405–423.

Vankar, P. S. Essential Oils and Fragrances from Natural Sources. *Resonance.* **2004,** *9*, 30–41.

Vinatoru, M. An Overview of the Ultrasonically Assisted Extraction of Bioactive Principles from Herbs. *Ultrason. Sonochem.* **2001,** *8*, 303–313.

Vinha, A. F.; Ferreres, F.; Silva, B. M.; Valentao, p.; Goncalves, A.; Pereira, J. A..; Oliveira, M. B.; Seabra, R. M.; Andrade, P. B. Phenolic Profiles of Portuguese Olive Fruits (*Olea europaea* L.): Influences of Cultivar and Geographical Origin. *Food Chem.* **2005,** *89*, 561–568.

Vita, J. A. Polyphenols and Cardiovascular Disease: Effects on Endothelial and Platelet Function. *Am J Clin. Nutr.* **2005,** *81*, 292S–297S.

Vorobiev, E.; Lebovka, N. Pulsed Electric Fields Induced Effects in Plant Tissues: Fundamental Aspects and Perspectives of Application. In *Electrotechnologies for Extraction from Food Plants and Biomaterials*; Vorobiev, E.; Lebovka N., Ed.; Springer: New York, 2008; pp 39–82.

Wang, J.; Sun, B.; Cao, Y.; Tian Y.; Li, X. Optimisation of Ultrasound-Assisted Extraction of Phenolic Compounds from Wheat Bran. *Food Chem.* **2008,** *106*, 804–810.

Wang, W.; Guo, J.; Zhang, J.; Peng, J.; Liu, T.; Xin Z. Isolation, Identification and Antioxidant Activity of Bound Phenolic Compounds Present in Rice Bran. *Food Chem.* **2015,** *171*, 40–49.

Watson, R. R. Functional Foods and Nutraceuticals in Cancer Prevention. John Wiley & Sons: New York, 2008.

Williams, L.; Bradley, L.; Smith, A.; Foxwell, B. Signal Transducer and Activator of Transcription 3 Is the Dominant Mediator of the Anti-Inflammatory Effects of IL-10 in Human Macrophages. *J. Immunol.* **2004,** *172*, 1567–1576.

Xia T.; Shi, S.; Wan, X. Impact of Ultrasonic-Assisted Extraction on the Chemical and Sensory Quality of Tea Infusion. *J. Food Eng.* **2006,** *74*, 557–560.

Yang, J.; Martinson, T. E.; Liu, R. H. Phytochemical Profiles and Antioxidant Activities of Wine Grapes. *Food Chem.* **2009,** *116*, 332–339.

Zabetakis, I.; Leclerc, D.; Kajda, P. The Effect of High Hydrostatic Pressure on the Strawberry Anthocyanins. *J. Agric. Food Chem.* **2000,** *48*, 2749–2754.

Zarnowski, R.; Suzuki, Y. Expedient Soxhlet Extraction of Resorcinolic Lipids from Wheat Grains. *J. Food Comp. Anal.* **2004,** *17*, 649–664.

Zderic, A.; Zondervan, E.; Meuldijk, J. Breakage of Cellular Tissue by Pulsed Electric Field: Extraction of Polyphenols from Fresh Tea Leaves. *Chem. Eng. Trans.* **2013,** *32*, 1795–1800.

Zhang, Y.; Sun, Y.; Xi, W.; Shen, Y.; Qiao, L.; Zhong, L.; Ye, X.; Zhou, Z. Phenolic Compositions and Antioxidant Capacities of Chinese Wild Mandarin (*Citrus reticulata* Blanco) Fruits. *Food Chem.* **2014,** *145*, 674–680.

Zhou, K.; Yu, L. Effects of Extraction Solvent on Wheat Bran Antioxidant Activity Estimation. *LWT—Food Sci. Technol.* **2004,** *37*, 717–721.

Zieliński, H.; Kozłowska, H. Antioxidant Activity and Total Phenolics in Selected Cereal Grains and Their Different Morphological Fractions. *J. Agric. Food Chem.* **2000,** *48*, 2008–2016.

Zougagh, M.; Valcarcel, M.; Rios, A. Supercritical Fluid Extraction: A Critical Review of Its Analytical Usefulness. *Trends. Anal. Chem.* **2004,** *23*, 399–405.

CHAPTER 3

CAROTENOIDS: TYPES, SOURCES, AND BIOSYNTHESIS

NIMISH MOL STEPHEN[1], GAYATHRI R.[2], NIRANJANA[2], YOGENDRA PRASAD K.[3], AMIT K. DAS[4], BASKARAN V.[5], and GANESAN P.[3*]

[1]*Department of Fish Processing Technology, Fisheries College and Research Institute, Tamilnadu Fisheries University, Ponneri - 601204, Tamilnadu, India*

[2]*SRM Research Institute, SRM University, Kattankulathur - 603203, Tamilnadu, India*

[3]*Department of Molecular Nutrition, CSIR-Central Food Technological Research Institute (CFTRI), Mysore - 570020, Karnataka, India*

[4]*Department of Grain Science and Technology, CSIR-Central Food Technological Research Institute (CFTRI), Mysore - 570020, Karnataka, India*

[5]*Department of Biochemistry and Nutrition, CSIR-Central Food Technological Research Institute (CFTRI), Mysore - 570020, Karnataka, India*

**Corresponding author, Tel.: +91 821 251 4876; fax: +91 821 251 7233; E-mail addresses: ganesanp@cftri.res.in; ganesan381980@yahoo.com.*

CONTENTS

ABSTRACT

Carotenoids are tetraterpenoid pigments, synthesized in all photosynthetic organisms including plants, algae, and microorganisms and are also synthesized in some non-photosynthetic bacteria, yeasts, and molds. Carotenoids are formed by the tail-to-tail linkage of the two C_{20} moieties (geranylgeranyl pyrophosphate [GGPP]) that results in the formation of the linear C40 hydrocarbon backbone which is susceptible to diverse structural modifications. In 1831, β-carotene was the first carotenoid isolated by Wackenroder. The empirical formula of β-carotene ($C_{40}H_{56}$) was established by Willstatter and Mieg. In the year 1937, Paul Karrer received Nobel Prize for elucidating the structure of β-carotene. It was first synthesized in 1950, and the commercial production of β-carotene was started by Roche in 1954. Currently, around 750 carotenoids are identified from natural sources. Based on the presence or absence of oxygen in their structure, the carotenoids are classified into oxygen-containing xanthophylls and oxygen-free carotenes. Carotenoids also vary structurally by the presence of other functional groups such as epoxy- and hydroxyl group. Some organisms produce partially degraded pigments called apocarotenoids or norcarotenoids, and few bacteria synthesize C_{45} or C_{50} carotenoids called homocarotenoids.

3.1 INTRODUCTION

Carotenoids are tetraterpenoid pigments, synthesized in all photosynthetic organisms including plants, algae, and microorganisms and are also synthesized in some non-photosynthetic bacteria, yeasts, and molds (Dufosse et al., 2005; Cazzonelli, 2011; Mata-Gomez et al., 2014). Carotenoids are formed by the tail-to-tail linkage of the two C_{20} moieties (geranylgeranyl pyrophosphate [GGPP]) that results in the formation of the linear C40 hydrocarbon backbone which is susceptible to diverse structural modifications. In 1831, β-carotene was the first carotenoid isolated by Wackenroder. The empirical formula of β-carotene ($C_{40}H_{56}$) was established by Willstatter and Mieg. In the year 1937, Paul Karrer received Nobel Prize for elucidating the structure of β-carotene. It was first synthesized in 1950, and the commercial production of β-carotene was started by Roche in 1954. Currently, around 750 carotenoids are identified from natural sources (Takaichi, 2011). Based on the presence or absence of oxygen in their structure, the carotenoids are classified into oxygen-containing xanthophylls and oxygen-free carotenes. Carotenoids also vary structurally by the presence of other functional groups such as epoxy- and hydroxyl group (Ganesan et al., 2011, 2013). Some organisms produce partially degraded pigments called apocarotenoids or norcarotenoids (Avalos & Cerda-Olmedo, 1986), and few bacteria synthesize C_{45} or C_{50} carotenoids called homocarotenoids (Baxter, 1960).

The major functions of carotenoids are light absorption in photosynthetic organisms and (photo)protection in all living organisms. During photosynthesis, carotenoids absorb radiant energy to chlorophyll molecules in a light harvesting function, dissipate excess energy via the xanthophyll cycle, and quench exited-state chlorophylls directly. Carotenoids are the first line of defense mechanism that quenches singlet oxygen ($^1O_2$), an ultimate factor responsible for photo-oxidative damage in plants, either physically by energy transfer mechanism or chemically by direct reaction with the radical (Amarowicz, 2011). Although carotenoids are not synthesized by humans and animals, their presence in the body is due to dietary intake of carotenoid-containing foods.

Carotenoids are considered as powerful food factors of great interest in many scientific disciplines because of their wide distribution and diverse functions. They are found to elicit profound effects on the maintenance of human health, and disease prevention. Today, this emerging class of nutrients has gained great attention in the nutritional supplementation industry and serves as a new frontier in the prevention of various lifestyle diseases like cancer, cardiovascular diseases, diabetes, and obesity through its powerful antioxidant behavior (Montonen et al., 2004; Maeda et al., 2007; Pashkow et al., 2008). The antioxidant efficiency of carotenoids in biological systems depends on their structures, site of action, partial pressure, and potential interaction with other bioactive molecules. Carotenoids also act as potential immunomodulators, and are reported to possess anti-inflammatory and antiallergic activity through modulating the expression of pro-inflammatory cytokines, and suppressing the action of IgE, respectively (Bellik et al., 2012). Some carotenoids (β-carotene, α-carotene, and β-cryptoxanthin) have pro-vitamin A activity which helps in promoting the growth and development of embryos, and vision enhancement (Bernstein et al., 2010; Costa et al., 2013). Carotenoids like lycopene, β-carotene, and fucoxanthin showed protective effect against UV-induced skin damage (Urikura et al., 2011; Offord et al., 2002). In addition to the health beneficial effects, carotenoids are used as food and feed additives. Astaxanthin is used as feed additive in aquaculture to provide the typical orange pigmentation to salmon and trout (Higuera-Ciapara et al., 2006) and also added to poultry feed to improve egg yolk coloration. An apocarotenoid derived from fungi, neurosporaxanthin is also used to improve pigmentation of egg yolks, skin, meat and legs of broilers, and flesh of fish (Paust et al., 1992). Few carotenoids like β-carotene, crocetin, and capxanthin are employed on an industrial scale as food colorant (Nells & De Leenheer, 1991; Jaswir et al., 2011). With these broad spectrums of background, this contribution mainly aimed to summarize the types, different sources, and the biosynthesis of carotenoids.

3.2 TYPES OF CAROTENOIDS

Although the trivial names of carotenoids were originated based on the biological source from which they have been isolated, they are generally classified on the basis of difference in their chemical structures, showed in Table 3.1. At the structural point of view, carotenoids are tetraterpenoids which contain eight isopentenylpyrophosphate units. The presence or absence of oxygen group mainly constitutes difference in the structure of carotenoids. Carotenoids which contain oxygen are termed as xanthophylls (zeaxanthin, neoxanthin, violaxanthin, and siphonaxanthin); on the other hand, carotenoids which lack oxygen are called as carotenes (α-carotene, β-carotene, and lycopene) (Fig. 3.1 and 3.2). Few carotenoids possess only the linear hydrocarbon backbone (acyclic carotenoids), which are susceptible to structural modifications, whereas most of the carotenoids contain cyclic end groups (cyclic carotenoids). Carotenoids also vary structurally by the presence of different functional groups such as epoxy- and hydroxyl group, and are termed as epoxy carotenoids and carotenols, respectively.

Some organisms produce specific carotenoids which distinctly vary with the number of carbon atoms and are known as apocarotenoids, norcarotenoids, homocarotenoids, and secocarotenoids. Apocarotenoids and norcarotenoids are the partially degraded pigments that contain less than 40 carbon atoms, which are formed by the oxidative cleavage of conjugated double bonds through chemical reaction, or by the enzymatic cleavage. Apocarotenoids play an important biological role as phytohormones, signaling molecules, and as shoot branching regulators. Retinol is the important apocarotenoid which is derived from central carbon double bond (15,15′) cleavage of β-carotene. Norcarotenoids are formed by the removal of CH_3, CH_2, or CH group from the carotenoid backbone, which

TABLE 3.1 Different Classes of Carotenoids Based on Their Structure, and the Presence of Functional Group.

Apocarotenoids	Chemical structure				Presence/absence of oxygen	
	Acyclic carotenes	**Cyclic carotenes**	**Carotenols/ Hydroxy carotenoids**	**Epoxy-carotenoids**	**Presence of oxygen**	**Absence of oxygen**
Retinol	ζ-Carotene	α-Carotene	α-Cryptoxanthin	Antheraxanthin	Antheraxanthin	α-Carotene
Bixin,	Phytoene	β-Carotene	β-Cryptoxanthin	Auroxanthin	Astaxanthin	β-Carotene
Crocin	Lycopene	γ-Carotene	Lutein	Luteoxanthin	Auroxanthin	δ-Carotene
Apo-8′-β-carotenal	Neurosporene	δ-Carotene	Rubixanthin	Neoxanthin	Canthaxanthin	γ-Carotene
Apo-8′-lycopennal	Phytofluene	α-Zeacarotene	Zeaxanthin	Violaxanthin	Capsanthin	Lycopene
Mycorradicin	Prolycopene	β-Zeacarotene	Zeinoxanthin	Fucoxanthin	Capsorubin	Neurosporene
Cachloxanthin	1,2-Dihydrolycopene	Tethyatene	Fucoxanthinol	Flavoxanthin	α-Crypoxanthin	Phytoene
Galloxanthin	Rhodopin	Torulene	Siphonaxanthin	Mutatoxanthin	β-Cryptoxanthin	Phytofluene
Sinensiaxanthin	Chloroxanthin	Renieratene	Alloxanthin	Cryptoflavin	Crocetin	α-Zeacarotene
Persicachrome	Lycoxanthin	Isorenieratene	Diatoxanthin	Latoxanthin	Lutein	β-Zeacarotene
Sinensiachrome	Spirilloxanthin	Chlorobactene	Parasiloxanthin	Salmoxanthin	Luteoxanthin	
Valenciaxanthin		Renierapurpurin	Nostoxanthin	Dinoxanthin	Lycophyll	
Cochloxanthin			Loroxanthin	Diadinoxanthin	Lycoxanthin	
			Saproxanthin	Lutein-5,6-epoxide	Neoxanthin	
			Caloxanthin	β-Carotene-5,6-epoxide	Rubixanthin	
			Crustaxanthin	β-Carotene-5,8-epoxide	Tunaxanthin	
			Nigroxanthin		Violaxanthin	
			Rhodopinol		Zeaxanthin	
			Lactucaxanthin		Zeinoxanthin	

Source: Britton et al. (2004), Carail and Caris-Veyrat (2006), Sugawara et al. (2006), Ganesan et al. (2010).

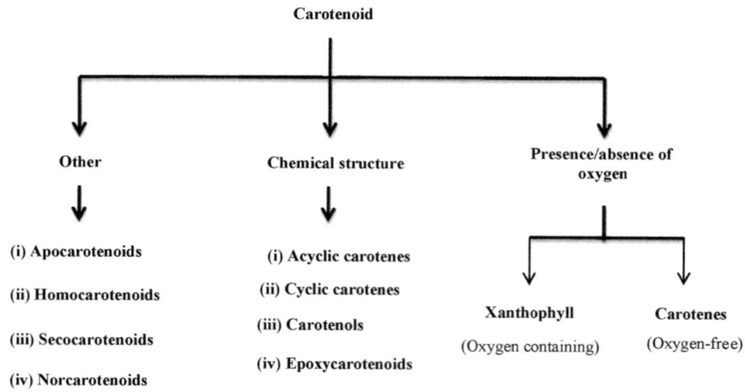

FIGURE 3.1 Classification of carotenoids based on the chemical structure.

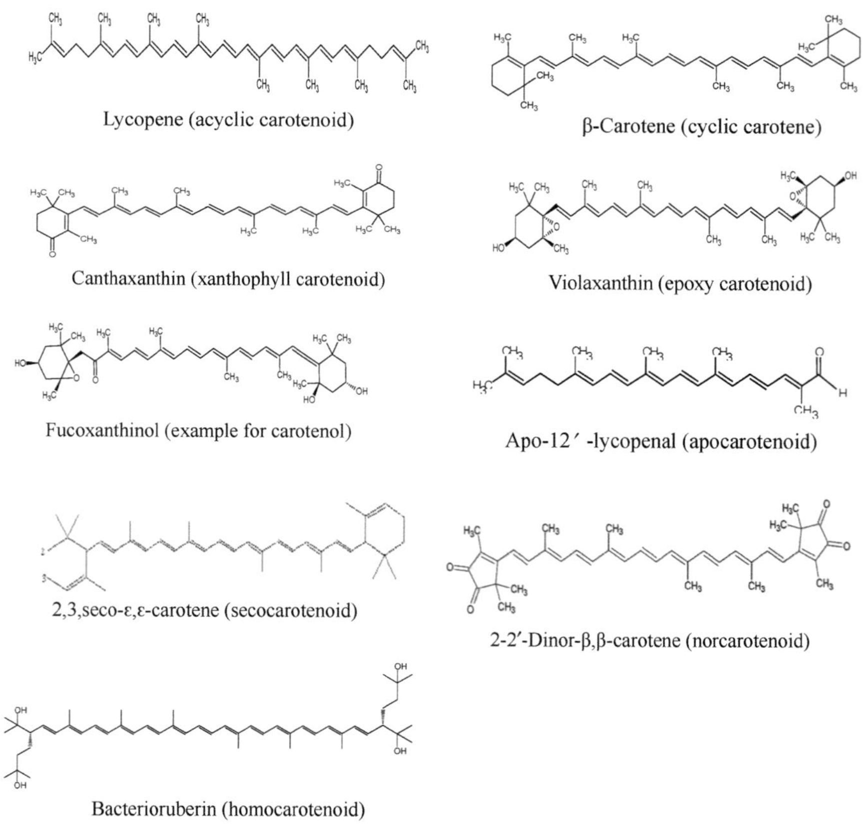

FIGURE 3.2 Chemical structures of different carotenoid classes.

include 2,2′-dinor-β,β-carotene, and 12,13,20-trinor-β,β-carotene. A few bacteria synthesize C_{45} or C_{50} carotenoids by further addition of isoprene to the C_{40} backbone called homocarotenoid. Secocarotenoids are formed by the fission of the bond between two neighboring carbon atoms along with the addition of one or more hydrogen atoms at each terminal ring structure. For example, 2,3-seco-ε,ε-carotene is formed by the fission of bonds between 2nd and 3rd carbon atom.

3.3 SOURCES OF CAROTENOIDS

Carotenoids are widely present in nature that provide leaves, fruits, vegetables, and flowers as well as marine microorganisms, algae, and vertebrates with distinctive yellow, orange, and reddish colors (Jaswir et al., 2011; Cazzonelli, 2011). Carotenoids are synthesized by all photosynthetic organisms, aphids, some bacteria, and fungi alike. They are not synthesized by humans and animals; however, their presence in the body is through dietary intake of carotenoid-containing foods. Numerous experimental and epidemiological studies have consistently demonstrated that individuals who eat more fruits and vegetables, which are rich in carotenoids, and who have higher serum carotenoid levels have a lower risk of cancer and cardiovascular diseases (Pashkow et al., 2008). Carotenoids are used as food colorant, and they are responsible for the color of birds like flamingo and canary, and certain insects and marine animals such as shrimp, lobster, and salmon (Zeb & Mehmood., 2004).

3.3.1 PLANTS

The major carotenoid pigments that occur in vegetables and fruits are β-carotene, α-carotene, β-cryptoxanthin, lycopene, lutein, zeaxanthin, neoxanthin, canthaxanthin, and capsanthin (Sangeetha & Baskaran, 2010; Mamatha et al., 2011). β-Carotene, α-carotene, and β-cryptoxanthin are the characteristic pigments of almost all yellow-orange vegetables and fruits. The major sources of β-carotene are carrot, mango, sweet potato, pumpkin, spinach, apricots, red peppers, and cantaloupe. Two or more servings of yellow-orange fruits and vegetables daily provide 3–6 mg of β-carotene. The nutritional content of 100 g carrot contains 9771 μg of β-carotene (Brill, 2009). Certain factors such as climate, genotype of the plant, and maturity of the plant play an important role in the determinant of carotenoid contents. Higher level of β-carotene is obtained when the carrots and squashes are grown under more sunlight. The concentration of carotenoid increases when the fruits undergo ripening; likewise, mature lettuces accumulate more carotenoids. It was reported that the ratio of β-carotene and β-cryptoxanthin changes in citrus fruits when they undergo ripening (Rodrigo et al., 2004). Food processing conditions can also affect the content of carotenoid (Mamatha et al., 2012). Almost 8–10% loss in the β-carotene and α-carotene content was observed when the yellow-orange vegetables like carrot, pumpkin, and sweet potato were subjected to mild heat treatment. It was reported that carotenoid content in fruits and vegetables is not altered to a great extent by common cooking methods such as steaming and boiling, but extreme heat induces oxidative destruction of carotenoids (Rao & Rao, 2007). Heat treatment can promote isomerization of the carotenoids from *trans*- to *cis*-isomeric forms in foods. The presence of dietary fat improves the bioavailability of β-carotene (Lakshminarayana et al., 2006; Mamatha & Baskaran, 2011; Nidhi et al., 2014). β-Cryptoxanthin is abundantly present in peach, papaya, pumpkins, orange, tangerine, and red sweet peppers. In orange juice, β-cryptoxanthin content is around 24 μg per 100 g of its nutritional content (Jaswir et al., 2011).

Lycopene is responsible for the red color in vegetables and fruits. Tomato and its products such as tomato sauce, tomato soup, and tomato juice are the predominant sources of lycopene. Tomatoes and tomato-based foods account for more than 85% of all the dietary sources of lycopene. The other sources of lycopene are red grapes, watermelon, pink grape fruit, papaya, guava, and apricot. The dietary intake of lycopene ranges from 0.5 to 27 mg/daily (Zeb & Mehmood, 2004). In tomato, lycopene content is around 3100 μg per 100 g of its nutritional content. The recommended level of lycopene for humans can be achieved by drinking two glasses of tomato juice a day. Previous study showed that the absorption of lycopene by the body is more efficient when the tomatoes are processed into juice, sauce, or paste (Zeb & Mehmood, 2004).

Lutein and zeaxanthin are the yellow pigment carotenoids which are present in spinach, broccoli, squash, kiwi fruit, grapes, orange juice, zucchini, yellow capsicums, persimmons, tangerines, mandarins, and abundantly in green leafy vegetables (Table 3.2) (Raju et al., 2007; Lakshminarayana et al., 2005). Lutein is more predominantly present than zeaxanthin in many of the fruits and vegetables. Egg yolk and maize contain highest concentrations of lutein and zeaxanthin. The highest quantity of lutein (60% of total carotenoids) was found in maize. Zeaxanthin was found to be rich in orange pepper with 37% of its total carotenoids (Sommerburg et al., 1998). It is also present abundantly in wolfberry. In green leafy vegetables, the amount of lutein was found to be 47% in spinach, 34% in stalks and leaves of celery, 27% in Brussels sprouts and scallions, 22% in broccoli, and 15% in green lettuce (Sommerburg et al., 1998). The content of zeaxanthin in dark green leafy vegetables ranges from 0–3%. In cooked spinach, the content of lutein and zeaxanthin was observed to be 7.0 mg per 100 mg of nutritional content (Rock, 1998). Zeaxanthin imparts coloration to the skin of birds and their egg yolk, and it also imparts skin pigmentation in pigs and some fishes (Sajilata et al., 2008).

TABLE 3.2 Carotenoids Content in Major Plant/Food Sources.

Carotenoid	Major sources	Amount—wet weight based (mg/100g)
β-Carotene	Carrot (raw)	18.30
	Carrot (cooked)	8.00
	Mango	2.15
	Mangos canned	13.10
	Sweet potato cooked	9.50
	Pumpkin canned	6.90
	Peppers red (raw)	2.40
	Peppers red (cooked)	2.20
α-Carotene	Carrot (raw)	5.00
	Carrot (cooked)	3.70
β-Cryptoxanthin	Mandarin oranges	1.77
	Orange juice	1.98
	Tangerine	1.60
	Papaya	0.47
Lycopene	Tomato (raw)	3.00
	Tomato (cooked)	4.40
	Tomato paste	29.30
	Tomato sauce	15.90
	Tomato soup	10.90
	Tomato juice	9.30
	Watermelon	4.90
	Papaya	3.40

TABLE 3.2 *(Continued)*

Carotenoid	Major sources	Amount—wet weight based (mg/100g)
Lutein	Spinach	6.26
	Broccoli	2.26
	Lettuce	1.25
	Green peas	1.84
	Watercress	10.71
Zeaxanthin	Maize	0.44
	Mandarin oranges	0.14
	Red pepper	0.60
Neoxanthin	Leek	1.00
	Arugula	1.00
	Lamb's lettuce	0.90
Violaxanthin	Yellow bell peppers	4.40
	Spinach	2.80
	Creamed spinach	2.50

Source: Rock (1998), Jaswir et al. (2011), Krinsky and Johnson (2005), Brill (2009), Rao and Rao (2007).

The xanthophyll epoxy-carotenoids such as neoxanthin and violaxanthin are also present in green leafy vegetables. Neoxanthin is present in leek (1.0 mg/100 mg), arugula (1.0 mg/100 g), and lamb's lettuce (0.9 mg/100 g). The major sources of violaxanthin are yellow bell peppers (4.4 mg/100 g), spinach (2.8 mg/100g), and creamed spinach (2.5 mg/100g). The daily intake of neoxanthin and violaxanthin was assessed as 0.5 mg and 1.2 mg, respectively (Biehler et al., 2011). Flowering plants such as marigold, little yellow star, and Mexican aster also contain significant amounts of neoxanthin and violaxanthin. Violaxanthin contributes 8% of the total carotenoids in the extracts of the flowering plant, *Canna indica* (Tinoi et al., 2006).

The sources of other carotenoids like canthaxanthin, capsanthin, capsorubin, crocin, and crocetin are *Cantharellus cinnabarinus* (mushroom plant), *Capsicum annum*, and *Crocus sativus* (saffaron plant) (Jaswir et al., 2011). The major source of canthaxanthin is edible mushroom, but it is also identified in some marine algae and crustaceans (Tanaka et al., 2012). Crocin and crocetin are produced in *Crocus sativus*, which belongs to the family *Iridaceae*, and its stigma contains various other carotenoids such as anthocyanin, carotene, and lycopene. Crocin and crocetin are widely used as food colorants (Bakshi et al., 2010). Capsanthin is present in the fruit part of *C. annum*, and it is used as a food colorant.

3.3.2 ALGAE

Algae are considered to be a rich source of various bioactive molecules including carotenoids. Carotenoids isolated from algae are reported to possess unique structures which represent specific functions (Ganesan et al., 2010, 2013). β-Carotene and zeaxanthin are commonly found almost in all algal divisions. But the acetylenic carotenoids such as alloxanthin, crocoxanthin, and monaxanthin are found only in some divisions or classes (Takaichi, 2011). Burczyk (1986) revealed the synthesis of ketocarotenoids, astaxanthin, canthaxanthin, and echinenone in 10 algal strains of *Chlorococcales*. Astaxanthin and its derivatives are present predominantly in some species of chlorophyta such as *Haematococcus pluvialis*, *Chlorococcum* sp., *Chlorella zofingiensis*, and *Chlorella vulgaris* (Del Campo et al., 2004; Hagen et al., 2000). The microalgal species like *Nannochloropsis oculata*, *Chaetoceros gracilis*, *Dunaliella salina* were reported to contain high amount of zeaxanthin. Zeaxanthin was found to be predominant in red algae such as *Porphyridium cruentum* and *Gracilaria damaecornis* (Grabowski et al., 2000; Schubert et al., 2011). The accumulation of zeaxanthin in brown alga *Macrocystis pyrifera* was observed by Garcia-Mendoza and Ocampo-Alvarez (2011). Lutein is most abundantly present in red (Rhodophyta) and green (Chlorophyta) algal classes. Some of the red and green algal species which are considered to be the major sources of lutein; they are *Muriellopsis* sp., *C. zofingiensis*, *Coccomyxa acidophila*, *Scenedesmus almeriensis*, and *Chlorella protothecoides* (Takaichi, 2011). *Eucheuma isiforme*, a red alga, is also considered as one among predominant species that accumulate lutein (Schubert et al., 2011). Violaxanthin and neoxanthin are mostly present in Chlorophyta (Takaichi & Mimuro, 1998). Diatoxanthin is mainly produced in Heterokontophyta, Haptophyta, Dinophyta, and Euglenophyta (Takaichi, 2011). *Undaria pinnatifida*, a marine brown alga, is considered as a major source of fucoxanthin (Terasaki et al., 2012). Noviendri et al. (2011) extracted fucoxanthin from two brown algae, *Sargassum binderi* and *Sargassum duplicatum*. Fucoxanthin is also present in Heterokontophyta. A significant amount of siphonaxanthin was found in marine green alga *Codium fragile* (Sugawara et al., 2014). Loroxanthin was identified in some classes of algae such as Euglenophyta, Chlorarachniophyta, and Chlorophyta (Yoshii et al., 2005; Takaichi, 2011). The factors which influence the production of carotenoids by algae are stress, size of the inoculum, concentration of inorganic phosphates, and intensity of light (Table 3.3). The production of astaxanthin in *H. pluvialis* increased by 2–3% (w/w on dry weight basis) when it was grown under stressed conditions. Under stressed condition, the green alga, *Botryococcus braunii* was found to produce higher amount of violaxanthin (6–9%), lutein (79–84%), astaxanthin (3–8%), zeaxanthin (0.32–0.78%), and β-carotene (1.75–2.14%) (Ambati et al., 2010). Enhanced carotenoids content was obtained in fresh water alga (*Cladophora* sp.) by the addition of inorganic phosphate into the culture medium (Khuantrairong & Traichaiyaporn, 2011). The mutant strain *Dunaliella salina zea1* is commercially used for the production of zeaxanthin. An increased production of β-carotene in *D. salina* was observed with the increased inoculum concentration (Zhu & Jiang, 2008). Red alga, *Gracilaria birdiae* was found

TABLE 3.3 Occurrence and Distribution of Carotenoids in Algae.

Carotenoid	Algae (species/classes)
Astaxanthin	*Haematococcus pulvialis*, *Chlorococcum* sp., *Chlorella zofingiensis*, *Chlorella vulgaris*, *Botryococcus braunii*
β-Carotene	*Botryococcus braunii*, *Dunaliella salina*, *Gracilaria birdiae*
Zeaxanthin	*Nannochloropsis oculata*, *Chaetoceros gracilis*, *Dunaliella salina*, *Porphyridium cruentum*, *Gracilaria damaecornis*, *Macrocystis pyrifera*, *Botryococcus braunii*, *Gracilaria birdiae*
Lutein	*Muriellopsis* sp., *Chlorella protothecoides*, *Eucheuma isiforme*, *Chlorella zofingiensis*, *Coccomyxa acidophila*, *Scenedesmus almeriensis*, *Botryococcus braunii*
Vialoxanthin	*Chlorophyta*, *Botryococcus braunii*, *Gracilaria birdiae*
Neoxanthin	*Chlorophyta*
Diatoxanthin	*Heterokontophyta*, *Haptophyta*, *Dinophyta*, *Euglenophyta*
Fucoxanthin	*Undari pinnatifida*, *Heterokontophyta*, *Sargassum binderi*, *Sargassum duplicantum*
Siphonaxanthin	*Codium fragile*
Loroxanthin	*Euglenophyta*, *Chlorarachniophyta*, *Chlorophyta*
Antheraxanthin	*Gracilaria birdiae*
Alloxanthin	*Gracilaria birdiae*

to produce high amount of violaxanthin under dark conditions, whereas in the presence of light, it produced high amount of zeaxanthin (Ursi et al., 2003). *G. birdiae* also accumulates other carotenoids such as antheraxanthin, alloxanthin, and β-carotene (Guaratini et al., 2012). Hii et al. (2010) observed that the production of fucoxanthin is more stable under dark conditions, and in alkaline pH.

3.3.3 FUNGI

Fungi produce only limited number of carotenoids which include β-carotene, astaxanthin, and neurosporaxanthin shown in Table 3.4. β-Carotene is predominant in the mucorales, *Blakeslea trispora*, *Phycomycus blakesleeanus*, and *Choanephora cucurbitarum*, and in the yeast, *Rhodotorula aurea* (Lampila et al., 1985). *B. trispora* is the major β-carotene producer employed on an industrial scale as a food colorant. It has been shown that mated culture of *B. trispora* exhibits increased production of β-carotene (Finkelstein et al., 1995). The major carotenoids present in *Rhodotorula* and *Rhodosporidium* species are β-carotene, γ-carotene, torulene, and torularhodin (Frengova & Beskova, 2009). Neurosporaxanthin is an amphipathic 35-carbon apocarotenoid, first discovered in the ascomycete, *Neurospora crassa*. Neurosporaxanthin has also been found in several species of the genera *Fusarium* (Avalos & Cerda-Olmedo, 1986; Bindl et al., 1970) and *Verticillium* (Valadon & Mummery, 1969; Valadon et al., 1982), as well as in *Podospora anserina* (Strobel et al., 2009). Lycopene synthesis in *B. trispora* was achieved by changing the environmental factors like pH, and addition of sodium carbonate and amines, which inhibit the enzyme catalyzing cyclization of lycopene to β-carotene. This results in increased yield of lycopene from 0.15 g/L to 0.7–1 g/L (Nells & De Leenheer, 1991). Induction of mutation in *B. trispora* produced up to 39 mg of β-carotene per gram of dry mass under standard laboratory conditions in which the original wild strains contained about 0.3 mg of β-carotene per gram of dry mass (Mehta et al., 2003). Blue light illumination stimulated the production of β-carotene in *Aspergillus giganteus* (213 μg/g) and *Rhodosporidium diobovatum* (700 μg/g), neurosporaxanthin in *Giberella fujikuroi* (159 μg/g), *Neurospora crassa* (152 μg/g), and *Phycomyces blakesleanus* (550 μg/g), and torulene in *Rhodotorula minuta* (150 μg/g) and *Verticillium agaricinum* (445 μg/g) (Cerda-Olmedo, 1989). A recent study reported that exposing 3-day-old mycelial cultures of *B. trispora* to ultrasonic treatment produced 173 mg/L of β-carotene and 82 mg/L of lycopene, which represented an increase of nearly 40.7% and 52.7%, respectively (Wang et al., 2014). Lycopene and β-carotene production were increased when oxygen vectors *n*-hexane and *n*-dodecane were added to culture of *B. trispora*. This treatment increased yield of lycopene from 51% to 78% and β-carotene from 44% to 65% (Xu et al., 2007). Span 20 at 0.2% increased the β-carotene production from 139 mg/L to 318 mg/L (Choudhari et al., 2008). Supplementation of medium with natural oils, amino acids, vitamin A, and antibiotics was found to increase the yield of β-carotene in *B. trispora* (Choudhari et al., 2008).

TABLE 3.4 Occurrence and Distribution of Carotenoids in Fungi.

Carotenoid	Fungal source
β-Carotene	*Blakeslea trispora, Phycomycus blakesleeanus, Choanephora cucurbitarum, Rhodotorula aurea, Rhodosporidium diobovatum, Aspergillus giganteus, Sporobolomyces roseus*
Torulene and torularhodin	*Rhodotorula minuta, Rhodosporidium* sp., *Verticillium agaricinum, Sporobolomyces roseus*
Astaxanthin	*Xanthophyllomyces dendrorhous, Peniophora* sp., *Phaffia rhodozyma*
Neurosporoxanthin	*Neurospora crassa, Fusarium* sp., *Verticillium* sp., *Podospora anserine, Giberella fujikuroi, Phycomyces blakesleanus*
Lycopene	*Blakeslea trispora*
Echinenone, phoenicoxanthin	*Phaffia rhodozyma*

The red-pigmented basidiomycete yeasts such as *Xanthophyllomyces dendrorhous* and *Phaffia rhodozyma* contain astaxanthin as principal carotenoid pigment (Andrewes et al., 1976; Zheng et al., 2006). *X. dendrorhous* was found to be a major astaxanthin producer at industrial scale (Johnson & Lewis, 1979; Johnson, 2003; Bhosale & Bernstein, 2005; Dominguez-Bocanegra et al., 2007). The biotechnological amenability of yeast made *X. dendrorhous* a good choice for large scale fermentation production of astaxanthin (Rodriguez-Saiz et al., 2010; Schmidt et al., 2011). Echinenone, 3-hydroxyechinenone, and phoenicoxanthin were also isolated and identified from *P. rhodozyma*. The overexpression of carotenogenic genes from *X. dendrorhous* in *Saccharomyces cerevisiae* produced high levels of β-carotene, up to 5.9 mg/g (dry weight)

(Verwaal et al., 2007). A food-grade yeast, *Candida utilis* has been engineered to confer a novel biosynthetic pathway for the production of carotenoids such as lycopene, β-carotene, and astaxanthin. The carotenoid biosynthesis genes were individually modified based on the codon usage of the *C. utilis* glyceraldehyde 3-phosphate dehydrogenase gene and expressed in *C. utilis* under the control of the constitutive promoters and terminators derived from *C. utilis*. The resultant yeast strains accumulated lycopene, β-carotene, and astaxanthin at 1.1, 0.4, and 0.4 mg per gram dry weight of cells, respectively (Miura et al., 1998). The striking reddish-pink color of the phylloplane yeast *Sporobolomyces roseus* was found to be due to the accumulation of three major carotenoid pigments β-carotene, torulene, and torularhodin (Davoli & Weber, 2002).

3.3.4 CYANOBACTERIA

Cyanobacteria are the primitive and diverse group of microorganisms which have been considered as rich source of biochemically active molecules represented in Table 3.5. Several bioactive compounds including carotenoids synthesized from cyanobacteria have emerged as templates for the development of anticancer drugs (Costa et al., 2012). Cyanobacterial pigments comprise the most colorful and attractive components among the microorganisms (Burja et al., 2001). Most common carotenoids synthesized by cyanobacteria are β-carotene, zeaxanthin, astaxanthin, echinenone, and myxoxanthophyll (Prasanna et al., 2010). Cyanobacteria also synthesize other carotenoids such as ε-carotene, γ-carotene, lycopene, canthaxanthin, oscillaxanthin (Takaichia & Mochimaru, 2007). Cyanobacteria possess two homologous desaturase genes, *CrtP* and *CrtQ* which mediate the formation of ζ-carotene and lycopene, respectively (Gombos & Vig, 1986). *Synechococcus* sp. (strain—PCC7942) was found to contain 52% of β-carotene, 38% of zeaxanthin, and minor amounts of caloxanthin, nostoxanthin, and cryptoxanthin (Cheng, 2006). *Thermosynechococcus elongatus* contains nostoxanthin in addition to β-carotene and zeaxanthin, while *Prochlorococcus marinus* contains β-carotene and zeaxanthin as well as α-carotene. A marine cyanobacteria, *Trichodesmium* sp., was found to synthesize a significant amount of β-carotene. All-*trans*-β-carotene, and a small amount of 9-*cis*-β-carotene were also identified in *Trichodesmium* sp. The protein sequence responsible for the biosynthesis of β-carotene in *Trichodesmium* sp. was identified as homologous to the key enzyme, retinyl palmitate esterase (Kelman et al., 2009).

TABLE 3.5 Occurrence and Distribution of Carotenoids in Cyanobacteria.

Cyanobacteria	Carotenoids
Synechococcus sp.	β-Carotene, zeaxanthin, caloxanthin, nostoxanthin, cryptoxanthin
Thermosynechococcus elongatus	β-Carotene, zeaxanthin, nostoxanthin
Prochlorococcus marinus	β-Carotene, zeaxanthin, α-carotene
Trichodesmium sp., *Calothrix elenkenii*, *Synechocystis* sp., *Lyngbya* sp.	β-Carotene

Certain environmental factors like light intensity and concentration of oxygen play a most critical role in enhancing the production of carotenoids. A previous study reported that higher light intensity (13–28 klx) stimulated the production of carotenoids in *Nostoc muscorum* and *Microcystis aeruginosa*. It was reported that *Calothrix elenkenii* accumulates high levels of β-carotene when it was grown in the presence of light–dark cycles (Prasanna et al., 2004). In a study, the synthesis of β-carotene was found to be higher under argon-enriched and air-tight environments. However, certain cyanobacterial species such as *Synechocystis* sp. and *Lyngbya* sp. synthesize considerable amount of β-carotene when they were grown under aerobic conditions (Prasanna et al., 2004; Prasanna et al., 2010). The carotenoid biosynthesis in cyanobacteria can be altered significantly by either over expression or deletion of particular genes. Over expression of *CrtR* gene in *Synechocystis* sp. was found to increase the accumulation of zeaxanthin by nearly 2.5-fold. Irradiation altered the ratio or the synthesis of canthaxanthin and β-carotene in terrestrial cyanobacterial species (Prasanna et al., 2010).

3.3.5 BACTERIA

In recent years, the interest in the production of natural carotenoids by microbes has been increased due to the consumer preferences for natural products and potential cost-effectiveness. Carotenoid production via biotechnological approaches through environmental and genetic manipulation of microorganisms has gained considerable attention, because it is sustainable and has no negative impact on the environment (Dufosse et al., 2005; Das et al., 2007; Yoona, et al., 2009; Wang et al., 2012; Naziri et al., 2014). Many bacterial species are considered as suitable host for the production of different carotenoids such as β-carotene, lycopene, canthaxanthin, zeaxanthin, astaxanthin, α-bacterioruberin, β-bacterioruberin, deinoxanthin, thermozeaxanthins, nostoxanthin, caloxanthin, sarcinaxanthin, and staphyloxanthin (Table 3.6) (Baxter, 1960; Asker & Ohta, 1999; Lutnaes et al., 2004; Marit et al., 2010; Tian & Hua,2010; Osawa et al., 2011; Ide et al., 2012).

TABLE 3.6 Occurrence and Distribution of Carotenoids in Bacteria.

Bacteria	Carotenoids
Paracoccus sp. strain DSM 11574	β-Carotene, echinenone, β-cryptoxanthin, canthaxanthin, astaxanthin, zeaxanthin, adonirubin, adonixanthin
Flavobacterium sp., *Paracoccus zeaanthinifaciens*	Zeaxanthin
Agrobacterium aurantiacum, *Paracoccus carotinifaciens*	Astaxanthin
Bradyrhizobium sp.	Canthaxanthin
Enterobacter sp. strain P_{41}	β-Carotene
Halobacterium salinarium, *Halobacterium sarcina*	α- and β-bacterioruberin
Staphylococcus aureus	Staphyloxanthin
Thermus thermophilus	Thermozeaxanthin
Deinococcus radiodurans	Deinoxanthin
Erythobacter sp.	Nostoxanthin
Micrococcus luteus	Sarcinaxanthin

A novel *Paracoccus* species strain DSM 11574 was found to secrete carotenoid-containing vesicles, which contain carotenoids such as β-carotene, echinenone, β-cryptoxanthin, canthaxanthin, astaxanthin, zeaxanthin, adonirubin, and adonixanthin (Hirschberg & Harker, 1999). Among bacterial species, *Escherichia coli* is the most convenient host for the production of carotenoids as it has a powerful genetic tool for metabolic engineering as well as a well-developed fermentation process (Misawa et al., 1990; Cunningham et al., 1993; Ruther et al., 1997). Metabolically engineered *E. coli* strains are found to produce high titer of carotenoids such as lycopene, β-carotene, zeaxanthin, and astaxanthin (Ruther et al., 1997; Sandmann, 2002; Alper et al., 2006; Yoon et al., 2006, 2007; Yuan et al., 2006; Cheng, 2006; Das et al., 2007). An increased production of isopentenyl diphosphate (IPP), a precursor compound of carotenoid biosynthesis in *E. coli,* was achieved through metabolic engineering in the endogenous 2-C-methyl-D-erythritol 4-phosphate (MEP) pathway or through introduction of mevalonate (MVA) pathway (Das et al., 2007; Yoona et al., 2009). Yoona et al. (2009) compared β-carotene production in different recombinant *E. coli* strains such as MG1655, DH5α, S17-1, XL1-Blue, and BL21. Among strains tested, DH5α was found to produce high titer of β-carotene (465 mg/L). In an experimental study, it was reported that a recombinant *E. coli* which harbored the whole MVA pathway of *Streptomyces sp. CL190* produced a high amount of lycopene (4.3 mg/L) (Vadali et al., 2005). Likewise, recombinant *E. coli* which uses the MVA pathway of *Streptococcus pneumonia* was found to produce lycopene at the level of 102 mg/L and β-carotene at 503 mg/L (Yoon et al., 2006, 2007). Novel and desirable carotenoids can also be produced in *E. coli* by expressing new carotenoids synthesis genes isolated from different microorganisms (Sandmann, 2002; Cheng, 2006). Introduction of seven bacterial carotenoid biosynthesis genes, *CrtE, CrtB, CrtI, CrtY, CrtZ, CrtX,* from *P. ananatis* and *CrtG* from *Brevundimonas* SD212 into *E. coli* synthesized a rare carotenoid, caloxanthin, and zeaxanthin and nostoxanthin (Tatsuzawa et al., 2000; Osawa et al., 2011).

Flavobacterium sp. and *Paracoccus zeaanthinifaciens* were found to synthesize significantly a higher amount of zeaxanthin (Shepherd et al., 1976; Hümbelin et al., 2002). *Flavobacterium* sp. was able to produce zeaxanthin at 190 mg/L of culture with the cell concentration of 16 mg/g of dried cellular mass. *Agrobacterium aurantiacum* and *Paracoccus carotinifaciens* were able to synthesize high concentration of astaxanthin, and *Bradyrhizobium* sp. was found to synthesize canthaxanthin. A Gram-negative yellow-pigmented, pleomorphic bacterial strain (TDMA-16^{T}), which was closely related to *Sphingomonas jaspsi* isolated from the fresh water sample at Tottori, Japan, was reported to produce carotenoids such as zeaxanthin and nostoxanthin (Asker et al., 2007).

A red-pigmented non-sulfur bacterium, identified as *Enterobacter* species P_{41}, was isolated from chicken feces which was found to synthesize

0.16 mg of β-carotene per gram of dried cell mass (Tanskul et al., 2013). A specific kind of carotenoids, α- and β-bacterioruberin were identified in halobacteria such as *Halobacterium salinarium* and *Halobacterium sarcina* (Baxter, 1960). *Lactobacillus plantarum* strain CECT7531 was one among 18 strains of *L. plantarum* observed to produce significant amounts of yellow C_{30} carotenoid, 4,4′-diaponeurosporene, which ranged from 1.8 to 54 mg/kg of dry cell mass (Garrido-Fernandez et al., 2010). A unique carotenoid, staphyloxanthin was considered as the best-recognized bacterial pigments that impart the eponymous golden color to the human pathogen *Staphylococcus aureus* (Liu, et al., 2005; Clauditz et al., 2006).

Few studies reported that thermophilic bacteria produce carotenoids which are rare in nature, and are considered to possess special mechanisms that stabilize cell membranes at high temperatures (Brock et al., 1984). Species belonging to the genus *Thermus* are known for their resistance to extreme temperatures, and are usually yellow or red pigmented due to their ability to synthesize carotenoids. A rare carotenoid, thermozeaxanthin, was isolated from the thermophilic eubacterium, *Thermus thermophilus* (Yokoyama et al., 1995; Tian & Hua, 2010). An extremophile, *Deinococcus radiodurans* is able to synthesize a unique carotenoid, deinoxanthin (Tian & Hua, 2010). Pigmented heterotrophic bacteria, *Erythobacter* sp. and *Micrococcus luteus*, isolated from the mid-part of the Norwegian coast, were reported to synthesize nostoxanthin and sarcinaxanthin, respectively (Stafsnes et al., 2010).

3.4 BIOSYNTHESIS OF CAROTENOIDS

The carotenoids biosynthesis in plant, algae, fungi, and bacteria generally involves a series of desaturation, cyclization, hydroxylation, and epoxidation commencing with the formation of phytoene (Fig. 3.3). Carotenoids are derived from the common 5-carbon (C5) building unit called IPP. It is the fundamental C5 biosynthetic unit from which all the terpenoids are constructed. This central precursor is formed through mevalonic acid (MVA) pathway and/or MEP pathway (Rohmer et al., 1993; Lichtenthaler et al., 1997; Rodriguez-Concepcion & Boronat, 2002).

Twenty-five carotenogenic (*Crt*) genes, whose gene products have different catalytic functions, have been identified as necessary to synthesize different carotenoids (Nishida et al., 2005). In the biosynthetic pathway, before chain elongation begins, the IPP is isomerized into dimethylallyl diphosphate (DMAPP). The immediate precursor for carotenoids biosynthesis, the C20-carbon GGPP, is formed by the sequential and linear addition of three molecules of IPP to one molecule of DMAPP. This reaction is catalyzed by one member of closely related family of prenyltransferase called GGPP synthase (*CrtE*). Then, the symmetrical 40-carbon phytoene is formed from the head-to-head condensation of two molecules of GGPP, which is catalyzed by the first carotenoid specific enzyme, phytoene synthase (*CrtB*).

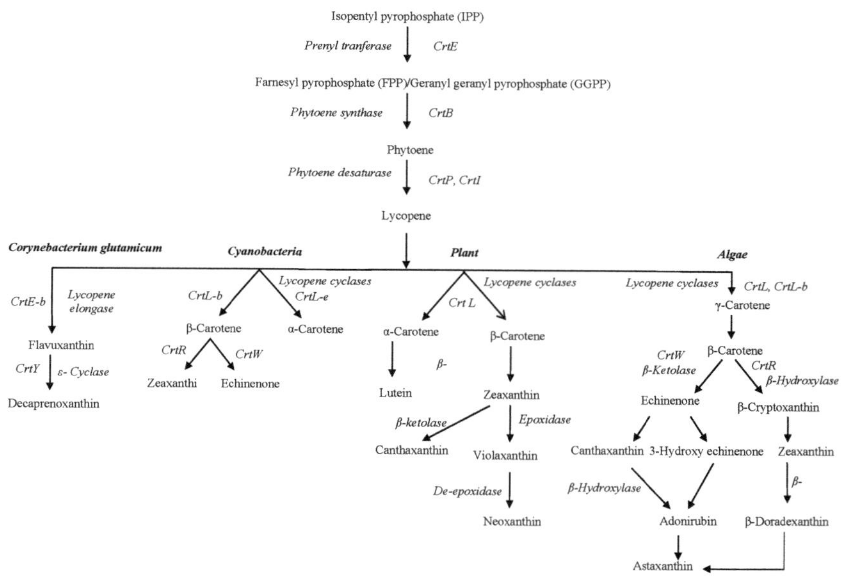

FIGURE 3.3 Biosynthetic pathways of carotenoids in plant, algae, cyanobacteria, and bacteria.

The phytoene undergoes a series of four desaturation reactions where two hydrogen atoms are removed (dehydrogenations) by *trans*-elimination from adjacent position to introduce a new double bond that contributes the chromophore in carotenoid pigments. The dehydrogenations are catalyzed by the phytoene desaturase which converts the phytoene to neurosporene in three desaturation steps or to lycopene in four steps. This is how colorless phytoene is transformed into the pink-colored lycopene. In plants and algae, the desaturation is carried out by phytoene desaturase (*CrtP*), ζ-carotene desaturase (*CrtQ*) and carotene isomerase (*CrtH*), whereas phytoene desaturase (*CrtI*) is the only enzyme involved in the formation of lycopene in bacteria and fungi (Giraud et al., 2004). However, the green sulfur bacteria possess *CrtP*, *CrtQ*, and *CrtH* (Frigard et al., 2004).

After the action of these three major enzymes (*CrtE*—GGPP synthase, *CrtB*—phytoene synthase, and *CrtI*—phytoene desaturase), the biosynthetic pathways diverge depending on the species leading to the accumulation of various different carotenoids. The acyclic lycopene undergoes modification by lycopene cyclases, ε-cyclase, and β-cyclase to produce α- and β-carotene, respectively. There are four specific genes identified for the lycopene cyclase activity, which include *CrtY*, *CrtL*, *CruA*, and *CruP* (Krubasik & Sandmann, 2000). The carotenes serve as substrate for the production of xanthophyll carotenoids such as lutein, zeaxanthin, and violaxanthin with the action of β-hydroxylase (*CrtG*; *CrtR*) and β-ketolases (*CrtO*—mono ketolase; *CrtW*). The conversion of zeaxanthin to violaxanthin involves addition of epoxide group by the activity of zeaxanthin epoxidase, which can be interconverted by the action of an enzyme called violaxanthin de-epoxidase (Cazzonelli, 2011).

Some organisms are able to synthesize modified form of carotenoids such as homocarotenoids and apocarotenoids. In case of *Corynebacterium glutamicum*, the lycopene (C-40) is modified to form a homocarotenoid (C-50) flavuxanthin by lycopene elongase, which is further modified to decaprenoxanthin by ε-cyclase. Apocarotenoids are derived from carotenoids mostly by oxidative cleavage. Neurosporaxanthin is an apocarotenoid synthesized by several fungi by the action of carotenoid oxygenase (Estrada et al., 2008).

KEYWORDS

- **carotenoids**
- **carotenes**
- **lycopene**
- **lutein**
- **zeaxanthin**

REFERENCES

Alper, H.; Miyaoku, K.; Stephanopoulos, G. Characterization of Lycopene-Overproducing *E. coli* Strains in High Cell Density Fermentations. *Appl. Microbiol. Biotechnol.* **2006,** *72*, 968–974.

Amarowicz, R. Lycopene as a Natural Antioxidants. *Eur. J. Lipid. Sci. Technol.* **2011,** *113*, 675–677.

Ambati, R. R.; Ravi, S.; Aswathanarayana, R. G. Enhancement of Carotenoids in Green Alga—*Botryococcus braunii* in Various Autotrophic Media under Stress Conditions. *Int. J. Biomed. Pharm. Sci.* **2010,** *4*, 87–92.

Andrewes, A. G.; Phaff, J. H.; Starr, P. M. Carotenoids of *Phaffia rhodozyma*, a Red-Pigmented Fermenting Yeast. *Phytochemistry* **1976,** *15*, 1003–1007.

Asker, D.; Beppu, T.; Ueda, K. *Sphingomonas jaspsi sp.* A Novel Carotenoid-Producing Bacterium Isolated from Misasa, Tottori, Japan. *Int. J. Syst. Evol. Microbiol.* **2007,** *57*, 1435–1441.

Asker, D.; Ohta, Y. Production of Canthaxanthin by Extremely Halophilic Bacteria. *J. Biosci. Bioeng.* **1999,** *88*, 617–621.

Avalos, J.; Cerda-Olmedo, E. Chemical Modification of Carotenogenesis in *Gibberella fujikuroi*. *Phytochemistry* **1986,** *25*, 1837–1841.

Bakshi, H.; Sam, S.; Rozati, R.; Sultan, P.; Islam, T.; Rathore, B.; Lone, Z;, Sharma, M.; Triphati, J.; Saxena, R. C. DNA Fragmentation and Cell Cycle Arrest: A Hallmark of Apoptosis Induced by Crocin from Kashmiri Saffron in a Human Pancreatic Cancer Cell Line. *Asian. Pac. J. Cancer. Prev.* **2010,** *11*, 675–679.

Baxter, R. M. Carotenoid Pigments of Halophilic Bacteria. *Can. J. Microbol.* **1960,** *6*, 417–424.

Bellik, Y.; Boukraâ, L.; Alzahrani, H. A.; Bakhotmah, B. A.; Abdellah, F.; Hammoudi, S. M.; Iguer-Ouada, M. Molecular Mechanism Underlying Anti-inflammatory and Anti-allergic Activities of Phytochemicals: An Update. *Molecules*. **2012,** *18*, 322–353.

Bernstein, P. S.; Delori, F. C.; Richer, S.; van Kuijk, F. J.; Wenzel, A. J. The Value of Measurement of Macular Carotenoid Pigment Optical Densities and Distributions in Age-Related Macular Degeneration and Other Retinal Disorders. *Vision. Res.* **2010,** *50*, 716–728.

Bhosale P.; Bernstein P. S. Microbial Xanthophylls. *Appl. Microbiol. Biotechnol.* **2005,** *68*, 445–455.

Biehler, E.; Alkerwi, A.; Hoffmann, L.; Krause, E.; Guillaume, M.; Lair, M.-L.; Bohn, T. Contribution of Violaxanthin, Neoxanthin, Phytoene and Phytofluene to Total Carotenoid Intake: Assessment in Luxembourg. *J. Food. Compost. Anal.* **2011,** 1–10.

Bindl, E.; Lang, W.; Rau, W. Untersuchungen Uber Die Lichtabhangige Carotinoidsynthese VI. zeitlicher verlauf der synthese der einzelnen carotinoide bei *Fusarium aquaeductu-um* unter verschiedenen Induktionsbedingungen. *Planta* **1970,** *94*, 156–174.

Brill, F. Literature Review: The Role of Carotenoids as Functional Foods in Disease Prevention and Treatment. *Nutrition* **2009,** *3*, 1–17.

Britton, G.; Liaaen-Jensen, S.; Pfander, H. Ed. Section II. Main list—Natural Carotenoids. In *Carotenoids Hand Book*; Springer Basel AG: USA, 2004; pp. 646.

Brock, T. D.; Genus, T. In *Bergey's Manual of Systematic Bacteriology*; Krieg, N. R. Ed.; Williams & Wilkins: Baltimore, MD, 1984, vol. 1; pp 333–337.

Burczyk, J. Cell Wall Carotenoids in Green Algae Which Form Sporopollenins. *Phytochemistry* **1986,** *26*, 121–128.

Burja, A. M.; Banaigs, B.; Abou-Mansour, E.; Burgess, J. G.; Wright, P. C. Marine Cyanobacteria—A Profile Source of Natural Products. *Tetrahedron* **2001,** *57*, 9347–9377.

Carail, M.; Caris-Veyrat, C. Carotenoid Oxidation Product: From Villain to Saviour?. *Pure Appl. Chem.* **2006,** *78*, 1493–1503.

Cazzonelli, C. I. Carotenoids in Nature: Insights from Plants and Beyond. *Funct. Plant Biol.* **2011,** *38*, 833–847.

Cerd-Olmedo, E. Production of Carotenoids with Fungi. In *Biotechnology of vitamin, pigments and growth factor*; Vandamme, E. J., Ed.; Elsevier Applied Science: New York, 1989; pp 27–42.

Cheng, Q. Structural Diversity and Functional Novelty of New Carotenoid Biosynthesis Genes. *J. Ind. Microbiol. Biotechnol.* **2006,** *33*, 552–559.

Choudhari, S. M.; Ananthanarayan, L.; Singhal, S. R. Use of Metabolic Stimulators and Inhibitors for Enhanced Production of β-carotene and Lycopene by *Blakeslea trispora* NRRL 2895 and 2896. *Biores. Technol.* **2008,** *99*, 3166–3173.

Clauditz, A.; Resch, A.; Wieland, K.-P.; Peschel, A.; Gotz, F. Staphyloxanthin Plays a Role in the Fitness of *Staphylococcus aureus* and its Ability to Cope with Oxidative Stress. *Infect. Immun.* **2006,** *74*, 4950–4953.

Costa, M.; Costa-Rodrigues, J.; Fernandes, M. H.; Barros, P.; Vasconcelos, V.; Martin, R. Marine Cyanobacteria Compounds with Anticancer Properties: A Review on the Implication of Apoptosis. *Mar. Drugs* **2012,** *10*, 2181–2207.

Costa, S.; Giannantonio, C.; Romagnoli, C.; Vento, G.; Gervasoni, J.; Persichilli, S.; Zuppi, C.; Cota, F. Effects of Lutein Supplementation on Biological Antioxidant Status in Preterm Infants: A Randomized Clinical Trial. *J. Matern. Fetal. Neonatal. Med.* **2013,** *26*, 1311–1315.

Cunningham, F. X.; Chamovitz, D.; Misawa, N.; Gantt, E.; Hirschberg, J. Cloning and Functional Expression in *Escherichia coli* of a Cyanobacterial Gene for Lycopene Cyclase, the Enzyme that Catalyzes the Biosynthesis of β-carotene. *FEBS Lett.* **1993,** *328*, 130–138.

Das, A.; Yoon, S.-H.; Lee, S.-H.; Kim, J.-Y.; Oh, D.-K.; Kim, S.-W. An Update on Microbial Carotenoid Production: Application of Recent Metabolic Engineering Tools. *Appl. Microbiol. Biotechnol.* **2007,** *77*, 505–512.

Davoli, P.; Weber, R. W. S. Carotenoid Pigments from the Red Mirror Yeast, *Sporobolomyces Roseus*. *Mycologist*. **2002,** *16*, 102–108.

Del Campo, J.; Rodriguez, H.; Moreno, J.; Vargas, M.; Rivas, J.; Guerrero, M. Accumulation of Astaxanthin and Lutein in *Chlorella zofingiensis* (*Chlorophyta*). *App. Microbiol. Biotecnol.* **2004,** *64*, 848–854.

Dominguez-Bocanegra, A. R.; Ponce-Noyola, T.; Torres-Munoz, J. A. Astaxanthin Production by *Phaffia rhodozyma* and *Haematococcus pluvialis*: A Comparative Study. *Appl. Microbiol. Biotechnol.* **2007,** *75*, 783–791.

Dufosse, L.; Galaup, P.; Yaron, A.; Arad, S. M.; Blanc, P.; Chidambara Murthy, K. N.; Ravishankar, G. A. Microorganisms and Microalgae as Sources of Pigments for food use: A Scientific Oddity or an Industrial Reality? *Trends. Food. Sci. Technol.* **2005,** *16*, 389–406.

Estrada, A. Z.; Mair, D.; Scherzinger, D.; Avalos, A.; Al-Babili, S. Novel Apocarotenoid Intermediates in *Neurospora crassa* Mutants Imply a New Biosynthetic Reaction Sequence Leading to Neurosporaxanthin Formation. *Fungal Genet. Biol.* **2008,** *45*, 1497–1505.

Finkelstein, M.; Haung, C. C.; Byng, G. S.; Tsau, B.-R.; Leach, J. *Blakeslea trispora* Mated Culture Capable of Increased β-carotene Production, U.S. Patent 5422247 A, June 6, 1995.

Frengova, G. I.; Beskova, D. M. Carotenoids from *Rhodotorula* and *Phaffia*: Yeasts of Biotechnological Importance. *J. Ind. Microbiol. Biotechnol.* **2009,** *36*, 163–180.

Frigard, N.-U.; Maresca, J. A.; Yunker, C. E.; Jones, A. D.; Bryant, B. A. Genetic Manipulation of Carotenoids Biosynthesis in Green Sulfur Bacterium *Chlorobium tepidum. J. Bacteriol.* **2004,** *186*, 5210–5220.

Ganesan, P.; Matsubara, K.; Ohkubo, T.; Tanaka, Y.; Noda, K.; Sugawara, T.; Hirata, T. Anti-Angiogenic Effect of Siphonaxanthin from Green Alga, *Codium fragile*. *Phytomedicine* **2010,** *17*, 1140–1144.

Ganesan, P.; Matsubara, K.; Sugawara, T.; Hirata, T. Marine Algal Carotenoids Inhibit Angiogenesis by Down-Regulating FGF-2-mediated Intracellular Signals in Vascular Endothelial Cells. *Mol. Cell. Biochem.* **2013,** *380*, 1–9.

Ganesan, P.; Noda, K.; Manabe, Y.; Ohkubo, T.; Tanaka, Y.; Maoka, T.; Sugawara, T.; Hirata, T. Siphonaxanthin, A Marine Carotenoid from Green Algae, Effectively Induces Apoptosis in Human Leukemia (HL-60) Cells. *Biochim. et Biophys. Acta* **2011,** *1810*, 497–503.

Garcia-Mendoza, E.; Ocampo-Alvarez, H. Photoprotection in the Brown Alga *Macrocystis pyrifera*: Evolutionary Implications. *J. Photochem. Photobiol.* **2011,** *104*, 377–385.

Garrido-Fernández, J.; Maldonado-Barragán, A.; Caballero-Guerrero, B.; Hornero-Méndez, D.; Ruiz-Barba, J. L. Carotenoid Production in *Lactobacillus plantarum. Int. J. Food Microbiol.* **2010,** *140*, 34–39.

Giraud, E.; Hannibal, L.; Fardoux, J.; Jaubert, M.; Dreyfus, B.; Sturgis, J. N.; Vermeqlio, A. Two Distinct crt Gene Cluster for Two Different Functional Classes of Carotenoid in *Bradyrhizobium. J. Biol. Chem.* **2004,** *279*, 15076–15083.

Gombos, Z.; Vig, L. Primary Role of the Cytoplasmic Membrane in Thermal Acclimation Evidenced in Nitrate-Starved Cells of the Blue-Green Alga, *Anacystis nidulans. Plant. Physiol.* **1986,** *80*, 415–419.

Grabowski, B.; Cunningham, F. X.; Gantt, E. Chlorophyll and Carotenoid Binding in a Simple Red Algal Light-Harvesting Complex Crosses Phylogenetic Lines. *PNAS* **2000,** *98*, 2911–2916.

Guaratini, T.; Lopes, N. P.; Marinho-Soriano, E.; Colepicolo, P.; Pinto, E. Antioxidant Activity and Chemical Composition of the Non Polar fraction of *Gracilaria domingensis (Kützing) Sonderex Dickie* and *Gracilaria birdiae (Plastino & Oliveira)*. *Rev. Bras. Farmacogn.* **2012,** *22*, 724–729.

Hagen, C.; Grünewald, K.; Schmidt, S.; and Müller, J. Accumulation of Secondary Carotenoids in Flagellates of *Haematococcus pluvialis* (*Chlorophyta*) Is Accompanied by an Increase in per Unit Chlorophyll Productivity of Photosynthesis. *Eur. J. Phycol.* **2000,** *35*, 75–82.

Higuera-Ciapara, I.; Felix-Valenzuela, L.; Goycoolea, F. M. Astaxanthin: A Review of its Chemistry and Applications. *Crit. Rev. Food. Sci. Nutr.* **2006,** *46*, 185–196.

Hii, S.-L.; Choong, P.-Y.; Woo, K.-K.; Wong, C.-L. Stability Studies of Fucoxanthin from *Sargassum binderi*. *Aust. J. Basic Appl. Sci.* **2010,** *4*, 4580–4584.

Hirschberg, J.; Harker, M. Carotenoid-Producing Bacterial Species and Process for Production of Carotenoids using Same. U.S. Patent 08,902,518, August 10, 1999.

Hümbelin, M.; Thomas, A.; Lin, J.; Jore, J.; Berry, A. Genetics of Isoprenoid Biosynthesis in *Paracoccus zeaxanthinifaciens*. *Gene*. **2002,** *297*, 129–139.

Ide, T.; Hoyab, M.; Tanakaa, T.; Harayama, S. Enhanced Production of Astaxanthin in *Paracoccus sp.* strain *N-81106* by using Random Mutagenesis and Genetic Engineering. *Biochem. Eng.* **2012,** *65*, 37–43.

Jaswir, I.; Noviendri, D.; Hasrini, R. F.; Octavianti, F. Carotenoids: Sources, Medicinal Properties and Their Application in Food and Nutraceutical Industry. *J. Med. Plants Res.* **2011,** *5*, 7119–7131.

Johnson, E. A. *Phaffia rhodozyma*: Colorful Odyssey. *Int. Microbiol.* ***2003,*** *6*, 169–174.

Johnson, E. A.; Lewis, M. J. Astaxanthin Formation by the Yeast *Phaffia rhodozyma*. *J. Gen. Microbiol.* **1979,** *115*, 173–183.

Kelman, D.; Ben-Amotz, A.; Berman-Frank, I. Carotenoids Provide the Major Antioxidant Defense in the Globally Significant N2-fixing Marine Cyanobacterium, *Trichodesmium*. *Environ. Microbiol.* **2009,** *11*, 1897–1908.

Khuantrairong, T.; Traichaiyaporn, S. The Nutritional Value of Edible Freshwater Alga *Cladophora sp. (Chlorophyta)* Grown under Different Phosphorus Concentrations. *Int. J. Agric. Biol.* **2011,** *13*, 297–300.

Krinsky N. I.; Johnson, E. J. Carotenoid Actions and Their Relation to Health and Disease. *Mol. Aspects Med.* **2005,** *26*, 459–516.

Krubasik, P.; Sandmann, G. Molecular Evolution of Lycopene Cyclases Involved in the Formation of Carotenoids with Ionone end Groups. *Biochem. Soc. Trans.* **2000,** *28*, 806–810.

Lakshminarayana, R.; Raju, M.; Krishnakantha, T. P.; Baskaran, V. Determination of Major Carotenoids in Few Indian Leafy Vegetables by High-Performance Liquid Chromatography. *J. Agric. Food. Chem.* **2005,** *53*, 2838–2842.

Lakshminarayana, R.; Raju, M.; Krishnakantha, T. P.; Baskaran. V. Enhanced Lutein Bioavailability by Lyso-phosphatidylcholine in Rats. *Mol. Cell. Biochem.* **2006,** *281*, 103–113.

Lampila, L. E.; Wallen, S. E.; Bullerman, L. B. A Review of Factors Affecting Biosynthesis of Carotenoids by the Order Mucorales. *Mycopathology* **1985,** *90*, 65–80.

Lichtenthaler, H. K.; Rohmer, M.; Schwemder, J. Two Independent Biochemical Pathway for Isopentenyl Di Phosphate and Isoprenoid Biosynthesis in Higher Plants. *Physol. Plant.* **1997,** *101*, 643–652.

Liu, G. Y.; Essex, A.; Buchanan, J. T.; Datta, V.; Hoffman. H. M.; Bastian, J. F.; Fierer, J.; Nizet, V. *Staphylococcus aureus* Golden Pigment Impairs Neutrophil Killing and Promotes Virulence Through Its Antioxidant Activity. *J. Exp. Med.* **2005,** *202*, 209–215.

Lutnaes, B. F.; Stramnd, A.; Pétursdóttir S. K.; Liaaen-Jensen, S. Carotenoids of Thermophilicbacteria *Rhodothermusmarinus* from Submarine Icelandic Hot Springs. *Biochem. Syst. Ecol.* **2004,** *32*, 455–468.

Maeda, H.; Hosokawa, M.; Sashima, T.; Miyashita, K. Dietary Combination of Fucoxanthin and Fish Oil Attenuates the Weight Gain of White Adipose Tissue and Decreases Blood Glucose in Obese/Diabetic KK-Ay Mice. *J. Agric. Food Chem.***2007,** *55*, 7701–7706.

Mamatha, B. S.; Sangeetha, R. K.; Baskaran, V. Provitamin-A and Xanthophyll Carotenoids in Vegetables and Food Grains of Nutritional and Medicinal Importance. *Int. J. Food Sci. Technol.* **2011,** *46,* 315–323.

Mamatha, B. S.; Arunkumar, R.; Baskaran, V. Effect of Processing on Major Carotenoid Levels in Corn (*Zea mays*) and Selected Vegetables: Bioavailability of Lutein and Zeaxanthin from Processed Corn in Mice. *Food Bioproc. Technol.* **2012,** *5,* 1355–1363.

Mamatha, B. S.; Baskaran, V. Effect of Micellar Lipids, Dietary Fiber and β-carotene on Lutein Bioavailability in Aged Rats with Lutein Deficiency. Nutrition, **2011,** *27,* 960–966.

Marit, H.; Stafsnes.; Kjell, D.; Josefsen.; Kildahl-Andersen, G.; Valla, S.; Trond, E.; Ellingsen.; Bruheim, P. Isolation and Characterization of Marine Pigmented Bacteria from Norwegian Coastal Waters and Screening for Carotenoids with Uva-Blue Light Absorbing Properties. *J. Microbiol.* **2010,** *8,* 16–23.

Mata-Gomez, L. C.; Montanez, J. C.; Mendez-Zavala, A.; Aguilar, C. N. Biotechnological Production of Carotenoids by Yeasts: An Overview. *Microb. Cell Fact.* **2014,** *13,* 12.

Mehta, B. J.; Obraztsova, I. R.; Cerda-olmedo, E. Mutants and Intersexual Heterokaryons of *Blakeslea trispora* for Production of β-carotene and Lycopene. *Appl. Environ. Microbiol.* **2003,** *69,* 4043–4048.

Misawa, N.; Nakagawa, M.; Kobayashi, K.; Yamano, S.; Izawa, Y.; Nakamura, K.; Harashima, K. Elucidation of the Erwiniauredovora Carotenoid Biosynthetic Pathway by Functional Analysis of Gene Products Expressed in *Escherichia coli*. *J. Bacteriol.* **1990,** *172,* 6704–6712.

Miura, Y.; Kondo, K.; Saito, T.; Shimada, H.; Fraser, P. D.; Misawa, N. Production of the Carotenoids Lycopene, β-Carotene, and Astaxanthin in the Food Yeast *Candida utilis*. *Appl. Environ. Microbiol.* **1998,** *64,* 1226–1229.

Montonen, J.; Knekt, P.; Järvinen, R.; Reunanen, A. Dietary Antioxidant Intake and Risk of Type 2 Diabetes. *Diabetes Care*. **2004,** *27,* 362–366.

Naziri, D.; Hamidi, M.; Hassanzadeh, S.; Tarhriz, V.; Maleki Zanjani, B.; Nazemyieh, H.; Hejazi, M. A.; Hejazi, M. S. Analysis of Carotenoid Production by *Halorubrum* sp. TBZ126: An Extremely Halophilic Archeon from Urmia Lake. *Adv. Pharm. Bull.* **2014,** *4,* 61–67.

Nells, H. J.; De Leenheer, A. P. Microbial Sources of Carotenoid Pigments Used in Foods and Feeds. *J. Appl. Bacteriol.* **1991,** *70,* 181–191.

Nidhi, B.; Mamatha, B. S.; Baskaran, V. Olive Oil Improves the Intestinal Absorption and Bioavailability of Lutein in Lutein-Deficient Mice. *Eur. J. Nutr.* **2014,** *53,* 117–126.

Nishida, Y.; Adachi, K.; Kasai, H.; Shizuri, Y.; Shindo, K.; Sawabe, A .; Komemushi, S.; Miki, W.; Misawa, N. Elucidation of a Carotenoid Biosynthesis Gene Cluster Encoding a Novel Enzyme, 2,2-hydroxylase, from *Brevundimonas* sp. strain *SD212* and Combinatorial Biosynthesis of New or Rare Xanthophylls. *Appl. Environ. Microbiol.* **2005,** *71,* 4286–4296.

Noviendri, D.; Jaswir, I.; Salleh, H. M.; Taher, M.; Miyashita, K.; Ramli, N. Fucoxanthin Extraction and Fatty Acid Analysis of *Sargassum binderi* and *S. duplicatum*. *J. Medic. Plant. Res.* **2011,** *11,* 2405–2412.

Offord, E. A.; Gautier, J.; Avanti, O.; Scaletta, C.; Runge, F.; Kramer, K.; Applegate, L. A. Photoprotective Potential of Lycopene, β-carotene, Vitamin E, Vitamin C and Carnosic Acid in UVA-Irradiated Human Skin Fibroblasts. *Free Rad. Biol. Med.* **2002,** *32,* 1293–1303.

Osawa, O.; Harada, H.; Choi, S.-K.; Misawa, N.; Shindo, K. Production of Caloxanthin 30-b-D-glucoside, zeaxanthin 3,30-b-D-diglucoside, and Nostoxanthin in a Recombinant *Escherichia coli* Expressing System Harboring Seven Carotenoid Biosynthesis Genes, Including *Crt X* and *Crt G*. *Phytochemistry* **2011,** *72*,711–716.

Pashkow, F. J.; Watumull, D. G.; Campbell, C. L. Astaxanthin: A Novel Potential Treatment for Oxidative Stress and Inflammation in Cardiovascular Disease. *Am. J. Cardiol.* **2008,** *101*, 58–68.

Paust, J.; Hoppe, P. P. M.; Kohler, W.; Lueddecke, E.; Schneider, J. U. Rheude, T. Feed Stuff Containing Neurosporaxanthine or its Ester(s)—Used to Improve Pigmentation of Egg Yolks, Skin, Meat and Legs of Broiler(s) and Flesh of Fish and Crustacea and their Shells. Patent DE 4100739 A1, June16, 1992.

Prasanna, R.; Pabby, A.; Singh, P. K. Effect of Glucose and Light–Dark Environment on Pigmentation Profiles in the *Cyanobacterium, Calothrix elenkenii. Folia Microbiol.* **2004,** *49*, 26–30.

Prasanna, R.; Sood, A.; Jaiswal, P.; Nayak, S.; Gupta, V.; Chaudhary, V.; Joshi, M. Natarajan, M. Rediscovering Cyanobacteria as Valuable Sources of Bioactive Compounds (Review). *Appl. Biochem. Microbiol.* **2010,** *46*, 119–134.

Raju, M.; Varakumar, S.; Lakshminarayana, R.; Krishnakantha, T. P.; Baskaran, V. Carotenoid Composition and Vitamin A Activity of Medicinally Important Green Leafy Vegetables. *Food Chem.* **2007,** *101*, 1598–1605.

Rao, A. V.; Rao, L. G. Carotenoids and Human Health. *Pharmacol. Res.* **2007,** *55*, 207–216.

Rock, C. L. Carotenoids: Biology and Treatment. *Pharmacol. Ther.* **1998,** *753*, 185–197.

Rodrigo. M. J.; Marcos. J. F.; Zacarias. L. Biochemical and Molecular Analysis of Carotenoid Biosynthesis in Flavedo of Orange (*Citrus sinensis* L.) During Fruit Development and Maturation. *J. Agric. Food Chem.* **2004,** *52*, 6724–6731.

Rodriguez-Concepcion, M.; Boronat, A. Elucidation of the Methylerythritol Phosphate Pathway for Isoprenoid Biosynthesis in Bacteria and Plastids. A metabolic Milestones Achieved through Genomics. *Plant. Physiol.* **2002,** *130*, 1079–1089.

Rodriguez-Saiz, M.; Dela Fuente, J. L.; Barredo, J. L. *Xanthophyllomyces dendrorhous* for the Industrial Production of Astaxanthin. *Appl. Microbiol. Biotechnol.* **2010,** *88*, 645–658.

Rohmer, M.; Knani, M.; Simonil, P.; Sutter, D.; Sahm, H. Isoprenoid Biosynthesis in Bacteria: A Noval Pathway for Early Steps Leading to Isopentenyl Diphosphate. *Biochem. J.* **1993,** *2*,295, 517–524.

Ruther, A.; Misawa, N.; Boger, P.; Sandmann, G. Production of Zeaxanthin in *Escherichia coli* Transformed with Different Carotenogenic Plasmids. *Appl. Microbiol. Biotechnol.* **1997,** *48*, 162–167.

Sajilata, M. G.; Singhal, R. S.; Kamat, M. Y. The Carotenoid Pigment Zeaxanthin—A Review. *Comprehen. Rev. Food. Sci. Food Safety*. **2008,** *7*, 29–49.

Sandmann, G. Combinatorial Biosynthesis of Carotenoids in a Heterologous Host: A Powerful Approach for the Biosynthesis of Novel Structures. *Chem. Biochem.* **2002,** *3*, 629–635.

Sangeetha, R. K.; Baskaran, V. Carotenoid Composition and Retinol Equivalent in Plants of Nutritional and Medicinal Importance: Efficacy of β-carotene from *Chenopodium album* in Retinol-deficient Rats. *Food Chem.***2010,** *119*, 1584–1590.

Schmidt, I.; Schewe, H.; Gassel, S.; Jin, C.; Buckingham, J.; Humbelin, M.; Sandmann, G.; Schrader, J. Biotechnological Production of astaxanthin with *Phaffia rhodozyma/ Xanthophyllomyces dendrorhous*. *Appl. Microbiol. Biotechnol.* **2011,** *89*, 555–571.

Schubert, N.; García-Mendoza, E.; Enríquez, S. Is the Photo-acclimatory Response of Rhodophyta Conditioned by the Species Carotenoid Profile? *Limnol. Oceanogr.* **2011,** *56*, 2347–2361.

Shepherd, D.; Dasek, J.; Suzanne, M.; Carels, C. Production of Zeaxanthin. US Patent 3,951-743, 1976.

Sommerburg, O.; Keunen, J. E. E.; Bird, A. C.; van Kuijk, F. J. G. M. Fruits and Vegetables that are Sources for Lutein and zeaxanthin: The Macular Pigment in Human Eyes. *Br. J. Ophthalmol.* **1998,** *82*, 907–910.

Stafsnes, M. H.; Josefsen, K. D.; Kildahl-Andersen, G.; Valla, S.; Ellingsen, T. E.; Bruheim, P. Isolation Characterization of Marine Pigmented Bacteria from Norwegian Coastal Waters and Screening for Carotenoids with UVA-Blue Light Absorbing Properties. *J. Microbiol.* **2010,** *48*, 16–23.

Strobel, I.; Breitenbach, J.; Scheckhuber, C. Q.; Osiewacz, H. D.; Sandmann, G. Carotenoids and Carotenogenic Genes in *Podospora anserina*: Engineering of the Carotenoid Composition Extends the Life Span of the Mycelium. *Curr. Genet.* **2009,** *55*, 175–184.

Sugawara, T.; Matsubara, K.; Akagi, R.; Mori, M.; Hirata, T. Anti-Angiogenic Activity of Brown Algal Fucoxanthin and its Deacetylated Product, Fucoxanthinol. *J. Agric. Food Chem.* **2006,** *54*, 9805–9810.

Sugawara, T.; Ganesan, P.; Li, Z.; Manabe, Y.; Hirata, T. Siphonaxanthin, a Green Algal Carotenoid, as a Novel Functional Compound. *Mar. Drugs.* **2014,** *12*, 3660–3668.

Takaichi, S. Carotenoids in Algae: Distributions, Biosyntheses and Functions. *Mar. Drugs.* **2011,** *9*, 1101–1118.

Takaichi, S.; Mimuro, M. Distribution and Geometric Isomerism of Neoxanthin in Oxygenic Phototrophs: 9-cis, a Sole Molecular form. *Plant Cell. Physiol.* **1998,** *39*, 968–977.

Takaichia, S.; Mochimaru, M. Carotenoids and Carotenogenesis in Cyanobacteria: Unique Ketocarotenoids and Carotenoid Glycosides. *Cell. Mol. Life. Sci.* **2007, 64,** 2607–2619.

Tanaka, T.; Shnimizu, M.; Moriwaki, H. Cancer Chemoprevention by Carotenoids. *Molecules*. **2012,** *17*, 3202–3242.

Tanskul, S.; Khoonchumnan, S.; Watanasit, S.; Oda, K. Application of a New Red Carotenoid Pigment-Producing Bacterium, *Enterobacter* sp. P41, as feed Supplement for Chicken. *Afr. J. Biotechnol.* **2013,** *12*, 64–69.

Tatsuzawa, H.; Maruyama, T.; Misawa, N. Fujimori, K.; Nakano, M. Quenching of Singlet Oxygen by Carotenoids Produced in *Escherichia coli*-attenuation of Singlet Oxygen-Mediated Bacterial Killing by Carotenoids. *FEBS. Lett.* **2000,** *484*, 280–284.

Terasaki, M.; Bhaskar, N; Kamokawa, H.; Nomura, M.; Stephen, M. N.; Kawagoe, C.; Hosokawa, M.; Miyashita. K. Carotenoid Profile of Edible Japanese Seaweeds: An Improved HPLC Method for Separation of major Carotenoids. *J. Aquat. Food. Prod. Technol.* **2012,** *21*, 468–479.

Tian, B.; Hua, Y. Carotenoid Biosynthesis in *Extremophilic Deincoccus*–Thermos Bacteria. *Trends. Microbiol.* **2010,** *8*, 11.

Tinoi, J.; Rakariyatham, N.; Deming, R. L. Determination of Major Carotenoid Constituents in Petal Extracts of Eight Selected Flowering Plants in the North of Thailand. *J. Sci.* **2006,** *33*, 327–334.
Urikura, I.; Sugawara, T.; Hirata, T. Protective Effect of Fucoxanthin against UVB-Induced Skin Photoaging in Hairless Mice. *Biosci. Biotechnol. Biochem.* **2011,** *75*, 757–760.
Ursi, S.; Pedersén, M.; Plastino, E.; Snoeijs, P. Intraspecific Variation of Photosynthesis, Respiration and Photoprotective Carotenoids in *Gracilaria birdiae* (*Gracilariales: Rhodophyta*). *Mar. Bol.* **2003,** *142*, 997–1007.
Vadali, R. V.; Fu, Y.; Bennett, G. N.; San, K. Y. Enhanced Lycopene Productivity by Manipulation of Carbon Flow to Isopentenyl diphosphate in *Escherichia coli. Biotechnol. Prog.* **2005,** *21*, 1558–1561.
Valadon, L. R. G.; Mummery, R. S. Biosynthesis of Neurosporaxanthin. *Microbios* **1969,** *1*, 3–8.
Valadon, L. R. G.; Osman, M.; Mummery, R. S.; Jerebzoff-Quintin, S.; Jerebzoff, S. The Effect of Monochromatic Radiation in the Range 350 to 750 nm on the Carotenogenesis in *Verticillium agaricinum. Physiol. Plant.* **1982,** *56*, 199–203.
Verwaal, R.; Wang, J.; Meijnen, J. P.; Visser, H.; Sandmann, G.; van den Berg, J. A.; van Ooyen, A. J. High-level Production of Beta-carotene in *Saccharomyces cerevisiae* by Successive Transformation with Carotenogenic Genes from *Xanthophyllomyces dendrorhous. Appl. Environ. Microbiol.* **2007,** *73*, 43–50.
Wang, B.; Lin. L.; Lu, L.; Chen, W. Optimization of β-carotene Production by a Newly Isolated *Serratia marcescens* Strain. *J. Biotechnol.* **2012,** *15*, 1–3.
Wang, H. B.; Xu, R. G, Yu, L. J.; Luo, J.; Zahang, L. W.; Huang, X. Y.; Zou, W. A.; Zhao, Q.; Lu, M. B. Improved Beta-carotene and Lycopene Production by *Blakeslea trispora* with Ultrasonic Treatment in Submerged Fermentation. *Z. Naturforsch. C.* **2014,** *69*, 237–244.
Xu, F.; Yuan, Q.-P.; Zhu, Y. Improved Production of Lycopene and β-carotene by *Blakeslea trispora* with Oxygen-Vectors. *Process. Biochem.* **2007,** *42*, 289–293.
Yokoyama, A.; Sandmann, G.; Hoshino, T.; Adachi, K.; Sakai, M.; Shizuri, Y. Thermozeaxanthins, New Carotenoid-Glycoside-Esters from Thermophilic Eubacterium *Thermos thermophilus. Tetrahedron Lett.* **1995,** *36*, 4901–4904.
Yoon, S. H.; Kim, J. E.; Lee. S. H.; Park, H. M.; Choi, M. S.; Kim, J. Y.; Lee, S. H.; Shin, Y. C.; Keasling, J. D.; Kim, S. W. Engineering the Lycopene Synthetic Pathway in *E. coli* by Comparison of the Carotenoid Genes of *Pantoea agglomerans* and *Pantoea ananatis. Appl. Microbiol. Biotechnol.* **2007,** *74*, 131–139.
Yoon, S. H.; Lee, Y. M.; Kim, J. E.; Lee, S. H.; Lee, J. H.; Kim, J. Y.; Jung, K. H.; Shin, Y. C.; Keasling, J. D.; Kim, S. W. Enhanced Lycopene Production in *Escherichia coli* Engineered to Synthesize Isopentenyl Diphosphate and Dimethylallyl Diphosphate from Mevalonate. *Biotechnol. Bioeng.* **2006,** *94*, 1025–1032.
Yoona, S.-H.; Leea, S.-H.; Dasa, A.; Ryua, H.-K.; Janga, H.-J.; Kima, J.-Y.; Ohb, D.-K.; Keaslingc, J.-D.; Kima, S.-W. Combinatorial Expression of Bacterial whole Mevalonate Pathway for the Production of β-carotene in *E. coli. J. Biotech.* **2009,***140*, 218–226.
Yoshii, Y.; Takaichi, S.; Maoka, T.; Suda, S.; Sekiguchi, H.; Nakayama, T.; Inouye, I. Variation of Siphonaxanthin Series among the Genus Nephroselmis (*Prasinophyceae, Chlorophyta*), Including a Novel Primary Methoxy Carotenoid. *J. Phycol.* **2005,** *41*, 827–834.

Yuan, L. Z.; Rouviere, P. E.; Larossa, R. A.; Suh, W. Chromosomal Promoter Replacement of the Isoprenoid Pathway for Enhancing Carotenoid Production in *E. coli*. *Metab. Eng.* **2006,** *8*, 79–90.

Zeb, A.; Mehmood, S. Carotenoids Contents from Various Sources and their Potential Health Applications. *Pak. J. Nutr.* **2004,** *3*,199–204.

Zheng, Y.-G.; Hu, Z.-C.; Wang, Z.; Shen, Y. C. Large-Scale Production of Astaxanthin by *Xanthophyllomyces dendrorhous*. *Food. Bioprod. Process*. **2006,** *84*, 164–166.

Zhu, Y.-H.; Jiang, J.-G. Continuous Cultivation of *Dunaliella salina* in Photobioreactor for the Production of β-carotene. *Eur. Food. Res. Technol*. **2008,** 227, 953–959.

CHAPTER 4

CARROT: SECONDARY METABOLITES AND THEIR PROSPECTIVE HEALTH BENEFITS

KAMLESH PRASAD[1*], RAEES-UL HAQ[1], VASUDHA BANSAL[2], MOHAMMED WASIM SIDDIQUI[3], and RIADH ILAHY[4]

[1]*Department of Food Engineering and Technology, SLIET, Longowal 148106, Punjab, India*

[2]*Center of Innovative and Applied Bioprocessing (CIAB), Phase-8, Industrial Area, S.A.S. Nagar, Mohali-160071, Punjab, India*

[3]*Department of Food Science and Post-Harvest Technology, Sabour, Bhagalpur 813210, Bihar, India*

[4]*Laboratory of Horticulture, National Agricultural Research Institute of Tunisia, Tunis, Rue Hнdi Karray 2049 Ariana, Tunisia*

**Corresponding author, E-mail address: dr_k_prasad@rediffmail.com*

CONTENTS

ABSTRACT

Fruits and vegetables are the major sources of plant bioactive components required for the proper functioning of body. The carrot consumption has cosmopolitan distribution throughout the globe with temperate regions as the major producers. Carrot is considered as a rich source of fibers, vitamins, minerals, and various secondary metabolites particularly terpenoids and phenolics. Carotenoids are the main class of the terpenoids and available in appreciable quantities. Presence of the carotenoids, a provitamin A, in carrot regard them as the cheap source. Color of the roots is an important characteristic emerging due to presence of the bioactive components associated with red, orange, yellow, and purple colors arising due to lycopene, carotene, lutein, and anthocyanins. The purple carrots have been said to be the phenolics-rich source among the other colored carrot varieties. Consequently, it has the highest antioxidant activities as compared to available other colored roots. The immense health-promoting benefits of the carrot are due to phytonutrients that have significantly increased their use in recent past. The well-known functions of carotenoids and phenolics as natural and potent antioxidants have concentrated their way of study as anticancer agents. The association of polyacetylene compounds and fiber compounds in carrot further contributes to the roots' functionality toward human health. The ideal storage temperature of the carrots is below the room temperature as higher temperatures decrease shelf life. The perishable nature of the roots has led to the development of new techniques of preservation of different processed form of carrot. The preservation of carrot in the different forms has significantly found to improve the marketability with minimum postharvest losses. The carrot pomace, a leftover after juice extraction in its dried form, have found to be of historical utilization among bakery products. The functionality imparted by the incorporation of different bioactive components of carrot to various processed products has further strengthened their use as a source-value addition. The present chapter is an effort of documenting the nature, composition, nutritional properties, and utilization of the carrots. The present chapter also describes the existence of different colorants, their extraction procedures, and various associated health benefits of carrot.

4.1 INTRODUCTION

Carrot (*Daucus carota* L.) is an important carotenoid-rich vegetable of *Apiaceae,* and previously known as *Umbelliferae* family, and grown as a fleshy edible root throughout the world (Janve et al., 2014). Fruits and vegetables are rich and cheap sources of nutrients, dietary fiber, and non-nutritive bioactive phytochemicals (Cordell et al., 2007). Increased consumption of the fruits and vegetables has significantly reduced the risk of various ailments including cancer (Block et al., 1992). Carrot as a temperate crop is being cultivated in India mainly during winter season. It is rich in several functional compounds like vitamins, minerals, and phytonutrients (Block, 1994; Torronen et al., 1996) and ranks seventh in overall contribution to nutrition among fruits and vegetables (Alasalvar et al., 2001). The significant health-promoting properties of carrots are because of their high carotenoid content, dietary fiber, and presence of appreciable amounts of vitamins (thiamine, riboflavin, niacin, B_5, pyridoxine, biotin, and cobalmine) and minerals (Manjunatha et al., 2003; Hanif et al., 2006; Pakin et al., 2004; Staggs et al., 2004). New carrot-derived products are found to have an increased consumption pattern in recent years (Alasalvar et al., 2001). Carotenoid pigments are not usually present among root vegetables except carrot, in which α-and β-carotenes predominate (Rodriguez-Amaya, 2001). The presence of carotene, fiber, and falcarinol compounds in carrots makes it a functional commodity. In fresh carrot, carotene is present in the chromoplasts and all-*trans*-α and β-carotenoids account for more than 90% of the total carotenoids (Simon & Wolff, 1987), which also contribute toward color of the carrot (Bushway & Wilson, 1982; Heinonen, 1990; Khachik et al., 1991; Mangels et al., 1993). The consumer interest has been shown on increased side during past years for the carrot carotenoids as nutritional supplements and food colorants (Simon et al., 1997; Rubatzky et al., 1999). Carrot being a veg-fruit, rich in various types of sugars, has paved the way for its use as table purposes and also as cooked or processed forms (Singh et al., 1999).

4.1.1 PRODUCTION

The total world production of carrot and turnip is 35.66 MMT with China, Russia, USA, and Uzbekistan as major producers contributing up to 57.5% of world's total production (FAOSTAT, 2013). India is the second largest producer of vegetable after China and ranked thirteenth in carrot and turnip production with an annual production of 1.15 MMT (Anon., 2012). The major carrot-producing states in India with their production and productivity data are shown in Table 4.1. Haryana, Andhra Pradesh, Punjab, Karnataka, Tamil Nadu, Uttar Pradesh, Assam, Bihar, and Jammu and Kashmir are the major carrot-producing states; Tamil Nadu leads with the productivity of 26.51 MT/Ha and Rajasthan 7.24 MT/Ha has the lowest productivity (Anon., 2012).

TABLE 4.1 Production and Productivity of Carrot in India.

State	2011–2012		2012–2013	
	Production*	Productivity**	Production	Productivity
Haryana	366.64	19.43	325.65	17.62
Andhra Pradesh	172.51	18.32	177.65	18.13
Punjab	115.02	20.47	120.03	20.47
Karnataka	94.10	94.10	97.90	18.47
Tamil Nadu	108.76	27.19	89.62	26.51
Uttar Pradesh	71.56	23.54	74.46	24.06
Assam	65.10	15.75	66.87	15.86
Bihar	59.35	12.23	60.45	12.33
Jammu and Kashmir	33.33	24.69	33.33	24.69
Delhi	23.01	17.11	27.97	25.55
Chhattisgarh	19.20	13.91	16.01	13.80
Rajasthan	4.26	3.80	12.04	7.24
Meghalaya	9.60	13.75	9.88	13.70
Jharkhand	–	–	7.11	11.47
Tripura	3.75	12.10	3.70	12.33
Sikkim	2.77	7.82	2.82	7.83
Nagaland	1.83	6.43	2.40	8.00
Mizoram	2.49	7.11	1.47	7.34
Odisha	–	–	1.03	12.88
Manipur	–	–	0.31	12.20
Total	1153.28	18.54	1130.70	17.97

*Production = '000 MT and **Productivity = MT/Ha.

4.1.2 BOTANY

The systematic taxonomic classification of plant carrot is presented in Table 4.2. The plant is a dicot with two cotyledons in embryo, diploid with nine chromosome pairs, and has fine hairs on its stem and leaves with a racemose inflorescence organized as umbel at the apex of the plant that grows during the second season as a distinctive characteristic feature of the family Umbelliferae (Bradeen & Simon, 2007).

TABLE 4.2 Taxonomy of *Daucus carota.*

Systematic Classification		Characteristics
Domain	Eukaryota	Have membrane-bound cell organelles
Kingdom	Plantae	Photosynthetic with cellolusic cell walls
Phylum	Magnoliophyta	Flowering plants
Class	Magnoliosida	Dicotyledons, possessing vascular tissue arranged in a ring and has tap root
Order	Apiales	Simple flowers, with the ability to self pollinate
Family	Apiaceae	Bisexual flower clusters arranged as umbel with five each sepals and petals with two locules in ovary and schizocarp fruit
Genus	Daucus	Wild carrots recognized by inflorescence and roots
Species	*D. carota*	Large conical tap root, colored red/orange/purple/black with sterile flowers

4.1.2.1 FOLIAGE OF CARROT

Foliage of carrot mainly includes the shoot system as leaves and flowers. The former doesnot take part in the propagation of plant but is actively involved in photosynthesis while the latter in the production of seeds through reproduction process.

4.1.2.1.1 Non-Reproductive Parts

The dorsoventrally flattened structures emerging from the nodes of stem are leaves. These photosynthetic parts are collectively known as phyllome of plant. The leaves of carrot are decompound (the compound leaf blade is pinnate more than thrice) or bi-pinnate/tri-pinnate or lobed (the leaflet-bearing axis is secondary or tertiary), linear or lanceolate (length more than width or wider at base than at the end), and dentate. The leaves are also opposite (leaf pairs at each node) and superposed (plane of occurrence is same) and with reticulate (dense network) venation. The stem is coniform, napiform, and fusiform, is epiterranean, reduced to disc from which the branches arise (Lawrence, 1951; Barclay, 2002).

4.1.2.1.2 Reproductive Parts

The presence of additional flowers on flowering axis is represented by inflorescence that can be racemose or cymose, depending on growth and development of flowering shoot. The umbel inflorescence has equal flower stalks (pedicel) equally originated from the regular points, and in carrot, there is a compound umbel with a secondary umbel arising from the tip of primary-branched umbel, which is closely related with corymbs in *Spiraea*.

The reproductive parts participate either directly (stamens and carpels) or indirectly (perianth) in the plant reproduction. These are present in an organized manner on the flower with basic four floral parts (calyx, corolla, stamen, and carpel) arranged in continuous spiral on the receptacle, which is an extended end of pedicel (stalk). The absence of any part from the above-described ones renders the flower as incomplete. The stalk of the flower is peduncle that is arising from the common origin with leaves (bract) to the receptacle. *D. carota* flowers have an extra internode between the bracteole (arising from top of the peduncle) and receptacle of each flower known as pedicel (Kingsley et al., 2008; Jones and Luchsinger, 1987; Sharma, 2009; Clarke & Lee, 2003).

4.1.2.1.2.1 Perianth

Calyx and corolla constitute the perianth, with the former present as green whorl of sepals that enclose the bud stage of the flower. The two parts of perianth are usually equal in number with corolla present inside the calyx and consists of the bright-colored or white petals that assist in pollination by attracting the pollinators. The absence of stamen or carpel in the flower produces unisexual imperfect flower that can be carpellate or staminate and occurs as monoecious (single plant) or diecious (different plants). Sepals can be aposepalous/polysepalous and synsepalous with the first one having separate distinguishable sepals and the later one with marginally fused sepals forming a tube-like structure. The calyx in the flower, if present, is reduced and aposepalous whereas hibiscus has fused sepals corresponding to gamopetalous or synsepalous. The petals in the carrot flower are apopetalous, in which petals are apart from each other. The other type of arrangement involves sympetalous, where they are fused fully or marginally to form a short tube as in *Foeniculum vulgare*. The regular or actinomorphic flower has equal-sized petals and any plane through center will divide it into two identical halves whereas the zygomorphic flower possesses bilateral symmetry because of varying size of sepals or petals. The flowers of Apiaceae family including carrot are having usually radial or actinomorphic symmetry and rarely zygomorphic or bilateral. The aestivation (arrangement of floral parts, particularly sepals and petals) in *D. carota* flowers is valvate, where petals are close to one another without contacting each other (Kingsley et al., 2008; Lawrence, 1951; Sharma, 2009; Clarke & Lee, 2003).

4.1.2.1.2.2 Reproductive Organs

Androecium (stamen) and gynoecium (carpel) contribute fertility to the flower, with the former one comprising thin anther-loaded filament and the latter one made of stigma, style, and ovary. The point of attachment of anther lobes with the filament can be from the base or back side known as basifixed or dorsifixed, respectively. The five stamens are apostemonous (separate) and basifixed in the carrot flower. The female ovarian structure has two locules present at the basal part of the carpel whose upper portions can be separated or fused designated as apocarpus and syncarpous, respectively. Carrot carpel has two styles attached to two locules each containing an ovary with three nuclei. The ovary in carrot is epigynous (upon gynoecium) in which the stamens and perianth parts merge together and forming hypanthium that is fused with the ovary and free parts of fused parts appear to be emerging from ovarian top forming an inferior ovary. The superior ovary is in hypogynous, where perianth and stamens are attached to receptacle below the gynoecium as in drupes (Sharma, 2009; Clarke & Lee, 2003).

4.1.2.1.2.3 Seed

The biennial cultivation of root crop develops flowers, and reproduction takes place in a similar way to the other flowering plants. Pollination is the transfer of viable pollen grains from anther lobes to stigma where the pollen grain germinates to form a pollen tube up to the ovary. The locular nuclei are fertilized by two pollens commonly referred as double fertilization. The fertilization yields a diploid zygote by fusion from single male nucleus with a female nucleus and triploid nucleus by combination of one male and two female nuclei. The fruit in carrots is a simple dry fruit known as cremocarp, which is a dry schizocarp fruit with two-seeded moieties each one present inside the mericarp (Sharma, 2009; Lawrence, 1951).

4.1.2.2 EDIBLE OR ROOT PORTIONS

The seed upon germination produces radicle that moves opposite to the shoot (phototropic and geotropic) deep inside the ground providing stability to the plant. These roots anchorage with the soil through primary root. The roots in carrot are tap and swollen adventitious forms, modified for food storage (Barclay, 2002). The secondary roots, given by the primary, bear minute root hairs for the uptake of minerals and water for the plant. The growth of the primary roots is governed by apical meristem encased within the root cap that secretes a mucous-like secretion for easier penetration. In case of the root crops, these primary roots are modified that store the synthesized plant matter resulting in their size enlargement and are classified as tubercular or tubular, napiform, fusiform, and coniform roots based on their shapes. The examples of the first-three root forms in their respective order are sweet potato, turnip, and radish, respectively (Sharma, 2009; Clarke & Lee, 2003).

The conical root is broad at base that gradually tapers at end, forming a conical structure and predominately associated with carrot. The central core, cambium, phloem, and periderm are the four distinctive tissues of carrot with the first two represented by a collective term "core" and the latter two by "cortex." The quality of the root depends upon these two proportions with good quality carrots containing high proportions of cortex in comparison to core (Nonnecke, 1989). The periderm part is rich in secondary plant metabolites (phenolics, terpenoids, and alkaloids) than the rest root portion and the edible portion is composed mainly of the saccharide-storing parenchymatic tissue associated with the phloem (Radovich, 2011). The vascular bundle systems are arranged in ring form with the growth of vascular cambium producing xylem and phloem tissue toward inward and outward, respectively, and their ratio is an important quality indicator that decreases with root growth (Sharma, 2009; Northolt et al., 2004).

4.1.3 CULTIVATION

Carrot is a biennial, subterranean, herbaceous root crop with alternate, compound leaves organized as a rosette and is grown worldwide, predominantly in temperate regions during spring, summer, and autumn seasons. The prevalence of high-temperature conditions in the tropics and subtropics limits carrot cultivation to winter season. The chilling treatment required by the European varieties for seed production restricts their farming to temperate regions only (Dev, 2009). The cultivation of carrot is carried through conventional farming (Woese et al., 1997). For better yield of carrot well-drained, loose, sandy soils rich in humus (Wrzodak et al., 2012) with four months to mature for the development of better flavor and color under the temperature range of 10–25°C. The morphological characteristics (length, weight, and diameter) are affected mainly by the photoduration, whereas the temperature affects chemical and sensory properties (Rosenfeld et al., 1998; Rosenfeld et al., 2002). Moderate temperatures of 15–21°C accumulate highest amounts of β-carotene in the root and is affected by variety, climate, type of soil, stage of maturity, harvest time, and post harvest storage parameters (Elkner, 2003; Vukasin et al., 2008). The deviation of temperature beyond the optimum conditions leads to the development of poor-colored, flavored, and shorter roots with thick central core in summer season. The lower temperature may retard the root growth producing slender, longer, and light-colored roots (Kelly et al., 2012), whereas the high temperatures produce higher yields of forked carrot (Nandal et al., 2008; Dahiya et al., 2007). Manosa et al. (2010) studied the root yields at two temperatures and concluded that the yield at 10°C was greater than at 18°C. The compact soils, uneven watering, excess nitrogen, and high temperatures in the soil result in the formation of forked, split, hairy, and tough roots, respectively. The forking limits the production of quality early season carrot in major parts (Singh,

1998). The carrot-terpenes mask its sweetness (Rosenfeld et al., 2004) and relatively more proportion of sugar can overcome this inhibition and is found in greater proportions in March-grown carrots (Marta et al., 2013). Nitrogen is regarded as the chief nutrient involved in the growth of carrots increasing the yield, size, and dry matter of both roots and leaves when applied at 100 kg/ha showing highest results (Moniruzzaman et al., 2013). The increase in the fertilizer resulted in increased mass and size of the leaves and roots with the combination of NPKS at 120–45–120–30 kg/ha elevated the yield up to 303% (Uddin & Hoque, 2004). The carrot yield and diameter decreased with the irrigation of saline waters, whereas the flavor, sodium content, and chlorophyll content increased (Unlukara et al., 2011). In northern plains and hills of India carrot seeds are sown during August to December and March to July, respectively. The seeds are covered uniformly with 20–30 mm of loose soil either by line sowing in flat beds or randomly scattering the seeds and then ridging or in ridges running parallel (Sakara et al., 2012). The highest quality and yields are obtained from seeds sown on ridges with broadcasting followed by ridge making method producing maximum yields of poor quality (Dahiya et al., 2007). The carrots sown in the early September in ridges yielded low-forked carrots and highest yield in Haryana (Nandal et al., 2008). The carrot needs regular irrigation with climate and soil type predicting the need of irrigation and mainly two irrigations are provided up to the seed germination (Burzo et al., 2005). The irrigation and rainfall directly affect the β-carotene and solid content in carrots (Fikselova et al., 2010). The ridges enhance the penetration of the delicate seedlings contributing toward production of better quality crop. The warm-weather conditions can lower the yields by damaging or killing the young plants (Kelly et al., 2012; Chetreanu & Atanasiu, 2011). The carrot root is enlarged hypocotyl with a prominent taproot and is developed by the plant from outside to inside (Nonnecke, 1989) with potassium and nitrogen increasing yield from 15.95% to 96.2%. The harvesting can be physical or mechanical with the latter method ideal for parallel row pattern of propagation in the farm. Harvesting is often started when the root sizes are 1.8 cm in diameter (Kotecha et al., 1998) and the immediate washing and cooling to <5°C is necessary to elevate the marketability of the carrots by maintaining the crispiness and freshness (Luo et al., 2014).

The central Asia is regarded as origin of *D. carota* L. (Simon et al., 2008) with its early use believed to be for its seeds used for medicinal aspects, and the archeological evidence of carrot seeds dates back 2000–3000 B.C. as reported by Banga (1957a). The assistance of nature in improving the carrot from tiny bitter root to a versatile root crop of modern time has taken years of human cultivation and domestication (Stolarczyk & Janik, 2011). Earlier, carrots were purple, yellow, or white and orange, which evolved from the yellow ones (Banga, 1957b). Eastern or Asiatic group (var. *altorubens*) and Western Group (var. *sativus*) are the two major groups of domesticated carrot as described by Vavilov (1951), with the first consisting of black-colored carrots and the second one the modern-cultivated ones. The roots of Eastern-group carrots may or may not be branched and are predominated by anthocyanin (purple, pink) pigments. The Himalayan–Hindu Kush region (Kashmir–Afghanistan) is recognized as the region where domestication principally started for color and flavor (Heywood, 1983; Mackevic, 1929; Stolarczyk & Janik, 2011). The carotenoid-pigmented Western roots are straight that evolved later in Turkey and Iran, regarded as the second hub of diversity (Vavilov, 1951; Stolarczyk & Janik, 2011). The cultivation of purple or yellow carrots dominated until 1600 A.D. in most parts of the world that followed the cultivation of orange varieties originating from Europe and America (Rubatzky et al., 1999).

4.1.4 CLASSIFICATION

The genus *Daucus* includes at least 20 species (Lain, 1981; Pimenov & Leonov, 1993), with *D. carota* the only important worldwide cultivated form of this genus. Based on root shape and size, carrot roots are classified into the following four major types (Decoteau, 2000; Nonnecke, 1989; Kelly et al., 2012; Simon et al., 2008).

4.1.4.1 CHANTENY

Chanteny (Fig. 4.1) carrots are tender, tapered, and shorter attractive roots than Danvers and Imperator which is having the blunt tip (12–18 cm long, 3–6 cm neck diameter) with color varying from medium to light orange. This type of carrot is primarily used for storage and processing worldwide due to low-waste production. The Chanteny along with Danvers are preferred for baby-food formulations.

4.1.4.2 DANVERS

The roots of these carrots are smooth, tapered, short, pointed, moderately wide at the top and longer than Chanteny types with length ranging from 12.5 to 21 cm and diameter from 2.5 to 4.5 cm. This is grown mainly for the consumption and marketing in its fresh form (Fig. 4.1).

4.1.4.3 IMPERATOR

These roots have a deep orange cortex with a lighter orange core and are 16–30 cm long, tapering to a pointed tip and are grown extensively for the marketing as fresh. This type of carrot is suitable for harvesting using mechanical means, as the attachment of the petiole is strong with the root (Fig. 4.1).

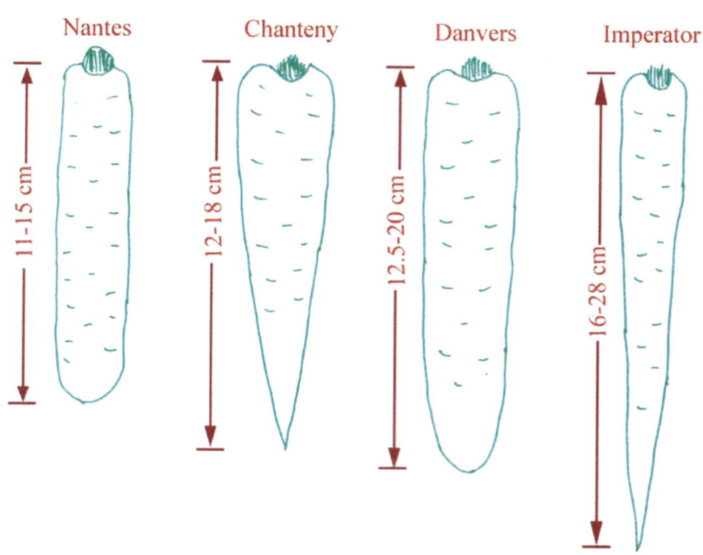

FIGURE 4.1 Different types of carrot on the basis of shape.

4.1.4.4 NANTES

European type carrot grown for marketing fresh with comparatively short length (11–21 cm), blunt end, and uniform diameter along length (cylindrical shape). This is grown in the plains of India. It has bright orange color with high sugar content making it less suitable for extended storage but is marked as the highest quality of all carrots due to very thick phloem. But presence of small petiole makes mechanical harvesting difficult for this carrot type (Fig. 4.1).

Hisar gairic, Ooty 1, Pusa kesar, Pusa yamdagni, Pusa meghali, Selection-223, Selection-21, Nantes, Arka suraj, Selection-223, Selection-29, and Chaman (Table 4.3) are the carrot varieties cultivated in India which are developed by different agricultural institutions. The improved varieties of carrot grown currently in India include Chanteny, Nantes, Danvers, Imperator, Hisar gairic, Pusa kesar, Pusa yamdagni, Selection-223, Selection-29, and Pusa meghali (Mohammad & Rehman, 2007). Pusa rudhira and Pusa asita are the two varieties with red and black-colored tap roots developed by IARI, New Delhi, and are found to be best suited for the cultivation under tropical climate.

TABLE 4.3 Improved Varieties of Carrot Developed by Various Agricultural Research Institutes in India.

Research Institute	Variety
IARI, New Delhi	Pusa yamdagni, Pusa meghali, Pusa kesar, Nantes, Pusa Nayanjyoti, Pusa Asita, Pusa Rudhira,
CCSHAU, Haryana	Hisar Gairic
PAU, Ludhiana	Sel No. 223, Sel No. 29
SKAUST(K), Kashmir	Chaman
Horticulture Research Center, TNAU, Ooty	Ooty 1
IIHR, Bangalore	Arka Suraj

4.2 NUTRITION

4.2.1 COMPOSITION

Carrots are among the nutritious foods available globally with the presence of substantial quantity of carotenoids, vitamin C, and phenolics. The composition aspects of carrot have been thoroughly investigated (Table 4.4). The leaf of the plant consists of both saturated fatty acids (SFA), monounsaturated fatty acids (MUFA), and polyunsaturated fatty acids (PUFA) with α-linolenic acid (ALA) as the chief form of PUFA (de Almeida et al., 2009; Leite et al., 2011).

TABLE 4.4 Chemical Composition of Fresh Carrot (*Daucus carota*).

Component	Composition		Component	Composition	
Moisture	84–95[1,2,3,4,5,6,7,9]		Zinc	0.2–0.24[3,6,9]	
Protein	0.6–2.0[1,2,3,4,5,6,9]		Copper	0.02–0.05[3,6,9]	
Carbohydrate	6–10.6[1,2,3,4,5,6,9]		Manganese	0.14[9]	
Fiber	0.6–2.9[1,2,3,4,5,6,7,8,9]		Iron	0.3–2.2[3,6,8,9]	
Fat	0.2–0.7[1,2,3,4,5,6,7,9]	g/100 g	Ascorbic acid	4–58[3,6,8,9,15]	
Total sugar	4.7–6.7[1,2,3,4,5,6,7,8,9,14]		Thiamine	0.04–0.1[3,6,9,13,15]	
Sucrose	3.68–4.54[8,10,14]		Riboflavin	0.02–0.06[6,8,9,13,15]	
Glucose	0.59–1.0[8,14]		Niacin	0.2–1.2[6,8,9,13,15]	mg/100 g
Fructose	0.55–1.0[8,14]		Pyridoxine	0.14[9]	
Calcium	2–80[2,3,6,8,9,15]		Tocopherol	0.66–0.87[9,12]	
Magnesium	9–12[3,6,9]		Carotenoids	5.33[6,8]	
Phosphorus	25–280[2,3,6,8,9]	mg/100 g	α-Carotene	3.5–10.7[8,9]	
Potassium	240–320[3,6,9,]		β-Carotene	6.85–18.3[8,9,16,17]	
Sodium	40–69[3,6,8,9]		Phylloquinone	0.0083–0.0132[9,11]	

1. Gill and Kataria (1974); **2**. Gopalan et al. (1996); **3**. Holland et al. (1991); **4**. Howard et al. (1962); **5**. Khanum et al. (2000); **6**. Kotecha et al. (1998); **7**. Sharma and Caralli (1998); **8**. Thomas (2008); **9**. USDA (2012); **10**. Lee et al. (1970); **11**. Damon et al. (2005); **12**. Chun et al. (2006); **13**. Gebhardt and Matthews (1991); **14**. Pennington and Douglass (2005); **15**. Hui (2006); **16**. Karnjanawipagul et al. (2010); **17**. Herrero-Martinez et al. (2006).

The chemical constitutes of carrot, as reported by Kotecha et al. (1998), include moisture 84–88.8%, protein 0.7–0.9%, fat 0.2–0.5%, carbohydrates 6–10.6%, total sugars 5.6%, crude fiber 1.2–2.4%, total ash 1%, calcium 0.34–0.80 mg/g, iron 0.004–0.02 mg/g, phosphorus 0.25–0.53 mg/g, sodium 0. 40 mg/g, potassium 2.40 mg/g, magnesium 0.09 mg/g, copper 0.0002 mg/g, and zinc 0.002 mg/g. The dietary fiber of fresh carrot includes pectin (7.41%), cellulose (80.94%), hemicellulose (9.14%), and lignin (2.48%) on dry weight basis (Nawirska & Kwasniewska, 2005). The carrot cell wall is composed of pectin, cellulose, lignin, and hemicellulose (Lineback, 1999). The sugars are present as glucose, fructose, and sucrose, whose composition in cooked form is 1.2%, 1.0%, and 2.8%, while in canned form is 0.8%, 0.6%, and 1.8%, respectively (Pennington & Douglass, 2005). The composition as given by Thomas (2008) includes protein 1%, sugars 8%, fiber 2%, sodium 40 mg/100 g, calcium 2 mg/100 g, phosphorus 280 mg/100 g, whereas Gopalan et al. (1996) reported 1.2% crude fiber, 1.1% total ash, calcium 80 mg/100 g, iron 2.2 mg/100 g, and phosphorus 53 mg/100 g in carrot. Gebhardt and Matthews (1991) published the composition of fresh carrot in *Nutritive Value of Foods* as carbohydrates (10%), proteins (0.9%) along with 30, 48, 0.6, 355, and 39 mg/110 g of mineral Ca, P, Fe, K, and Na, respectively.

The nutritional composition of carrots on dry weight basis as reported by Singh et al. (2001) is 9.8% protein, 7.7 mg/100 g iron, 0.8 mg/100 g copper, 1.8 mg/100 g manganese, and 3.2 mg/100 g zinc. The dietary fiber (23.76 g/100 g) and carotene content (60.21–79.47 mg/100 g) in carrots are reported (Anderson & Bridges, 1988; Rakcejeva et al., 2012). The carrot pomace contains moisture 84.23%, protein 6.21%, carbohydrate 32.22%, ash 5.78%, iron 0.12 mg/g, phosphorus 2.94 mg/g, and Ca 2.91 mg/g (Shymala & Jamuna, 2010), whereas the carrot powder composition ranged for moisture from 4.77% to 8.78%, total ash 5.05% to 6.93%, crude fat 2.43% to 3.45%, crude protein 6.16% to 14.8%, and crude fiber 17.05% to 24.66% (Pandey et al., 2003; Gazalli et al., 2013). The total dietary fiber on dry weight basis in carrots is 39.5%, containing 11.2% soluble and 28.3% insoluble fiber, whereas as the composition of dietary fiber on fresh weight basis is 5.7%, including 1.6% as soluble and 4.1% as insoluble fiber (Khanum et al., 2000).

The crude protein (g/100 g), crude fat (g/100 g), crude fiber (g/100 g), total ash (g/100 g), carbohydrate (g/100 g), and β-carotene (mg/100 g), as reported by Singh et al. (2013), in powder and grits of carrot are 7.51, 7.91; 12.69, 7.51; 1.39, 0.95; 7.42, 9.67; 2.69, 7.07; 68.30, 64.88; and 36.94, 33.48, respectively. The total carotenoids, β-carotene and vitamin C of carrot pomace are 4.0, 3.92, and 3.53 mg/100 g, respectively (Shymala & Jamuna, 2010), whereas Pandey et al. (2003) found carotene content as 6.91 mg/100 g in fresh carrot and 30 mg/100 g in carrot powder.

4.2.2 SECONDARY METABOLITE

The phytochemicals meaning the "plant chemicals" (Fig. 4.2) are the recognized secondary metabolites which are having lowering action toward

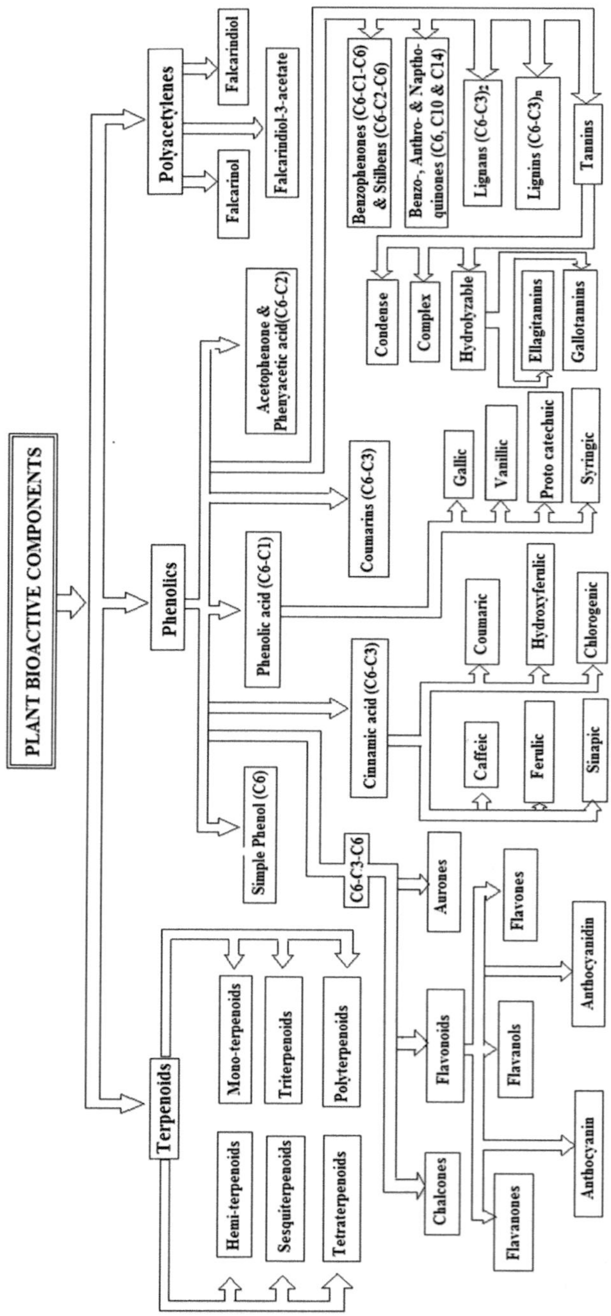

FIGURE 4.2 Classification of the plant secondary metabolites.

the incidence of chronic diseases. A significant number of phytochemicals have been reported to be present in the fruits, vegetables, and cereals but the major percentage is still unknown and the estimated individual number of the identified phytochemicals exceeds 5000 (Liu, 2003). The secondary metabolites isolated from carrot plant include steroids, volatile oils, terpenoids, tannins, flavonoids, carotenoids, and hydrocarotene (Vasudevan et al., 2006). The major phytochemical present in carrots is carotenoids that are among nature's most wide spread pigments with antioxidant and provitamin roles. The chemical components present in the *D. carota* are quercetin, pyrrolidine, daucine, daucosterine, tiglic acid, and essential oils like carotol, daucol, copaenol, gerenol, limeonene acid, pinene acid, and cineole acid with the first three essential acids being the major edible oils present in seeds with a respective percentage of 30.55, 12.60, and 0.62 (Karakaya & Sedef, 1999; Thomas, 2008; Ozcan & Chalchat, 2007). Ceska et al. (1986) reported the presence of two coumarin compounds (8-MOP and 5-MOP) distributed in the whole carrot plant. The seeds of the carrot plant contain petroselenic, palmatic, stearic, oleic, linoleic, arachidic, vaccenic, asaron, bisabol, and tiglic acids along with protein, fiber, and ash (Ozcan & Chalchat, 2007; Mosig & Schramm, 1955). The carrot leaves are also the excellent source of precious omega-3 and omega-6 fatty acids, whose low ratio {omega-6–omega-3 (n-6/n-3)} in diet is essential to diminish the malnutrition-related disorders (de Almeida et al., 2009; Simopoulos, 2002). Carrots are also known to contain vitamin-like components that include myoinositol, choline, total choline, and betaine with composition in raw and cooked ones as 12 and 52, 6.82 and 0.44, 8.79 and 8.77, and 0.34 and 0.11 mg/100 g, respectively (Clements & Darnell, 1980; Zeisel et al., 2003). The carrot seed extract is significantly involved in lowering the biochemicals, serum-glutamate oxaloacetate transaminase, serum-glutamic pyruvic transamiinase, and alkaline phosphatase to normal whose increase is mediated by the hepatotoxin thioacetamide (Singh et al., 2012).

The existence of different vitamins in carrot as reported by different studies include thiamine 0.04 mg/100 g, riboflavin 0.02 mg/100 g, niacin 0.2 mg/100 g, provitamin A 5.33 mg/100 g, and ascorbic acid 4 mg/100 g (Kotecha et al., 1998); provitamin A 270 mg/100 g, vitamin C 10 mg/100 g, α-carotene 10.65 mg/100 g, and β-carotene 18.25 mg/100 g; 4.9 mg/100 g vitamin C, 2.2 mg/100 g β-carotene, 4.9 mg/100 g vitamin C, and 2.2 mg/100 g β-carotene (Singh et al., 2001). Tocopherol content decreased from 0.87 to 0.71 mg/100 g in fresh and frozen blanched carrots, respectively, with major type represented by α-tocopherol (Chun et al., 2006). The presence of different vitamin B-complex in carrots have been reported: B1, B2, and B3 (Hanif et al., 2006), B5 (Pakin et al., 2004), and B7 (Staggs et al., 2004).

The carrot is ranked sixth in context of the total phenolic content among the consumed vegetables (Chu et al., 2002) and about 67% of phenolics in diet are dominated by flavanoids (Liu, 2004). Lignans are present mostly in fiber-rich plants including vegetables like carrots, asparagus, and broccoli (Saleem et al., 2005). Several studies have revealed that carrots contain sterols, that include campesterol 2.2 mg/100 g, stigmasterol 12.8 mg/100 g, and β-sitosterol 0.08 mg/100 g (Normen et al., 1999); liginin, 93 mg/100 g secoisolarici–resinol (Milder et al., 2005); phenolic acids including 10 mg/100 g chlorogenic acid, 0.1 mg/100 g caffeic acid and 0.46 mg/100 g protocatechuic acid (Mattila & Hellstro, 2007); 86.5 mg/100 g dry weight xylitol (Wang & van Eys, 1981). The chlorogenic acid, caffeic acid, ferulic acid, and *p*-hydroxybenzoic acid are the main phenolics present in different colored (orange, red, purple, and white) carrots (Sun et al, 2009) and the purple form contains the highest amounts of phenolics (Alasalvar et al., 2001) on an average about nine times more than other colors (Leja et al., 2013).

Carrots contain terpenoids (Fig. 4.2), and more than 90 entities have been recognized with the major mono- and sesquiterpenoids including *p*-cymene, limonene, β-mycrene, sabinene, terpinolene, γ-terpinene, β-caryophylene, β-bisabolene, γ-bisabolene, α-humulene, and α-pinene that provide aroma to the root (Kjeldsen et al., 2003; Christensen et al., 2007; Kreutzmann et al., 2008a; Soria et al., 2008). The carotenoids and terpenoids are derivatives of the basic isoprenoid structure with mono and sesqui forms bearing two and three basic structures scoring the carbon atoms to 10 and 15, respectively. The volatile compounds are dominated by mono- and sesquiterpenoids up to a range of >97% with former in higher concentration than latter. The distinctive flavor of carrots is dominated mainly due to terpenoids and the cooking time of 10 min results in loss (>85%) of carrot volatiles (Alasalvar et al., 1999; Simon et al., 1980; Lund and Bruemmer, 1992; Kjeldsen et al., 2003). The key nonterpenoid volatile compounds present are 3-hydroxy-2-butanone, ethanol, hexanal, acetic acid, and erythro- and thero-2,3-butanediol (Soria et al., 2008).

Carrots also contain aliphatic polyacetylene (C17) alcoholic compound called falcarinol earlier known as carotatoxin. Falcarindiol and falcariniol-3 acetate are other types of falcarniol compounds (Fig. 4.3) present throughout in root with major amount within the peel (Acworth et al., 2011; Koidis et al., 2012) and the composition ranges from 0.70 to 4.06 mg/100 g FW and 4.57 to 27.06 mg/100 g DW, respectively (Pferschy-Wenzig et al., 2009; Christensen & Kreutzmann, 2007). Falcarindiol is present in greater amounts in carrot root than falcarniol and the former along with di-caeffic acid contributes more to the bitterness of the carrot (Kreutzmann et al., 2008a). The falcarniol compounds' thermal stability is more at elevated temperatures (50–100°C) than processing below the mentioned temperature (Rawson et al., 2010).

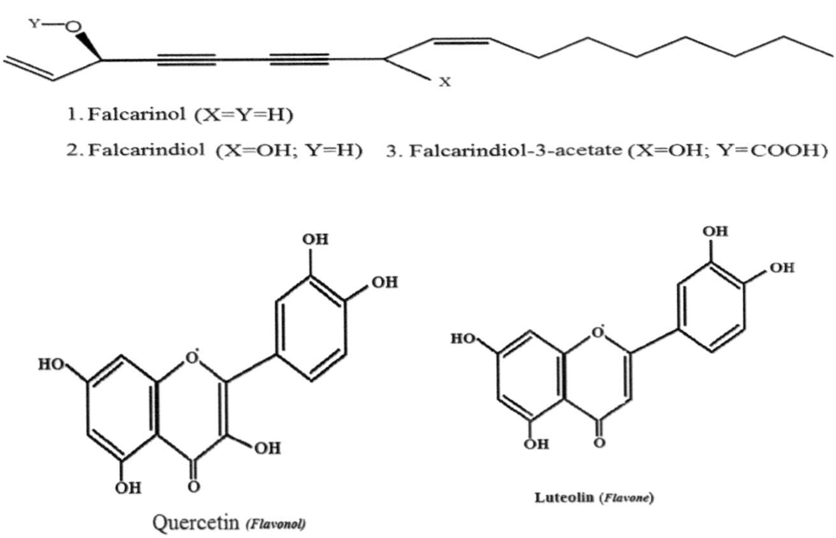

FIGURE 4.3 Polyacetylene compounds (a), flavonol (b), and flavone (c) in carrot.

4.3 THERAPEUTIC VALUE AND COLORANTS OF CARROT

4.3.1 MEDICINAL VALUES OF CARROT METABOLITES

Carrot is one of the highly nutritious root crops containing considerable amounts of vitamins, minerals, and carotenoids, with nitrogen balancing, diuretic properties, and is effective in elimination of produced uric acid from the body (Anon., 1995). As for the therapeutic effects, carrot (Fig. 4.4) is employed in lowering the blood sugar, preventing cancer, and to treat diabetes, dyspepsia, heart disease, gout, carcinomatous ulcers, amenorrhea, angina, asthma, diarrhea, increased blood pressure, elevated cholesterol, liver and skin problems, and wrinkles (Hartwell, 1971; Duke, 1997, 2006). Carrots along with cabbage, soybeans, garlic, ginger, and other edible members of Apiaceae family are considered as having the highest anticancerous activities (Caragy, 1992). Carrots are not a significant source of calories and dietary nutrients. It ranks seventh in overall contribution to nutrition among fruits and vegetables (Alasalvar et al., 2001). The carrot root and its extracts are often used for the prevention of constipation, inflammation particularly in tonsillitis, and liver-associated dysfunctions, and also as antihelmintic and antiseptic (Ahmed et al., 2005). Abdulaali (2009) reported the effectiveness of alcoholic carrot extracts on inhibiting the *in vitro* growth of *Pseudomonas aeruginosa*. Kanji, a fermented carrot product produced by fermentation of carrot by *Lactobacilli*, is effective in controlling the growth of *Staphylococcus aureus*, the causative agent of the food poisoning in human gut. The bacteriocins synthesized by the *Lactobacilli* inhibit the growth of microorganisms including *S. aureus*, *Bacillus* and spores of *Clostridium* within canned foods (Sowani & Thorat, 2012). The chemotherapy-resistant cancers could be treated using black carrot extracts either solely or synergistically in combination with anticancer drugs (Sevimli-Gur et al., 2013). Among the different colored carrots, the red variety shows higher antioxidant activity and develops elevated phenolic levels during lean-rainy season (Leja et al., 2013). The carcinogens and free radicals in the cigarette cause lymphocyte DNA damage which could significantly be decreased by consumption of carrot juice and β-carotene supplements (Lee et al., 2011).

Several investigations have shown the effectiveness of secondary metabolites (Table 4.5), including carotenoids, in combating the free radicals and also for the singlet oxygen, thereby inhibiting carcinogenesis, cardiovascular diseases (CVD), monocyte adhesion, platelet activation in the cell, rheumatism, cataracts, and aging (Duke, 1997; Edge et al., 1997; Krinsky & Johnson, 2005; Awad & Fink, 2000; Giovannucci et al., 1995; Krinsky et al., 2004; Mayne, 1996; Sies & Stahl, 1995; Steinmetz & Potter, 1996; Franceschi et al., 1998; Filotheou et al., 2010); providing hemoprotective, photoprotective, and proper immune functioning properties (Young & Woodside, 2001; Deshpande et al., 1995); regulating

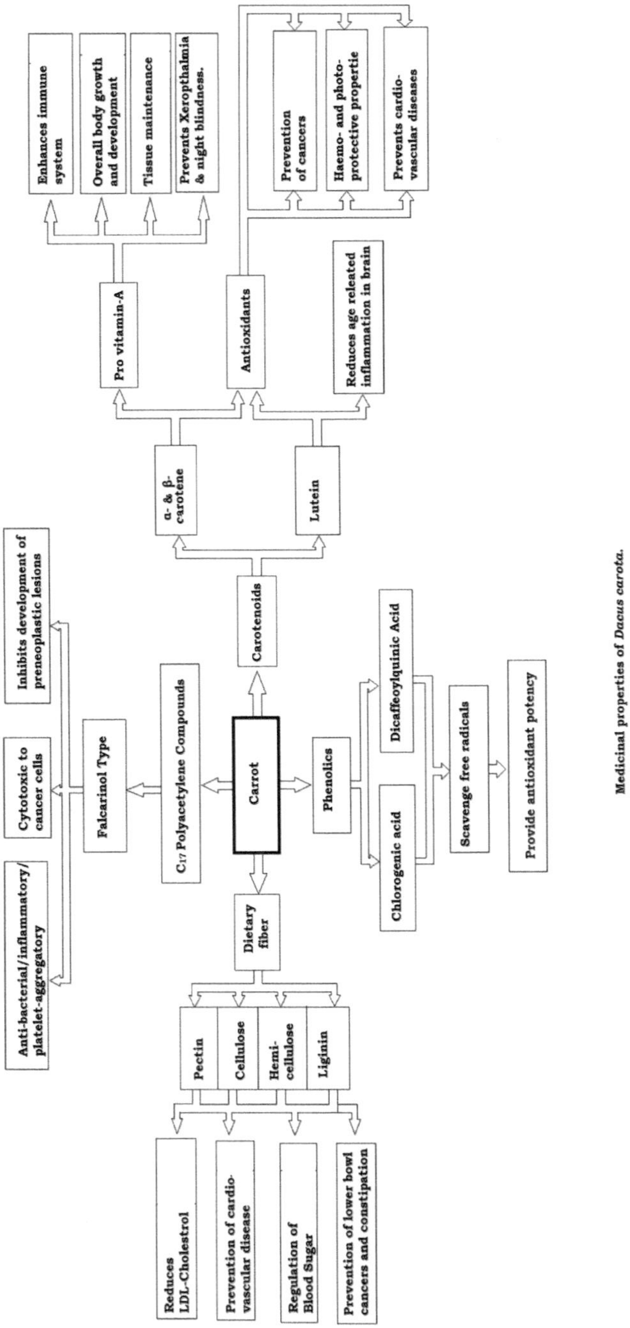

FIGURE 4.4 Therapeutic uses of carrot.

TABLE 4.5 Classification of Some Active Components and Their Functions.

Compound			Disease	Reference
Polysaccharide	Dietary fiber		GI-tract, heart health, anti-cancer, diverticulitis, GERD	Yun (1999), Weisburger (2000), Anderson and Hanna (1999), Knauf and Facciotti (1995), Schneeman (1994), Sontag (1999), O'Keefe (1996)
Terpenoids	Monoterpenoid	Limonene	Anticancer, antioxidant	Steinmetz and Potter (1996), Jaga and Duvvi (2001), Makris and Watson (2001), Hirsh et al. (2001), Weisburger (2000)
	Tetraterpenoid	Carotenoids	Eye health, heart health, anti-cancer, antioxidant	Giugliano (2000), Eastwood (1999), Steinmetz and Potter (1996)
Polyacetylenes	Falcarinol compounds		Anticancer, anti-inflammatory, antiplatelet-aggregatory, antibacterial	Lund and White (1990), Zidorn et al. (2005)
Phenolics	Flavonoids	Quercetin	Anticancer, antioxidant, anti-inflammatory, osteoporosis, heart health, detoxification, prevents asthma	Weisburger (2000), Mazur (1998), Borek (2001), Knekt et al. (2002)
	Hydroxycinnamic acids (phenolic acid)	Chlorogenic acid	Antioxidants, anticancer	Cheng et al. (2007)
	Coumarins		Antibacterial, vasorexalant anti-inflammatory, antioxidant	Hoult and Paya (1996)
	Tannins		Antioxidants, antimicrobial, Blood pressure, modulate immune responses	Chung et al. (1998), Hayashi et al. (2008)
	Lignans		Antioxidant, anti-estrogenic, anticancer	Adlercreutz and Mazur (1997)

TABLE 4.5 *(Continued)*

Compound		Disease	Reference
Antioxidant	Ascorbic acid	Cataract, cardiovascular diseases, cancers, antioxidant, healing wounds	Combs (1998), Carr and Frei (1999)
Phytosterols	Campesterol Stigmasterol β-Sitostanol	Anticancer, inhibits cholesterol absorption, anti-inflammation, antioxidant	Hartmann (1998), Ostlund (2002), Dutta (2003)

the oxidation or reduction enzyme activity and hormone metabolism, and impact of antiviral and antibacterial effects (Sun et al., 2001; Sun et al., 2002; Chu et al., 2002; Dragsted et al., 1993; Waladkhani & Clemens, 1998). The increased action of phytochemicals play a key role against the cancers associated with mouth, phyranx, GI tract, lungs, bladder, and breast (Anon., 1997). Liu (2004) reported that the enhanced effect of phytochemicals is due to their synergistic and combined effect on each other which cannot be leveled by the expensive dietary supplements. The reactive and unstable entity with an unpaired electron is free radical that can be a derivative of oxygen (hydroxyl, superoxide, or hydroperoxyl) or *in vivo* produced reactive oxygen species (ROS) like nitric oxide (Prior & Cao, 2000) initiating chain reaction which ultimately results in the cell disruption which can be terminated by the dietary antioxidants (Halliwell et al., 1995; Ames, 1983; Kaur & Kapoor, 2001). The carcinogenic consequences of few synthetic antioxidants have limited their application in food systems, thereby increasing the interest of using natural phytochemicals obtained from fruits and vegetables (Yen et al., 1998; Gazzani et al., 1998). The high intake of plant products among vegetarians and nonvegetarian individuals have considerably lowered down their blood-pressure levels (Ascherio et al., 1992).

4.3.1.1 CARROT LEAVES

The PUFA/SFA ratio indicates whether food is healthy or unhealthy with reference of 0.45, above this is considered healthy and below as unhealthy. Carrot leaves are among the balanced sources of ALA and linoleic acid (LA) that have cancer, asthma, thrombosis, artherosclerosis, and coronary heart disease (CVD)-lowering properties upon the consumption of dietary food source in low ratios. The ALA/LA ratio in carrot leaves has been observed in the range of 0.36–0.57 (Leite et al., 2011; de Almeida et al., 2009). The proposed mechanism of phytochemical action on cancer has been demonstrated by Liu (2004). The normal cellular processes of the body produce chemicals that are ROS-like hydrogen peroxide and the superoxide anion-free radical that cause secondary oxidative stress whenever there is imbalance between antioxidants and oxidants (Butt & Sultan, 2011). The oxidizing agents cause continuous oxidative stress to body cells which adversely affects the cell components leading to the risk of CVD and cancer (Ames & Gold, 1991; Liu & Hotchkiss, 1995; Ames et al., 1993), which can be retarded by the consumption of dietary antioxidants in sufficient quantities. The single- and double-strand breakage of DNA in addition with the base mutation and chromosomal breakage is further enhanced if the affected DNA strand is left unrepaired (Ames et al., 1993).

4.3.1.2 CARROT ROOT

4.3.1.2.1 Terpenoids

4.3.1.2.1.1 Sterols

Plant sterols, like campesterol, stigmasterol, and β-sitosterol, present in carrots regulate permeability and fluidity of phospholipid bilayers, inhibit cholesterol absorption, possess anticancer, antiatherosclerosis, antiinflammation, and antioxidant properties (Hartmann, 1998; Awad & Fink, 2000; Ostlund, 2002; Dutta, 2003).

4.3.1.2.1.2 Carotenoids

Carotenoids (Fig. 4.5) are regarded as the natural light-collecting pigments in photosynthesis that protect the cells from photosensitization (Demmig-Adams et al., 1996). There is difficulty in consuming very high amounts of β-carotene from carrots as the body is not able to convert excess of it to vitamin A. The carrot consumption in surplus amounts is associated with providing an orange tone to the skin, particularly to hand, palm, and feet (Small & Catling, 1999). The combined effect by the joint interaction between phenolic acid, β-carotene, and ascorbic acid as well as between flavonoids and tocopherols has been verified by the studies (Trombino et al., 2004; Marinova et al., 2008).

Carotenoids function beyond the natural coloring matter as antioxidants that are chemoprotective agents with health-promoting or disease-preventing properties and consumption of its lower levels is related with elevated incidence of cancer and heart disease in persons (Thomas, 2008). The carotenoid's role in photoprotection and in photosynthesis is attributed to its quenching and deactivating ability of oxygen atom forming on light exposure (Britton, 1995). The conjugated double bonds in carotenoids function efficiently as scavengers of oxygen, thereby preventing

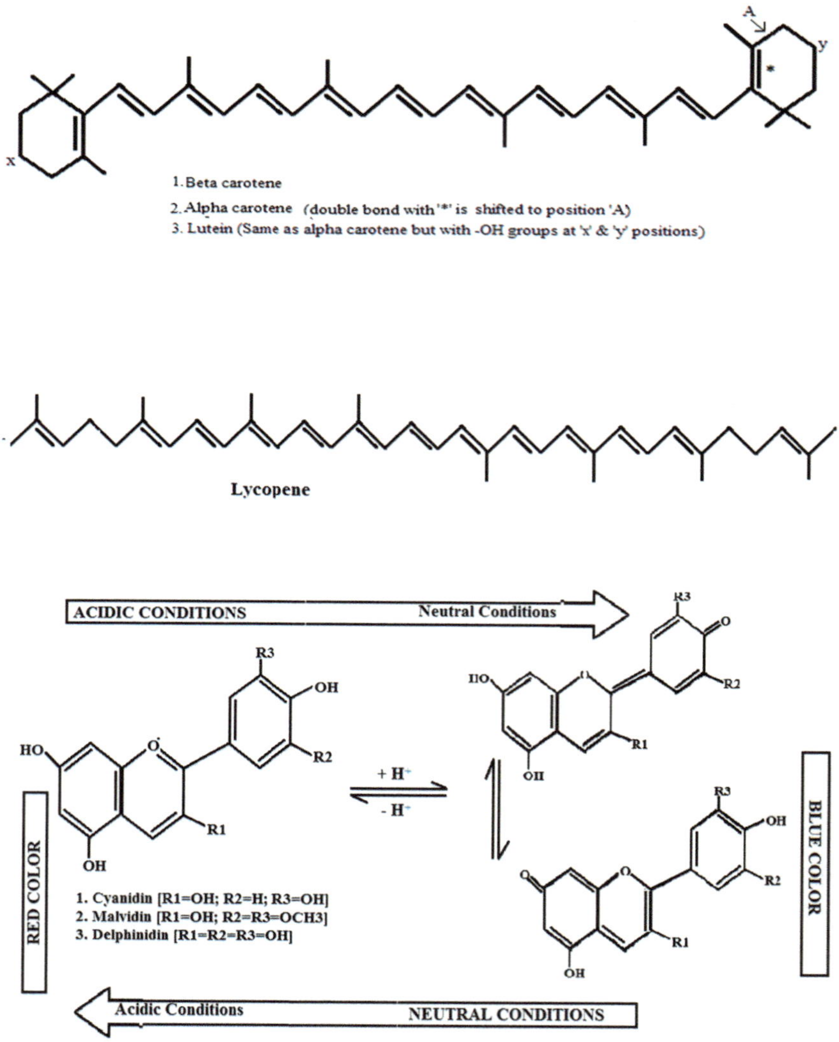

FIGURE 4.5 Carotenoids (a and b) and anthocyanin skeletal structure (c).

peroxidation of lipid molecules within a cell (Krinsky, 1998; Stahl & Sies, 1996; Burton & Ingold, 1984). Most investigations on carotenoid for its antioxidant property are for β-carotene, which is known as antioxidant vitamin along with vitamin C and E (Kaur & Kapoor, 2001). The consumption of natural fresh food with ample carotenoids (carotene, lycopene, and lutein) has been found to improve the immune system and significantly lower down the risk of CVD and cancer (Tanaka et al., 2012; Kohlmeier & Hastings, 1995; van Poppel & Goldbohm, 1995). Carotenoids possess the immunomodulatory effects that could enhance immune system to tumorigenesis and cell communication resulting in restricting the expansion of tumor-initiated cells (Johnson, 2002). The susceptibility of conjugated bonds with the peroxy radicals (ROO$^{\bullet}$) forms a carbon-centered radical (β-car$^{\bullet}$) that changes reversibly into chain-carrying peroxy radical (β-car-OO$^{\bullet}$) upon reacting with oxygen. The antioxidant activity of β-carotene is conferred by its reactivity toward peroxy radicals and the relative stability of β-car$^{\bullet}$ structure (Burton, 1989).

The biological antioxidants like carotenes have a net positive effect in delaying the immune-related disorders by stretching the time period between infection and symptoms, namely, antioxidants lower down the drug-resistant HIV strains during the infection of HIV and delay its surfacing as AIDS (Baranowitz et al., 1996). Kaur and Kapoor (2001) described that the high percentage of carotenoids in diet has lowering down action of breast and lung cancer incidence in smokers. β-Carotene is effective in preventing the autooxidation of lipids in addition to inhibiting the photoxidation by neutralizing the peroxyl radicals (Britton, 1995). The higher level of total carotene in blood plasma significantly lowered the DNA damage (Southon, 2000). The hydrophobic nature of carotenoids is responsible for their interaction with lipids and other hydrophobic structures like internal core of membranes rich in lipids. Carotenoids occur with proteins as carotenoproteins necessary for the existence, transport, and function of hydrophobic protein. The lipoproteins serve as the possible transport vehicles for distribution of carotenoid molecules to whole body all the way through blood stream with HDL and LDL mainly associated with carotenes and xanthophylls, respectively (Britton et al., 2008). β-Carotene as the major form of provitamin A enhances immune system, tissue maintenance, cell growth and differentiation, synthesis of glycoproteins and hormones, mucus secretion from the epithelial cells, protects heart-related aliments, urinary tract infections, cataracts, arthritis, high blood pressure, bronchial asthma, overall growth, and development of bones (Wolf, 1980; Combs, 1998; Faulks & Southon, 2001; Bender, 2003; Seo & Yu, 2003). α- and β-carotenes are the chief carrot carotenoids with physiological activity of 50% and 100%, respectively, in human body for the provitamin A (Panalaks & Murray, 1970; Simpson, 1983) with one molecule of β-carotene forming two retinol molecules inside the human system. The preventive effect of α-carotene in retarding tumorigenesis in liver, lung, skin, and colorectum has been observed to be greater than the β-carotene (Murakoshi et al., 1992; Donaldson, 2004; Narisawa et al., 1996). Some studies have questioned the preventive role of carotenoids on derms and respiratory tract oncogenesis (Lambert et al., 1990; Wolterbeek et al., 1995). β-Carotene constitutes major form of carotenoids present in carrots (60–80%), followed by α-carotene (10–40%), lutein (1–5%), and other minor carotenoids (0.1–1%) (Sun & Temelli, 2006).

The carotenoid molecules lacking provitamin A function are seen more active than their provitamin A counterparts in increasing the cell-mediated immune system (Chew & Park, 2004). The β-carotene may serve as more important nutrient than the vitamin A as the latter is unable to scavenge singlet oxygen and works as poor antioxidant (Bendich, 1989). Central cleavage and excentric cleavage pathway are the two metabolic pathways involved in the formation of vitamin A from β-carotene with the former pathway as the main mechanism of action (Tanaka et al., 2012).

4.3.1.2.2 Polyacetylenes

Carrots contain falcarinol, falcarindiol, and falcarindiol-3-acetate (Fig. 4.3) that are C17-polyacetylene compounds with both positive and detrimental properties (Table 4.5). The falcarinol and the falcarindiol-3-acetate types are more cytotoxic over falcarindiol that collectively induce apoptosis on the cells of lymphoid leukemia, whereas the isolated carotenoids did not show any appreciable affect on apoptosis, signaling the formers role. These are the natural pesticides of plant that possess cytotoxicity toward several lines of cancer cells and provide additionally the roots with anti-inflammatory, antibacterial, and antiplatelet aggregatory properties (Hansen et al., 2003; Lund & White, 1990; Bernart et al., 1996; Zidorn et al., 2005; Christensen & Brandt, 2006; Zaini et al., 2012). Falcarindiol and acetyfalcarinol are the other two types of polyacetylene compounds available in carrots that along with myristicin (phenyl propanoid) provide fungal, bacterial, and disease resistance to roots along with antiplatelet and anti-inflammatory effects (Yates et al., 1983; Christensen & Brandt, 2006).

4.3.1.2.3 Vitamins

The presence of biological antioxidant in form of vitamin C is actively involved in the synthesis of neurotransmitters, steroid hormones, and collagen; conversion of cholesterol into bile pigments; absorption of minerals (iron and calcium); prevention of cataract, certain cancers, and CVD; healing of wounds and burns; strengthening the capillary walls and preventing bruising and clotting of blood in vessels (Combs, 1998, Carr & Frei, 1999). Vitamin A deficiency is among the world's three major micronutrient problems and results in blindness of nearly 20,000 children every year, with night blindness and xerophthalmia as initial symptoms (Jiang et al., 2008; Rathore, 2001). In the developing countries like India, the deficiency of retinol is a major cause of blindness in children (van den Berg et al., 2000; Kumar et al., 2001) and administration of the foods rich in provitamin A, such as carrots, can serve as beneficial alternative in combating this problem (El-Arab et al., 2002).

4.3.1.2.4 Phenolics

Phenolics are another important secondary metabolites characterized by an aromatic ring derived from tyrosine and phenylalanine. Phenolic compounds (Fig. 4.2) are biologically active ingredients having one or more phenolic residues formed by shikimate and acetate pathways. The phenolics are classified into simple mono and polyphenols (Harborne, 1989) with flavones, flavonols, and their glycoside derivatives as major subclasses. Phenol, thymol, and cresol are some simple phenols thoroughly distributed in plants (Laura, 1998). Phenolic acids, the nonflavonoid polyphenolics containing carboxylic group attached to phenol with the basic structure C_6–C_1, are subdivided into two major classes as hydroxyl cinnamic and benzoic acids, whereas polyphenols contain at least two phenolic rings that confer both desirable and undesirable attributes to plant produce (Babic et al., 1993; Liu, 2004; Marinova et al., 2005; Bravo, 1998). Polyphenols were considered as antinutritional due to the property of precipitating with macromolecules but consideration has changed recently from antinutritional to nutritional as reported by several investigations supporting their desirable role in health promotion (Laura, 1998). The polyphenols are known to possess more radical-inhibition activity than vitamin C and E (*in vitro*) within the body (Rice-Evans et al., 1996).

The antioxidant properties of phenolics can serve as possible replacers of synthetic antioxidants (BHT, BHA) in food systems (Temelli et al., 2012).

The major phenolics present in plant (Fig. 4.2) are phenols, phenolic acids, phenylacetic acids, hydroxycinnamic acids, coumarines, condensed tannins, flavonoids, xanthones, stilbenes, lignans, melanins, and naphtoquinones (Goodwin and Mercer, 1983) with the first seven present in major concentrations in vegetables (Sinha & Hui, 2011). Alasalvar et al. (2001) reported different types of hydroxycinnamic derivatives of phenolic acids in carrots as 3′-feruloyquinic acid, 5′-feruloyquinic acid, 4′-feruloylquinic acid, 3′,4′-diferuloylquinic acid, 3′,5′-diferuloylquinic acid, 3′-caffeoylquinic acid, 5′-caffeoylquinic acid, 3′,4′-dicaffeoylquinic acid, 3′,5′-dicaffeoylquinic acid, 3′-*p*-coumaroylquinic acid and 5′-*p*-coumaroylquinic acid. These compounds are derived through phenylpropanoid metabolism from phenylalanine (Dixon & Paiva, 1995) to scavenge the generated free radicals, thereby providing antioxidant potency (Hollman et al., 1996; Nagai et al., 2003). Although phenolics are distributed throughout the carrot root, the periderm tissue is highly concentrated than any other part of the root (Mercier et al., 1994) with chlorogenic acid as the main derivative of hydroxycinnamic acid present, representing 42.1–61.8% of total phenolic compounds (Liu, 2004; Zhang & Hamauzee, 2004). The hydroxycinnamic acids are strong antioxidants that combat the ROS and inhibit the lipid oxidation (Cheng et al., 2007; Sroka & Cisowski, 2003). The individual phenolic content decreases from the exterior (peel) to the interior (xylem) (Goncalves et al., 2010) with peel accounting for 54.1%, phloem 39.5%, and xylem only 6.4% of the phenolics (Zhang & Hamauzee, 2004). The higher concentrations of phenolics in the carrot peel can be utilized as a source of neutraceutical value addition for food products.

Coumarins are another group of phenolics widely present in the Apiaceae family, including carrot, celery, fennel, and parsley (Zobel, 1997; Cherng et al., 2008). These compounds possess antibacterial, anti-inflammatory, antioxidant, vasorelaxant, and immunomodulatory activities (Hoult & Paya, 1996). The blood-clotting property of coumarins is well established (Makris & Watson, 2001; Hirsh et al., 2001) and its administration can prevent the excessive loss of blood on wounding.

Lignans are biphenolic plant phenolics with phenylpropane ($[C_6–C_3]_2$) as basic structure and are diversely reported in vegetables like carrot, kale, broccoli, cabbage, leek, and sweet pepper (Milder et al., 2005). These compounds are known to possess antioxidant, anti estrogenic properties and lower down the possibility of cancer (Kurzer & Xu, 1997; Ju et al., 2001; Adlercreutz & Mazur, 1997; Webb & McCullough, 2005). The application of lignin in cancer, chemotherapy, antimicrobial, and other pharmacological activities has lead to its increased utilization interest in foods (Saleem et al., 2005). Seco-isolariciresinol has been found to be active against colon adenocarcinoma (Caco-2) cell lines in clonogenic and MMT assay. The potent cytotoxicity of matairesinol on promyleocytic leukemia HL-60 cells has been studied (Kang et al., 2003) whose dosage reduced the viability of HL-60 cells by apoptosis (programmed to death) and DNA damage (Saleem et al., 2005). Lariciresinol and pinoresinol have been quoted to possess anti-inflammatory effects that inhibit the cytokine (TNF-α) production from macrophages. The cytokines are a group of pro-inflammatory moieties secreted by macrophages and lymphocytes upon activation by inflammatory signals (Cho et al., 2001; Cho et al., 1998; Duan et al., 2002). Pinoresinol is also said to have antioxidant property based on oxygen-radical absorbance capacity (Kikuzaki et al., 2004).

The phenolic polymer of aromatic alcohol sharing basic structure with lignan is denoted as lignin that is lipophilic component of plant cell wall. The dominance of guaiacyl liginin in carrots, radish, and spinach denotes these as guaiacyl-rich vegetables (Bunzel et al., 2005).

Lignins do not undergo hydrolysis and are thought to absorb cadmium, copper, and bile acids with the absorption of first two entities producing reactive oxygen and assisting the latter components' excretion apart from its absorption (de Vries, 2003; Eastwood, 2005; Guo et al., 2008; Valko et al., 2006).

Flavonoids are the plant phenolics mainly associated with sugars (aglycones) rendering them a very low reactivity toward free radical that contributes them with low-antioxidant properties (Hollman & Arts, 2000; Ross & Kasum, 2002; Rice-Evans et al., 1997). The basic structure comprises of aromatic rings two in number denoted as "A" and "B" joined together by three carbon "C" ring (C_6–C_3–C_6) (Fig. 4.3) and the distinction changes in oxygenated heterocylic C_3 ring classifies them into subtypes with major ones including flavonols, flavanols (3-ol), flavones, flavanones, isoflavanoids, and anthocynadins (Liu, 2004, Rice-Evans et al., 1997). The sugars associated in glycoside formation can be galactose, xylose, and rhamanose but the major involved sugar is glucose (Markham, 1982). The flavonol ring A originates from three molecules of malonyl CoA, whereas aromatic B ring with chromanol C ring from phenyalanine (Fatland et al., 2004; Tsao & McCallum, 2009).

The major flavonoid taken daily through the diet is quercetin (~70%) found also in carrots that may prevent the precise rise of enzyme inhibition in the proliferation of cancer cells (Hertog et al., 1993). Quercetin (Fig. 4.3) possesses strong antioxidant properties (Ruzic et al., 2010) and has been linked with lowering of the death rates due ischemic heart disease, reduction in lung cancer incidence, lowering down the risk of type-2 diabetes, and prooxidant effect on plasma proteins (Knekt et al., 2002; Young et al., 1999). Polyphenols particularly flavanoids prevent oxidation of LDL cholestrol (C) that decreases the incidence of atherosclerotic plaque formation inside the walls of arteries and hence develop CVD. An early occurrence of atherosclerotic plaque formation is initiated by taking up the oxidative modified LDL-C by the macrophages resulting in formation of foam cells that eventually lead to formation of atherosclerosis. The ability of caffeic and chlorogenic acids to decrease LDL-C oxidation is more than the simple (monomeric) phenolics.

Flavanoids are also potent to inhibit lipid peroxidation, scavenge free radical, modulate cell-signaling pathways, detoxify the cancerous substances by enzyme stimulation, and reduce the inflammation associated with the development of free radicals (Esterbauer et al., 1992; Heim et al., 2002; Rice-Evans & Packer, 2003; Hollman & Katan, 1999; Miura et al., 1995; Terao et al., 1994; Sanchez-Moreno et al., 1998).

The inhibitory effect of anthocyanins on HMG-CoA reductase diminishes the cholesterol synthesis, thereby preventing from the diseases associated with it (Winston, 1996). The role of isoprenoids to decrease the inhibition of HMG-CoA reductase is an efficient way to inhibit the synthesis of cholesterol usually dominated by the malignant cells. The terpenoids are also known to reduce the LDL-C levels and decrease tumor proliferation (Elson & Yu, 1994; Pearce, 1992).

Luteolin (Fig. 4.3) another secondary metabolite available in carrot has been cited to help in reducing the age-mediated brain inflammation and memory deficits (Jang et al., 2010). Kahkonen et al. (1999) reported that carrot flesh extracts did not show antioxidant activity toward oxidation of pure methyl lineolata at 40°C as compared to carrot peel and leaves extracts.

Tannins are plant phenolics that include hydrolysable, condensed, and complex tannins widely found in vegetables. Condensed tannins reported in carrots act as antioxidants, inhibit growth of many microorganisms, accelerate blood clotting, reduce blood pressure, decrease serum lipid levels, and control influence of immune responses (Gu et al., 2004; Serrano et al., 2009; Chung et al., 1998; Hayashi et al., 2008).

Carrot consumption in diet may be useful toward the neurodegenerative diseases, such as Parkinson, Alzheimer, and Huntington disease (Okuda, 1992). In addition to enhancing the nutritional value of foods, carrot in various forms (juice, pomace, or whole) is used for the fortification of foods in order to incorporate the therapeutic properties. Shikimic acid has antiviral properties and is found to fight against flu. The above-mentioned neurodegenerative diseases result in brain damage with the latter one due to genetic disorder and beyond one's control. The cause ascertained to Alzheimer involves the small plaques of protein deposited upon different parts of brain leading to the loss of memory, whereas exhaustion of dopamine from the basal ganglia of brain leads to rigidity, stiffness, and tremors in key muscles causing Parkinson. The developed method for getting extract from grated carrot rich in phenolics compound and shikimic acid have been used to treat the neurodegenerative, cardiovascular, and cancer diseases. The cut stress (grate) method starts the carrot's metabolism and a herbicide (glycophosphate) provides the carbon for its functioning involving inhibition of enzymes and accumulation of phenolic compound and shikimic acid in hefty quantities (Velazques, 2013).

4.3.1.2.5 Minerals

Carrots contain calcium, magnesium, and potassium that regulate muscle relaxation and contraction (Sheng, 2000; Rude, 2000). The production of phytochemicals takes place in secondary metabolic pathways with the help of minerals. The presence of calcium is effective in preventing osteoporosis and certain cancer lines particularly colon cancer (Boik & Jungi, 1996).

4.3.1.2.6 Dietary Fiber

Dietary fiber is the most abundant phytochemical of vegetables (Thomas, 2008) and consists of polysaccharides, oligosaccharides, and lignin that form the structural components of plants and it is indigestible by human digestive enzymes. Figure 4.4 presents a few health benefits related to the dietary fiber and its association with lower part of gastrointestinal tract (GI tract) protects against colorectal disorders (Calixeto, 1998) with distinguished dietary fibers acting differently on cancers (Smith et al., 1998). Fibers are classified into two main types depending on their aqueous solubility as soluble and insoluble fibers that have different as well as same physiological effects with fruits and vegetables as the principal rich and cheap sources of dietary fiber. Both the types of fibers hold water that creates bulk in the human gastrointestinal tract and its high intake can interfere with the absorption of the certain minerals (Sinha & Hui, 2011). The fiber of carrot pomace has been reported to provide hypocholesterolemic effect with reducing blood glucose levels signaling its possible use in the diets of the obese persons (Afify et al., 2013). Diverticulitis (inflammation of small pouches formed along GI tract) and gastroesophageal reflux disease (GERD), the main cause of indigestion that causes the pain known as "heart burn," are the two common diseases associated with the insufficient ingestion of dietary fiber (Sontag, 1999; O'Keefe, 1996). Noncellulosic polysaccharides like pectin, gums, and mucilage compose the soluble fibers whereas insoluble fibers mainly consist of cellulose, hemicellulose, and lignin (Yoon et al., 2005). The calorific value of dietary fibers is almost zero as body is unable to absorb it from the GI tract but the potential health benefits associated with the intake of dietary fiber have significantly led to an increase in consumption of fiber-rich products throughout the world. The high dietary fiber content in carrot (Bao & Chang, 1994) imparts vital benefits toward human health; the fiber-rich diets are found to decrease, hinder, and treat cardiovascular and diverticular disease (Gorinstein et al., 2001; Villanueva-Suarez et al., 2003). The health benefits of either soluble or insoluble dietary fibers are immense including decrease in the blood cholesterol levels, regulation of blood-sugar levels, enhanced weight control, glycemic control, and prevention of the lower bowl cancers, constipation, and heart diseases (Anderson et al., 1994).

The overall nutrient absorption rate is affected by the intake of the soluble fibers that can be helpful for the people suffering with the high levels of blood glucose and cholesterol (Jensen et al., 1997). Sugars, mainly sucrose, raise the triglyceride and cholesterol levels than starch while dietary fiber may have lowering triglyceride effect (Anderson, 1980). The LDL-C and high-density lipoprotein cholesterol (HDL-C) have, respectively, direct and inverse relationship toward CVDs (Chobanian et al., 1991). Soluble dietary fiber has been regarded as an efficient hypocholesterolemic agent that binds with the LDL-C in the intestines hindering its absorption that results in overall decrease of bad cholesterol within the body and some findings have shown that insoluble fibers could also be effective in decreasing the cholesterol levels (Erkkila et al., 1999; Knopp et al., 1999; Rosamond, 2002; Chau et al., 2004a; Chau et al., 2004b). Hsu et al. (2006) reported hypocholesterolemic and hypolipidemic effects of insoluble fiber-rich fraction obtained from the carrot pomace whereas the pectin present as calcium pectate in the cells of carrot imparts cholesterol-reducing properties (Kaur et al., 2012). Physicochemical properties such as water-holding capacity and cation-exchange capacity might be related to the cholesterol-lowering property of water insoluble fibers (Chau & Cheung, 1999). Consumption of soluble fibers lowers down the emptying rate of stomach particulates into the intestines resulting in the proper digestion of food particles and extends the feeling of satiety past the meal (French and Read, 1994; Spiller, 1994). Fruits and vegetables have long been used toward the production of dietary fiber (Figuerola et al., 2005; Grigelmo-Miguel & Martin-Belloso, 1999), whereas the fiber source from carrots is yet to be industrialized. There is a significant effect of carrot fiber on weight gain, reduction of LDL-C level, triglycerides, and fat, whereas there is no effect of carrot fiber on serum HDL-C level (Parveen et al., 2000).

4.3.2 CARROT COLORANTS

The root crop has a large diversity in terms color with the major ones as orange, red, black, purple, yellow, and white (Table 4.6). The differences among color in the carrot involve the presence of different colorants as carotenoids and anthocyanins (flavanoids) that constitute a vast group of compounds. The red carrots have been shown to contain higher amounts

TABLE 4.6 Different Types of Carrot on the Basis of Color.

Carrot color	Image		Colorants
	Whole root	Transverse cut	
Black	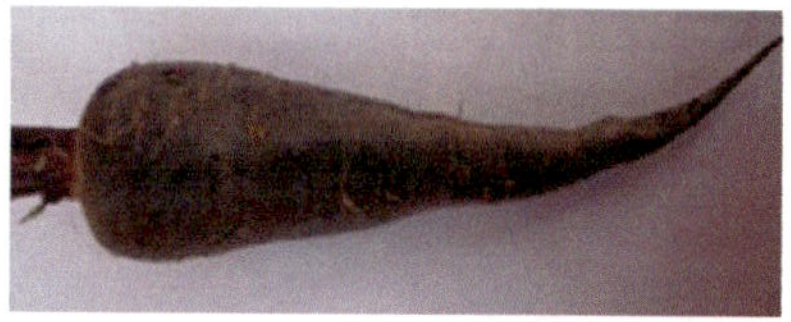		• Contains high amounts of anthocyanins mainly acylated cyanidin • Acts as a powerful antioxidant
Purple			• Rich sources of anthocyanins • Contains oxygenated carotenoid (lutein) and non-oxygenated forms (α- and β-carotene)
Orange	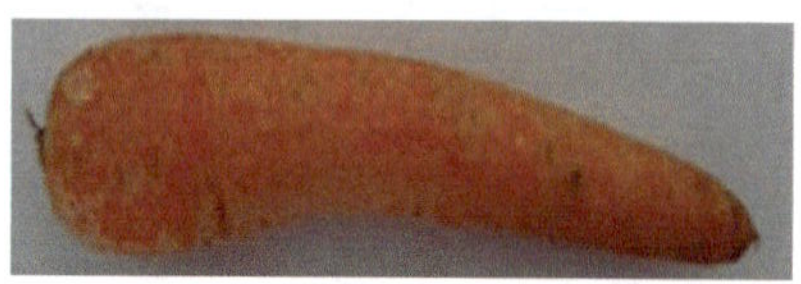		• Carotenoid-rich sources especially β-carotene. • Has antioxidant and provitamin A activity
Red			• Presence of lycopene imparts red color • Contains considerable amount of β-carotene and lutein

TABLE 4.6 *(Continued)*

Carrot color	Image		Colorants
	Whole root	Transverse cut	
Yellow			• About half of carotenoids exist as lutein • Lycopene and α-carotene absent
White			• Devoid of α-carotene and lycopene; with good amounts of vitamin C • Contains minute quantities of lutein and β-carotene

of terpenoids, whereas the purple one contains the least among the colored carrots. The volatile components of yellow are comparable with the orange variety (Kreutzmann et al., 2008b; Kebede et al., 2014).

4.3.2.1 CAROTENOIDS

The carotenoids are basically C40 tetra-terpenoids (Fig. 4.5) built from the basic C_5 isoprenoid units joined head to tail with hydrocarbon carotenoids known as carotenes (β-carotene or lycopene) and those containing oxygen as xanthophylls (lutein or zeoxanthin); these are present in chromoplasts and chloroplasts without or with chlorophyll, respectively (Rodriguez-Amaya, 2001; Vickie & Elizabeth, 2007). The *trans*-form of carotenoids is dominant in nature that may transform into *cis*-form upon processing, cooking, and exposure to light. The nutritional significance of *trans*-carotenoids is more as compared to their *cis*-forms with the latter type exhibiting less potency due to the significant decrease in the vitamin A activity (Clydesdale et al., 1991, Rodriguez-Amaya, 1989; Sweeney & Marsh, 1971). The *cis*-isomers possess low biological activity and readily absorb UV radiations resulting in their fader color (Fraser & Bramley, 2004).

The vitamin A activity is expressed either as international unit (IU) or retinol equivalent with 10 units of former equivalent to 1 unit of latter. The 1.2 and 0.6 μg of α- and β-carotene are equivalent to 1 IU (Bauernfeind, 1972; NAS-NCR, 1980).

The number of characterized carotenoids exceeds 600 and about 40 carotenes are reported to be present in the carrots with the major forms including α- and β-carotene, lycopene, cryptoxanthin, lutein, violaxanthin, neoxanthin, and antheraxanthin as reported by Grassmann et al. (2007), Heinonen et al. (1989), Muller (1997), and Gross (1991). The carrots are imparted with red and orange color by lycopene and carotene (α- and β) pigments, respectively. Black or purple color in carrot is due to the presence of anthocyanins, whereas oxygenated carotenoids (lutein) are responsible for yellow color and the absence of the mentioned pigments result in the white color (Surles et al., 2004; Nicolle et al., 2004; Molldrem et al., 2004).

The presence of β-carotene content in fresh carrot peels is almost similar to the carrot flesh, which supports the fact that its presence is concentrated in the peels (Negi & Roy, 2001). The long series of conjugated double bonds in carotenes contributes to their color with more/less alternate unsaturation providing the deeper/fader color. The α-carotene is fader in color than its β-counterpart due to the occurrence of one less conjugated double bond. As compared to β-carotene, lycopene has two more double bonds (2 nonconjugated and 11 conjugated) imparting it a bright red color. The precursors of vitamin A carotenes are known to exceed more than 40 in number and chiefly include α-carotene, β-carotene, and β-cryptoxanthin, which are enzymatically converted to vitamin A in liver and intestines of the human body (Simpson, 1983; Vickie & Elizabeth, 2007).

The oxygenated derivatives of carotenes, xanthophylls are yellow-orange in color whose presence is noticed when chlorophyll is degraded during the autumn season and exist as cryptoxanthin and lutein in corn and leaves, respectively (Gross, 1991).

4.3.2.2 ANTHOCYANINS

Anthocyanins (of the Greek *anthos*—flower and *kianos*—blue) are regarded as the largest group of phenolics containing water-soluble vacuolar pigments (Kammerer et al., 2004) imparting red, purple, or blue color to plant and plant products present mainly in fruits. The anthocyanidin or aglycone (Fig. 4.5) is the fundamental moiety of anthocyanin structure consisting of heterocyclic oxygenated ring "C" joined with two "A" and "B" rings. The association of anthocyanidins with sugars forms anthocyanins (Konczak & Zhang, 2004; Delgado-Vargas & Paredes-Lopez, 2003). The anthocyanidin in nature occurs mainly in association with sugar molecules, anthocyanins (Castaneda-Ovando et al., 2009) and are distributed in certain vegetables, roots, and cereals (Mazza & Miniati, 1993). The glycosylation pattern and pH of the system are responsible for the color change (Fig. 4.5) of these pigments (Fossen et al., 1996; Li et al., 2012). The color form is favored in acidic conditions, and at $pH < 4$, these pigments are fully colored (Delgado-Vargass et al., 2000). The glycosided anthocyanidins referred to as anthocyanins belong to a class of phenolic flavonoids that are based mainly on cyanidin, pelargonidin, malvidin, peonidin, petunidin, and delphinidin (de Pascual-Teresa & Sanchez-Ballesta, 2008; Andersen & Jordheim, 2006). The 3-OH and 5-OH positions are more commonly glycosylated than the 7-OH and differences in methoxylation and hydroxylation in ring-B differentiates anthocyanins (Francis & Markakis,

1989). The presence of anthocyanins is markedly distributed among the plant kingdom mainly in fruits and flowers as red, blue, purple, or blue-red colors with the acylated derivatives of cyanidin as the major anthocyanin pigments present in black and purple root colors (Schwarz et al., 2004). The presence of various phenolics, vitamin C, and E in the black carrot has increased the attention of scientific community toward its use (Singh et al., 2011; Alasalvar et al., 2001). The higher levels of phenolics increase the antioxidant property of purple-haze carrot than other black varieties (Algarra et al., 2014). The use of pectinase in enhancing black-juice extraction also elevated its antioxidant composition by doubling the anthocyanin content and raising the yield levels of total phenolics and total flavonoids by 27% and 46%, respectively (Khandare et al., 2011). The stability of anthocyanins is affected by temperature, pH, oxygen, light, metallic ions, and enzymes. This has led recent studies to focus on enhancing the chemical stabilization (Rein, 2005). The study by Koley et al. (2014) on different varieties of Indian carrot reported absence of anthocyanins in red and orange varieties of carrot but anthocyanins were detected in black carrot along with total phenolics in abundant quantities.

Anthocyanins are flavonoids and differ from the carotenoids by possessing positively charged oxygen in the molecule and water solubility. The presence of black/purple color in carrot is the result of anthocyanin pigments that have relatively low levels of toxicity and thus can be used as the source of colorants in foods. However, the glycosylation of anthocyanins occurs with different sugars but glucose is the main glycosidic sugar attached with anthocyanins (Li et al., 2012). The major anthocyanins associated with the black carrot are acylated with ferulic, sinapic, and *p*-coumaric acids, whereas peonidin exists without acylation (Kammerer et al., 2004; Dougall et al., 1998; Kammerer et al., 2003).

The black carrot color extracts can be employed as food colorants in juices, soft drinks, preserves, confectionaries, and jellies due to formation of bright strawberry red color in acidic pH system (Downham & Collins, 2000) offering a healthier option to consumer against various diseases (Khandare et al., 2011). The fermented carrot beverage, *Kanji,* is prepared from the black carrot. β-D-Glucopyranosyl-(1>6)-[β-D-xylopyranosyl-(1>2)]-β-D-galactopyranosyl-(1>O3)-cyanidin is the nonacylated anthocyanin whereas the 2-methoxycinnamic acid and 4-hydroxy-3,5-dimethoxycinnamic acid are acylated with 6-*O*-acyl-β-D-glucopyranosyl-(1>6)-[β-D-xylopyranosyl-(1>2)]-β-D-galactopyranosyl-(1>O3)-cyanidins, in black carrot (Elham et al., 2006; Baker et al., 1994). Algarra et al. (2014) and Montilla et al. (2011) reported the acylated forms of anthocyanins in black carrot varieties and detected cyanidin 3-xylosylgalactoside, cyanidin 3-xylosylglucosylgalactoside, and the latter's ferulic, sinapic, and coumaric acids derivatives. The generally regarded as safe (GRAS) status of the anthocyanins obtained from black carrot extracts removes the hindrances in its commercialization (Khandare et al., 2011). Kammerer et al. (2004) detected seven anthocyanin compounds in black carrots, five of them acylated with hydroxybenzoic and hydroxycinnamic acids. The acylated anthocyanins, ranging from 55% to 99% in black carrots (Kammerer et al., 2004), are known for their higher color stability than other types of anthocyanins when used in food systems. Narayan and Venkataraman (2000) observed the absence of acylated anthocyanin components in Nantes cell culture and only two anthocyanin pigments, cyanidin-3-lathyroside and cyanidin-3-β-D-glucopyranoside, were isolated. The anthocyanin in purple carrot constitutes 33.65% and 29.85% of cyanidin-3-xylosyl-glucosyl-galactoside acylated with ferulic and comaric acid, respectively, and 28.7% cyanidin-3-xylosyl-galactoside as the third compound as detected by high-performance liquid chromatography (HPLC) (Assous et al., 2014). The antioxidant activity of purple-colored carrots due to the presence of high levels of phenolics is by far at par than the other colored varieties of the *genus* (Sun et al., 2009; Leja et al., 2013).

4.4 EXTRACTION OF BIOACTIVE COMPONENTS

The color of the carotenoids is presented by the alternate unsaturated double bonds that constitute the light-absorbing chromophore, also responsible for its instability with light, oxygen, and heat pressing the need of using antioxidants in the extraction process. The light absorption of the chromophore has a direct relation with the number of alternate double bonds in the electron-rich polyene chain that serves as the preferable sink to electrophiles (e.g., free radicals), the reason for carotenoids' antioxidant property (Takaichi, 2000; Furr, 2004). Xanthophyll are the oxygenated forms of hydrocarbon carotenes containing one or more oxygenated groups such as β-cryptoxanthin and canthaxanthin with hydroxy and keto-groups, respectively. The natural occurrence of the oxygen in xanthophylls imparts it with partial or more polar characteristics and it may be esterified with fatty acids (astaxanthine dipalmitate) (Bijttebier et al., 2014).

Carotenogenic root crop (e.g., carrot) mainly contains carotenes, and xanthophylls are abundant in maize.

4.4.1 CAROTENOID EXTRACTION

The primary extraction process of carotenoid uses an organic solvent (either polar or the combination of both polar and apolar) that transfers the pigments from the sample into the organic phase following their subsequent quantification. The other methods of extraction involve use of supercritical fluids (SCFs), microwaves, and others, and are briefly discussed here along with the traditional extraction technique.

4.4.1.1 CONVENTIONAL SOLVENT EXTRACTION

The colored matter extracted by water-miscible solvents (ethanol, methanol, acetone, or their mixes) with more nonpolar solvents (hexane, diethyl ether, and methylene chloride) from the macerated or powdered sample is transferred into a separating flask added with measured amounts of distilled water. The grounded sample is extracted until it becomes colorless followed by the separation of the two phases (Pinheiro-Santana et al., 1998).

The water concentrates the pigments into apolar solvent and the flask is agitated followed by discarding the lower water phase upon separation. The process goes on 2–5 times till the discarded layer is colorless and the extract is desiccated with anhydrous Na_2SO_4 to remove aqueous residue. Another separation method is of partitioning layers by the use of acetone extracts in petroleum ether in small proportions with gentle addition of water without shaking or agitation and leaving it for discarding lower aqueous layer followed by washing of extract as in the former process.

In case of dried samples, use of water-immiscible solvents for extraction and rehydration prior to extraction with water-miscible solvents yields good extracts (Rodgureiz-Amaya & Kimura, 2004). The extract can be either used as bioactive source or as coloring matter for foods and its estimation is achieved by measuring the absorbance at 400–550 nm in a UV–vis spectrophotometer with 450 nm used predominately in studies (Pinheiro-Santana et al., 1998). The predominantly absorbed wavelength

(λ_{max}) of most carotenoids generates three peaks and greater the conjugation greater is the absorption with red color lycopene absorbing the longest wavelengths in UV–vis spectrum. The perception of color from carotenoids is due to presence of at least seven alternate double bonds (Rodgureiz-Amaya & Kimura, 2004).

The solvent mixture and extraction procedure vary in different foods as Park (1987) and Koca et al. (2007) used 3:2 and 7:3 hexane:acetone mixture for extraction of carotenes from carrot, whereas Mustapha and Babura (2009) employed 95% alcohol to the macerated sample at 70–80°C with addition of petroleum ether in separating flask. Solvents with low boiling point (e.g., petroleum ether 30–60°C) should be preferred to reduce the heat damage as these eventually are reduced or removed by evaporation. The ground samples were extracted with 80% acetone in a vortex shaker for about 1 h and later centrifuged at 8000 × *g* whose supernatant was used for determining the bioactive components including carotenoids by Hernandez-Ortega (2013).

4.4.1.2 SUPERCRITICAL FLUID EXTRACTION

The fluids over their critical pressure and temperature constitute SCFs and are used as solvents in supercritical fluid extraction (SCFE) process. The viscosity and density of SCFs behave like gas and liquid, respectively (Sihvonen et al., 1999). The different SCFs include carbon dioxide, methanol, ethane, ethylene, hexane propane, toluene, water, and pyridine (Brignole, 1986) with CO_2 as the major and frequently used solvent in food applications for extracting process due to its nontoxicity, easy availability, incombustibility, liquid-like density, inertness toward extracted components and it is "Generally Regarded as Safe" (Lang & Wai, 2001; Brunner, 2005; Shi et al., 2007; Shilpi et al., 2013). The properties like low surface tension, higher diffusion coefficient, and liquid-like densities of SCFs enable their easy penetration into porous matrix to extract the solutes (Shilpi et al., 2013). The bed of solid substrate powdered sample matrix is placed in stainless steel basket (extractor vessel) operated at required parameters of temperature, pressure, and SCFs flow rate. The SCF flowing through the matrix solublizes the components of interest and transfers them to separator where depressurization changes the SCF into gas and the extraction of components is facilitated through an attached ball valve

(Sihvonen et al., 1999; Temelli et al., 2012; Sandra et al., 2014). The ideality of CO_2 as SCF for extraction of heat-sensitive components is enhanced by its low values of critical temperature (31.1°C) and pressure (73.8 MPa) (Reverchon et al., 1993). The β-carotene extraction from carrots at different temperature and pressure with varying contact periods was studied by Kaur et al. (2012). The extraction was carried out from 5 to 360 min in the loaded vessel held in heat-controlled convective oven with β-carotene concentration in the extract increasing with the increase in temperature, pressure, and extraction time. The maximum extracted yield was reported to be at 45°C and 35 MPa by Kaur et al. (2012). Vega et al. (1996) and Goto et al. (1994) studied extraction of carotenoids from carrots by SCFE with latter study concluding that disruption of cell membranes is necessary for complete extraction from freeze-dried carrots. Ranalli et al. (2004) extracted the crude oil from carrots at different combinations of pressure and temperature. The carotenoid presence in extracted oil was described to be 339.3–745.5 μg/g identified and quantified using HPLC. The highest extraction of essential oil from carrots was achieved at 313 K and 10 MPa and was separated by GC–MS (Sandra et al., 2014). The comparison between SCFE and solvent extraction of carotenoid molecules from the carrot root has been studied (Barth et al., 1995) whose findings reported an increase in (7%) provitamin A activity with carotenoids extracted by SCFE.

The high moisture content in tissues negatively affects the extraction of lipophilic components by SCFE (Wang & Weller, 2006; Shi et al., 2013) and particle size needs to be decreased for efficient extraction of carotenoids. The major factors affecting SCFE process are presented here.

4.4.1.2.1 Particle Size and Moisture

The particle size plays a significant role in carotenoid extraction by SC-CO_2 with smaller size (up to a certain size 0.5–1.0 mm) (Sun & Temelli, 2006) yielding higher amounts due to the higher exposure of surface area and decreasing the matrix length through which solvent has to pass (Goto et al., 1994; Eggers, 1996). The very small particles cause channeling effect resulting in increased internal mass transfer due to their compaction property (Reverchon & de Marco, 2006). The extraction values significantly increased from 1109.9 to 1369.6 μg/g dry carrot when the size was decreased from 2 to 1 mm, and upon further reduction of the particle size to 0.25 mm the extraction yield increased to 1503 μg/g dry carrot (Goto et al., 1994; Sun and Temelli, 2006).

The extraction of α- and β-carotene increased on lowering down of the moisture ratio, whereas a reverse trend was seen for lutein. Lowering down the moisture content from 84% to 17.5% in carrots increased α- and β-carotene extraction from 184 to 442 and 354 to 668 μg/g dry carrot, respectively. The lycopene extraction at a moisture content of 50–60% yielded trace amounts. The relatively polar components, namely, lutein, utilize water as a cosolvent during extraction process, whereas its presence is not favored in relatively nonpolar compounds like lycopene and carotenes (Sun & Temelli, 2006; Vasapollo et al., 2004; Martinez, 2008). The formation of ice crystals from the available moisture in the samples leads to value clogging ultimately affecting the extraction (Liu et al., 1994). The removal of moisture is necessary for improving the extraction yield of lipid-soluble molecules by SC-CO_2 (Durante et al., 2014).

4.4.1.2.2 Temperature and Pressure

Increase in the temperature greatly affects the extraction process above the crossover pressure in SC-CO_2 extraction. There is respective increase and decrease in the solvent density at constant temperature and pressure up on the former's increase in pressure and latter's temperature increase. The density plays a vital role in extraction process with its increase directly increasing the solubility power leading to high extraction rates. The decrease in SCFs density at lower pressures decreases the extraction power whereas increased pressure elevates the extraction yield due to greater penetration of low viscous solvent into solid matrix (Reverchon & de Marco, 2006; Shi et al., 2009; Sahena et al., 2009). Increased pressure and temperature increase the carotenoid extraction (Shi et al., 2013). However, on increasing the temperature more than 80°C has resulted in the carotenoid isomerization (Lenucci et al., 2010; Shi et al., 2008). The extraction of lycopene increased both at the pressure of 33.5–45 MPa maintained at 66°C and at the temperature of 45–65°C with a pressure of 45 MPa. Lycopene recovery of 100% and 96% was achieved at 100°C for 50 and 40 min, respectively, from the tomato skin, whereas the yield was 20% at 60°C for 80 min (Ollanketo et al., 2001; Vasapollo et al., 2004).

4.4.1.2.3 *Use of Cosolvent or Modifiers*

CO_2 dissolves apolar or somewhat polar compounds efficiently and has limitations for the extraction of high molecular weight and polar compounds. The extraction by pure SC-CO_2 solvent is favored more for nonpolar and slightly polar components (Brunner, 2005) and the use of cosolvents, namely, ethanol, methanol, water, canola oil, and others, in small volumes (<20%) can extend the range of extraction toward polar and high molecular weight components by enhancing the solvent affinity toward poorly soluble solutes (Durante et al., 2014; Shilpi et al., 2013). The enhanced extraction by SC-CO_2 up to 599 and 892 μg/g dry carrots for α- and β-carotene, respectively, was achieved upon further decrease in moisture content to 17.5%. Acetone, methanol, ethanol, hexane, and water are cosolvents used with all of them increasing the lycopene extraction except the last one. Vegetable oils used as a cosolvent in carotene extraction include canola oil and hazel-nut oil, the former when used in carrot extraction yielded twice the extract than without its use (Vasapollo et al., 2004; Sun & Temelli, 2006). The yields increased more than twice for carotenes (α- and β-carotene) and greater than four times in case of lutein as compared to the results of SC-CO_2 alone (Sun & Temelli, 2006). The use of ethanol and 5% chloroform as modifiers for the carotenoid extraction from carrots has been reported by Marsili and Callahan (1993) and Chandra and Nair (1997), respectively.

4.4.1.3 *MICROWAVE EXTRACTION*

Microwave extraction or microwave-assisted extraction (MAE) uses electromagnetic waves that cause alterations in the cell structure resulting in the accelerated extraction of cellular components within a very short period of time. The frequency and wavelength of microwaves range from 300 MHz to 300 GHz and 1 cm to 1 m, respectively, with a frequency of 2450 MHz commonly used in commercial microwaves (Kingston & Jassie, 1988). The basis for microwave extraction involves conversion of absorbed microwaves into thermal energy upon passing through a medium (Zhang et al., 2011), and the use of microwaves for the first time in MAE was reported by Ganzler et al. (1986). The MAE works just opposite to the conventional extraction processes involving simultaneous transfer of heat and mass toward the same direction compared to conventional extraction that causes opposite transfer of heat and mass within the sample (Chemat et al., 2009). The dipole rotation and ionic conduction property of polar molecules in presence of microwaves produce significantly high temperature and pressure inside the sample resulting in the lysis of cells, releasing their components into solvent phase and hence enhance the extraction rate due to its constant mixing with the solvent (Gill-Chavez et al., 2013; Mandal et al., 2007; Jain et al., 2009; Routray & Orsat, 2011). The use of MAE for carotenoids and phenolics has been reported by different studies (Zhao et al., 2006; Choi et al., 2007; Pasquet et al., 2011; Li et al., 2011; Moreira et al., 2012; Simsek et al., 2012). The dielectric constant, solubility, and dissipation factor (efficiency of microwaves to increase temperature of different solvents) of solvents are considered before their selection for MAE. The ratio of dielectric loss factor, efficiency of absorbed microwave conversion into heat, and dielectric constant is dissipation factor (Jain et al., 2009). The high dielectric constant (polar molecules) is compulsory for MAE as the polar molecules readily absorb microwaves (Wang & Weller, 2006) and the dissipation factor influences the yield of extraction as reported by Ajila et al. (2011) for the extraction of phenolics by utilization of the solvents of high dissipation factor (ethanol or methanol). The solvent with low dielectric constant (hexane) are used to form mixed solvents as they absorb microwaves at a very low energy level used in the extraction of heat-labile and nonpolar components in carrot (Eskilsson & Bjorklund, 2000).

MAE has both its pros and cons over traditional extraction with higher extraction rates in relatively shorter time, low solvent consumption, and low environmental pollution as its advantages (Garcia-Ayuso et al., 1999, 2001), whereas modification in the basic chemical structure (at high power > 1000 W) and difficulties in recovery of nonpolar components serve as the limitations of the process (Young, 1995; Gill-Chavez et al., 2013). The open-vessel and closed-vessel systems are the two main types of MAE extraction systems with the latter system commonly employed for extraction purposes operating at controlled temperature and pressure. The temperatures can be raised to 100°C above than the atmospheric boiling temperature and are mainly used for digestion or extraction under harsh conditions (Jassie et al., 1997). The open system looks like the Soxhlet extraction unit with a simple design of quartz cell equipped with a vapor condenser operating at atmospheric pressure. The open system offers a relative safe extraction than the closed and heated solvent is refluxed continuously through sample (Letellier et al., 1999). MAE is used for various cell metabolites, namely, alkaloids and steroids extracted by both (open and closed) systems, whereas terpenes, alkaloids, carotenoids, essential oils, pyrimidine glycosides, and gossypol by closed system (Kaufmann & Christen, 2002). Hiranvarachat et al. (2013) employed different pretreatments (soaking in citric acid, blanching) to the carrot tissue and examined the improvement of carotenoid extraction with acetone, ethanol, and hexane as mixed solvent. The extraction yields at optimal conditions showed significant improvement in the pretreated samples (58.04–61.62 mg/100 g dry weight basis) than the untreated ones (51.79 mg/100 g dry weight basis) but the values were low as compared to Soxhlet extraction (61.13 mg/100 g dry weight basis). Hiranvarachat and Devahastin (2014) extracted carotenoids from peels of carrot using intermittent microwaves at different intermittency ratios (fraction of microwave radiation time to total processing time) and low power (180–300 W). MAE may result in the reduction of extraction efficiency for the matrix by its prolonged use due to the significant degradation of carotenoids mainly due to the applied heat treatment (Hiranvarachat & Devahastin, 2014).

4.4.1.4 SAPONIFICATION

It is an important pretreatment step before HPLC analysis for removing unwanted pigments and esterified fatty acids from carotenoids that may otherwise hinder its chromatographic separation. The process involves addition of methanolic potassium hydroxide plus hexane or petroleum ether to the carotenoid extract at room temperature for 16 h or at 56°C for 20 min followed by its washing to remove alkali with water (AOAC International, 1995). The obtained organic phase is passed through the bed of anhydrous sodium sulphate (Na_2SO_4) to obtain dry pigments (Oliver & Palou, 2000). The esterified carotenols (fatty acids esterified with the oxygen of xanthophylls) are hydrolyzed into free carotenols that are easy to separate, identify, and quantify on the chromatograph than their esterified forms (Fisher & Kocis, 1987; Oliver et al., 1998). This step involves loss of some carotenoids and is skipped when there are alkaline-sensitive pigments (fucoxanthin, lutein, and violaxanthin) or absence of carotenol esters (carrot, tomato, leafy vegetables) in the carotenoid extract (de Quiros & Costa, 2006). The studies related to saponification of carrot extracts use methanolic KOH for 30 min in darkness (Granado et al., 1992; Howard et al., 1999).

4.4.1.5 CHROMATOGRAPHIC SEPARATION

The separation of mixed components by differential absorption between the stationary and mobile phases is chromatographic separation. Column chromatography (CC), thin layer chromatography (TLC), and HPLC are included in liquid chromatography involving two immiscible liquids and separation technique based on solubility (Ahuja, 2006). All the three techniques separate the components of mixture by the interaction of stationary phase (column) with the mobile phase and this process widely finds application in the separation of pharmaceutical compounds and plant/animal metabolites on different stationary phases.

4.4.1.5.1 CC and TLC

The CC is a reliable method for separation of groups with same polarity in large quantities, whereas TLC is a rapid-quality analysis where components of mixture can be isolated, separated, identified and quantified from the extract (Grillini, 2006; Touchstone & Dobbins, 1978). These two chromatographic techniques have already been used for the carotenoid separation as reported (Rodriguez-Amaya et al., 1988; Premachandra, 1985; Mercadante, 1999). The polarity and arrangement of the molecular structure in addition to the number of alternate double bonds or conjugation drive the separation process on stationary phases. The stationary phase is packed within a glass column in CC, whereas in TLC this phase (aluminum oxide and silica) is coated onto glass, plastic, or metallic plate (Stahl, 1956; Frey, 1993).

A spot or streak of sample material to be separated is positioned on the adsorbent plate that is positioned in the developing chamber saturated with the solvent vapors and a little amount of moving phase. The capillary forces move the mobile phase on the stationary phase and form a developed plate that is removed from the chamber when the mobile phase reaches near its top (solvent front) and is visualized for compounds. The colored components are easily differentiated, whereas the colorless ones are examined by detection techniques, namely, under the UV light or use of derivatizing reagent that enables the visualization of the bands which may fluoresce or appear as dark spots and are indexed for references by calculating their rates of flow (R_f) value. Polar compounds are found at lower regions while nonpolar components at higher regions of plate due to more and less strong attraction to adsorbent, respectively. The R_f of the component is determined by the ratio of distance of solvent front on the developed plate upon distance of component from the origin of spot up to two decimal places (Grillini, 2006; Jork et al., 1990, 1994; Hahn-Deinstrop, 2007). The choice of mobile phase selection is reported by Zweig and Sharma (1984). The isomerization, oxidation, and degradation of carotenoids on the adsorbed plate limit its use mainly for identification purposes. The low separation efficacy along with the greater time consumption limits TLC's role in the preliminary detection and identification of carotenoids present in a sample (Karnjanawipagul et al., 2010).

In case of CC the stationary phase, mainly silica, is mixed with some solvent (hexane or ethylacetate) and poured into the column lined with cotton wool plug from the base that prevents the exit of silica during development of column. The column is packed by slapping the column walls to remove the air bubbles and gently pouring the separation mixture followed by the sufficient solvent without disturbing the column. The distinguishable colored samples are separated by virtue of their color, whereas the invisible ones are collected in a rack of test tubes followed by their examination on TLC. The activated magnesia and diatomaceous earth were used as column packings in the ratio of 1:3 by Aalbersberg (1991) for the separation of carotenoids with 10% and 20% of acetone in hexane as solvent. The stationary phase also acts as an adsorbent for certain components and holds them back with hydrogen bonding, van der Waals forces, and so on, until a desired component is isolated and then the adsorbate can be removed by solvent changes (polarity, pH, ionic strength, concentration, etc.). The change and variation of the phase and solvent extends CC's scope in separation and isolation process (Pavia & Russell, 1999). The use of high pressure to enhance the movement of solvent is the distinguishing feature of HPLC over CC (Ahuja, 2006).

4.4.1.6.2 HPLC Separation

The analytical chemical method for the qualitative and quantitative analysis of carotene involves use of HPLC that consists of four major components a sampler, pump, column, and detector. The sampler brings the sample into solvent phase that is pressurized by pumps to pass it through the adsorbent column with a mobile phase, leading to the separation of components owing to their different flow rates on the porous adsorbent column similar to the TLC. The mobile phases most commonly used for the carotenoid analysis include acetonitrile (ACN) or methanol (Me-OH) or its mixtures and for the enhanced separation of nonpolar compounds small quantities of organic solvents (water, hexane, propanol, acetone, etc.) are added (van den Berg et al., 2000; de Quiros & Costa, 2006; Mathiasson et al., 2002; Chandrika et al., 2003; Rozzi et al., 2002; Englberger et al., 2003). The granular materials most commonly used to pack the columns include silica, alumina or any other type of high polar material, namely, polymers (C_{30}, C_4, C_{18}) with minute size (2–50 μm) are typically spherical coated with nonpolar alkyl chains that separate analytical samples on the basis of polarity with nonpolar coated silicon particles. The components coming out of the column are received by a detector which is mostly a mass spectrophotometer; the other types of detectors include charged aerosol, conductivity, fluorescence, electrochemical, refractive index, conductive, evaporative light scattering detectors, and others (Ahuja, 2006). The components absorbing UV light in detector is proportional to the amount of component present and the chromatographic data systems convert the results into usable data. The time taken by each component to come out (elute) of the column denotes its retention time that can be used in the detection of components. The liquid chromatographic columns involving polymeric C_{30} significantly improved the carotenoid separation of polar as well as apolar carotenoid isomers in HPLC (Sander et al., 2000). Careri et al. (1999) used acetonitrile–methanol (0.1 M ammonium acetate)–dichloromethane mobile phase in C_{18} column, whereas Lacker et al. (1999) used methanol–methyl *tert*-butyl ether mobile phase in C_{30} column for the carotenoid separation. The solvent and column used by Aalbersberg (1991) involve acetonitrile–methanol–ethyl acetate (EtOAc) (80:180:2) and C_{18}, respectively, for the carotenoid separation. The quantitative analysis of the sample is provided by the detector (UV–vis and Photodiode array/PDA) that generates an electrical signal proportional to its amount coming out of the column. Supelcosil LC-18 (150 × 46 mm, 5 μm) column along with methanol as solvent A and acetonitrile (90:10, v/v) as solvent B were used in carotenoid separation of carrots with constant concentration of solvent throughout the process. The injected sample is dissolved in dichloromethane/methanol (1:1 v/v) and the mobile phase flow rate at 1 mL/min.

The separated α-carotene, β-carotene, and lutein were identified by using UV detector at 450 nm (Sun & Temelli, 2006). The analysis time significantly reduces whereas sensitivity increases in ultra-high-performance liquid chromatography (UHPLC) due to use of column packings with less than 2 µm diameter (Rivera and Canela-Garayoa, 2012; Swartz, 2005).

The lesser selectivity of UHPLC (due to absence of C_{30} ligand column) in carotenoid separation in spite of its quick analysis (1/4 time of HPLC) restricts its use in carotenoid separation process, whereas HPLC C_{30} column is so far the most effective column employed for the separation of carotenoids (Bijttebier et al., 2014). Ligor et al. (2013), Rojas-Garbanzo et al. (2011), Hiranvarachat et al. (2013), Chumnanpaisont et al. (2014), and Sancho et al. (2010) have also used C_{30} column for the carotenoid separation.

4.4.1.6.2.1 Identification and Quantification

As already stated, the conjugated unsaturation in case of carotenoids contributes to their color by absorbing the particular regions of the UV–vis spectrum with maximum absorption of three peaks/wavelengths denoted as λ_{max}.

The minimum presence of seven conjugated double bonds imparts the color with maximum 11 unsaturated bonds in lycopene developing red color due to the absorbance of the farther regions in spectrum. The spectrum shape and maximum absorption is unique to a chromophore that serves as basic clue for their identification. The steric hindrance in the cyclic carotene results in the decrease of λ_{max} toward shorter wavelength (hypsochromic shift), decrease in absorbance (hypochromic effect), and less-defined spectral peaks, which are responsible for yellow-orange color of β-carotene in spite of the equivalent number of conjugated double bonds present as in lycopene. The maximum absorption wavelength (λ_{max}) is specific to a chromophore along with spectrum shape and the increase in double-bond conjugation results in the absorption of light with greater wavelength. The majority of carotenoids show maximum absorption at three different wavelengths with α-carotene (λ_{max} 424, 448, 476 nm) losing conjugation in one of the rings resulting in a slightly fader color (yellow) than β-carotene (λ_{max} 429, 452, 478 nm) and lycopene (λ_{max} 448, 474, 505 nm), despite the same number of unsaturated bonds with a one less conjugated bond (Rodgureiz-Amaya & Kimura, 2004). The chromatography of reference standards is used along with the UV–vis spectrophotometry in confirming the carotenoid peaks.

The different fractions obtained with solvents were evaporated to give an adequate absorbance in the spectrophotometer (Siong & Lam, 1992).

The Beer–Lambert law states that the absorbance of light is directly proportional to the concentration and is obeyed by the carotenoid solution also. The information about the absorption coefficients of carotenoids in different solvents is necessary for their spectrophotometric quantification. The absorption coefficient denoted as $A_{1\,\mathrm{cm}}^{1\%}$ indicates the absorbance at a particular wavelength of 1% solution in 1 cm light-path cuvette and is involved in calculating the concentration.

4.4.1.6.3 Gas Chromatography

This is an organic analytical technique used in the research and quality control for separating the fairly volatile mixtures (nonpolar and semipolar) using a gaseous mobile phase and a tubular stationary phase packed with solid or coated with high boiling point viscous liquid encased in a temperature controlled oven.

The GC has a higher separation power and sensitivity than LC whose main components are carrier gas (hydrogen or helium), injector, column, detector, and integrator (Kaklamanos et al., 2012). The basic steps of GC analysis include extraction of sample, clean-up step, determinative step, and lastly confirming the results. The samples must be volatile and heat stable (<400°C). Samples are injected into the injector which causes their rapid vaporization and are swept out with the gas mix to the column where they separate on the basis of volatility. The temperature-controlled column is initially at a lower temperature that causes low boiling point components to separate first and reach detector and the process goes on upon gradual increase of the temperature in the column (Hajslova & Cajka, 2006, 2008). The time each component takes to reach the detector is called retention time, that is, the sum of adjusted retention time and gas hold-up time. The basic measure in identifying components by GC involves the use of retention time (Ahuja, 2006). The detectors are mass spectrometers and hence the name GC–MS. It isolates ionized atoms or molecules on the basis of variation in their mass to charge ratio (Kaklamanos et al., 2012).

The detectors can be selective toward a specific component or nonselective with a wide range of detection that converts some physical and chemical properties of components into an electrical signal. The flame-ionizing detector (FID) is a classical nonselective detector that crucifies the components coming out of column into their respective ions by burned hydrogen gas. The electrical conductivity of flame is measured for the detection of components that contribute to a peak on the integrator. The data analysis of the components by integrator gives the chromatogram that is a graphical representation of the sample elusion and response. The height and area within each peak are employed for determining the quantity of each component and its identification in the sample mix. The mono- and sesquiterpenes are the main volatiles accounting for ~99% total volatiles in stored carrot as identified by GC–FID, –mass spectrometry (MS) and –olfactory (O). The fused silica capillary column (CP-Wax 52CB) with FID operating at 250°C and chiral column were used in GC–FID and GC–MS separation with both techniques involving helium as carrier gas. GC–O involved the use of above two columns mounted onto the sniffing port (Kjeldsen et al., 2003). The head space–solid phase microextraction–gas chromatography–mass spectrometry (HS–SPME–GC–MS) analysis of thermally and high pressure high temperature (HPHT) processed colored (red, yellow, orange, and purple) carrots showed a visible decrease in the terpenes (associated with harsh flavor [Howard et al., 1995] in large quantities) than the blanched ones. The static headspace analysis/GC/MS developed by Alasalvar et al. (1999) concluded detection of 35 different volatiles in carrot out of which mono- and sesquiterpenes constitute about 97% of the total volatiles. The volatiles appreciably didnot change over a storage period of 28 days but showed a significant loss (88.6–95.5%) upon cooking for 10–30 min.

Kebede et al. (2014) hypothesized that HPHT enhances the concentration of terpene alcohols at the expense of terpenes resulting in the latter's decrease in concentration. The gas chromatograph attached with a mass spectrometer has been employed by Fukuda et al. (2013) for volatile profiles and aroma characteristics of 14 carrot varieties. The helium gas was used as carrier with a flow rate of 1.0 mL/min and the oven temperature was gradually increased from 40°C to 220°C at the rate of 8°C/min and identification of 43 volatile components with 4 unidentified ones was reported.

4.4.2 EXTRACTION OF OTHER BIOACTIVE COMPONENTS

The difference in the solubility of components within different solvents (whether polar or apolar) is the major differentiating step in the extraction of various metabolites. The different extraction techniques, as described in the carotenoid extraction, also apply for the extraction procedures of the other metabolic components and need not to be elaborated again. The extraction of some other bioactive components in carrot is briefly documented below.

4.4.2.1 POLYACETYLENE

The mechanical operation, especially peeling, affects the polyacetylene content in carrots (Koidis et al., 2012). The time–temperature combinations significantly affected the polyacetylene composition in carrots with the lower temperatures (50–60°C) and higher temperatures (50–60°C) showing a respective decrease and increase in concentration (Rawson et al., 2010). Quick identification: the carrot roots were minced to purée and mixed with water (2 times the sample wt.) and extracted 3–4 times with dichloromethane in a separation flask. The Na_2SO_4 crystals are used for removal of moisture from the extract that is concentrated by heating (30°C) at low pressure. N_2 gas is used for further removal of solvent and the obtained sample is redissolved in ethanol for examination in GC–MS (Pferschy-Wenzig et al., 2009).

4.4.2.1.1 Extraction

The frozen or freeze-dried samples are used for falcarinol extraction by accelerated solvent extraction (ASE). The freeze-dried sample or frozen sample is macerated to purée and is added with EtOAc, stirred and left overnight and filtered. The removed residue is once again extracted with EtOAc for 3 h and the combined extracts are added with Na_2SO_4 to remove the water (Kreutzmann et al., 2008a). The cartridges of ASE were filled with 1 g freeze-dried powder and diatomaceous earth to mark up the volume (Rawson et al., 2011; Aguilo-Aguay et al., 2014). The extraction of sample under optimized conditions by different solvents, namely, EtOAc (Pferschy-Wenzig et al., 2009; Aguilo-Aguay et al., 2014) and ACN (Rawson et al., 2010) yielded the extracts that were dried at 37°C under N_2 (Koidis et al., 2012). The dried sample is redissolved in the corresponding solvents, filtered, and transferred into the HPLC vials. The samples are analyzed by RP-HPLC maintained at the flow rate of 0.65–1 mL/min and injection volume of 5–20 μL. The separation column C_{18} is held at constant temperature of 40°C with varying solvent gradient ratios changing with the time.

The mobile phase used for the chromatographic separation includes solvent mixture of ACN-H_2O and MeCN-H_2O used by Rawson et al. (2011), Koidis et al. (2012), and Kreutzmann et al. (2008a), respectively. The ratio of said solvents employed as different gradients along with their respective timings include 20:80 (0–5 min), 50:50 (10 min), 53:47 (30 min), 65:35 (45–50 min), 75:25 (70–72 min), 95:5 (90–95 min), and 100:0 (55–105 min).

The elutes coming out from the chromatograph column are analyzed by using a UV–vis or DAD detector operating at 205 nm and CAD that has the advantage over UV–vis in detecting a nonchromophore compound. The CAD detected greater amounts of impurities than the UV–vis detector due to low response of these compounds with UV and varying coefficients of extinction (Acworth et al., 2011).

4.4.2.2 ANTHOCYANIN EXTRACTION

The relatively hydrophilic or polar nature of phenolic compounds is achieved efficiently by the utilization of more polar solvents like methanol, ethanol, water, acetone, and others. Polar nature of anthocyanin pigment necessitates use of polar solvents for extraction and is mainly extracted by solvent extraction (Castaneda-Ovando et al., 2009). Algarra et al. (2014) used ethanol:water (1:1 v/v) mixture for extraction of anthocyanins from peeled and chopped carrot roots in presence of 0.01% HCl, whereas the freeze-dried paste of black carrots grounded into fine powder was subjected to extraction with acidified aqueous methanol (0.1% HCl in 80% methanol v/v) (Longo et al., 2007).

The assistance of water in aiding enzyme degradation of phenolics favors the use of dried or freeze-dried or frozen samples (Stalikas, 2007). The acidic or low pH stability of the phenolic compounds aids them to remain neutral in acidic polar solvents and enhances their extraction (Tsao, 2010). The acidified methanol proved to be a superior extraction solvent than the aqueous acetone as reported by Lee et al. (2004). The use of mild acidified solvent is necessary for prevention of anthocyanin degradation (Longo & Vasapollo, 2006), whereas strong acid solutions may result in degradation of acylated anthocyanins (Kapasakalidis et al., 2006).

The anthocyanin degradation accelerates with heating upon changing the pH toward alkalinity (from pH 4.3 to 6.0) (Kirca et al., 2007). The centrifugation of the mixture at 2000–4000 rpm for some minutes allows the separation of residue that is again extracted with the solvent (Benakmoum et al., 2008; You et al., 2011; Toor et al., 2006). The crude extract is heated in vacuum at <40°C (Assous et al., 2014; Koley et al., 2014) or in shaking water bath at 90°C (Li et al., 2012) to dryness.

The samples are redissolved in the solvent and analyzed by HPLC for identification of different components. The nonpolar components that may interfere with the separation process are usually removed by extracting the sample with petroleum ether and EtOAc. Different studies analyzing anthocyanins have employed the use of C_{18} column encased in thermostablized jacket with DAD (Tsao, 2010; Li et al., 2012) and UV–vis detector (Koley et al., 2014; Assous et al., 2014). The combination of formic acid, water, and other polar solvents (methanol, acetonitrile) is the main choice of solvents used as mobile phases during a dynamic gradient of elution (Li et al., 2012; Montilla et al., 2011).

The HPLC associated with UV–vis and PDA detectors are mainly employed for the separation, identification, and quantification purpose of anthocyanin pigments (Mikanagi et al., 2000; da Costa et al., 2000; Hong & Wrolstad, 1990). The intricacy to acquire anthocyanin reference compounds and their similar spectral analysis limit the use of above-mentioned detectors and pave the way for advanced detectors, like nuclear magnetic resonance and MS as the preferred devices for identification (Castaneda-Ovando et al., 2009).

4.5 CARROT PRODUCTS

Fruits and vegetables are the protective foods, essential for the normal functioning of body systems by providing vitamins, minerals, and bioactive components. The use of carrots both in the fresh and processed forms has increased recently due to a lot of positive health benefits. The carrot juice is a rich source of carotenoids. The blending of carrot juice with other juices or yoghurt has been attempted (Roy et al., 1996; Hassanin, 1999; Profir & Vizireanu, 2013; Karangwa et al., 2010; Mishra & Das, 2003; Cliff et al., 2013; Salwa et al., 2004) and found high acceptability levels. Several methods of juice extraction of carrot have been tried and found that the juice extracted using hydraulic pressing yields high-quality juice and pomace (Haq et al., 2013a). Alklint et al., (2004) showed that carbonation of acidified carrot juice increases the shelf life due to reduction of pH, whereas the juice with high levels of phenolics and vitamins was yielded from organic grown carrots (Hallmann et al., 2011).

The black carrots are specifically used in formation of carrot Kanji, a popular health drink of northern India. The carrot Kanji serves as probiotic drink with an effective action against bacterial toxins (Sowani & Thorat, 2012). The dehydrated carrots have been used in the development and fortification of different foods involving beverage, soups, snacks, bakery products, and so on (Prasad et al., 2014; Lin et al., 1998; Suman & Kumari, 2002; Jain et al., 2012; Gayas et al., 2012; Baljeet et al., 2014; Dar et al., 2012; Kumar & Kumar, 2011). The dehydrated carrot fractions are found to have the potential to be used as a source of prebiotic in different foodstuffs (Haq et al., 2013b). The process mapping for the production of function fractional carrot powder has been provided (Fig. 4.6). The high fiber fraction in dehydrated carrot serves as possible edible source of fiber for the development of fiber-rich food products.

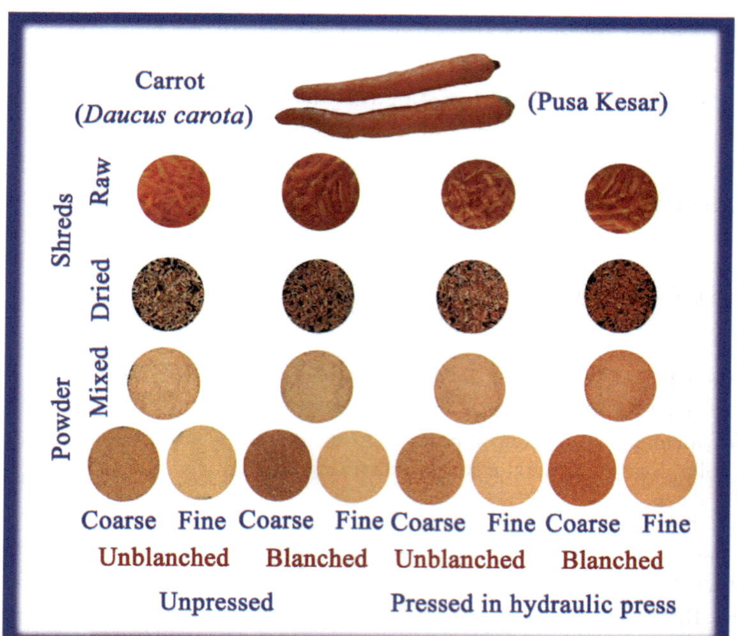

FIGURE 4.6 Pictorial mapping of carrot powder production.

The lactose-free milk obtained from the soya beans contains appreciable amounts of proteins, carbohydrates, fats, vitamins, and minerals. Madukwe et al. (2013) fortified soy milk with dehydrated carrot powder, whereas Singh and Katiyar (2014) developed a low-cost supplement for children from carrot, papaya, black gram, and groundnut powder. The nature of carrot to fit in the fruit category differentiates it from other vegetables and also extends its use both as processed vegetable and fruit products (Fig. 4.7). The carrots are popular for their use in different sweet dishes, namely, gazar ka halwa and kheer mix (Sampathu et al., 1981; Manjunatha et al., 2003; Suman & Kumari, 2002). "Kanwal Carrot Dessert" has been recently commercialized under the brand name of "Kanwal" developed from powdered milk, sugar, coconut, carrot, and dry fruits (Mansoor et al., 2013). The utilization of carrots in processed food products functions as a low-cost source of provitamin A in addition to value addition. The sweet Indian product commonly known as gazrella or sweet meat is prepared from the carrot shreds, sugar, condensed milk, and oil (Basantpure et al., 2003; Hatan & Malhotra, 2012; Singh et al., 2006). The attempts for the utilization of carrot in candy development have been reported by Durrani et al. (2011) and Madan and Dhawan (2005). The former study involved the development of honey-based carrot candy, whereas the latter study used sugar, coconut powder, and jaggery for carrot candy development. The preserve is the intermediate moisture food and contains the whole fruit pieces in the heavy sugar syrup and is locally termed as "murrabba." The carrot preserve involves boiling the pieces in sugar solution until 55–70% total soluble solids (Lal et al., 1986) and has been studied (Sethi & Anand, 1982; Singh et al., 1999).

4.6 CONCLUSIONS

Carrots are among the rich sources of nutrient and the plant bioactive components like dietary fiber, phenolics, and more importantly carotenoids imparting it various functional and medicinal properties but the perishable nature and seasonal availability of the crop makes its availability difficult for the whole year. The health benefits of the carrot along with its nutrition profile presses the need for the processing of root crop into various products. A sizable population lives well below the poverty line in developing countries like India; the low-cost food sources of vitamin A need to

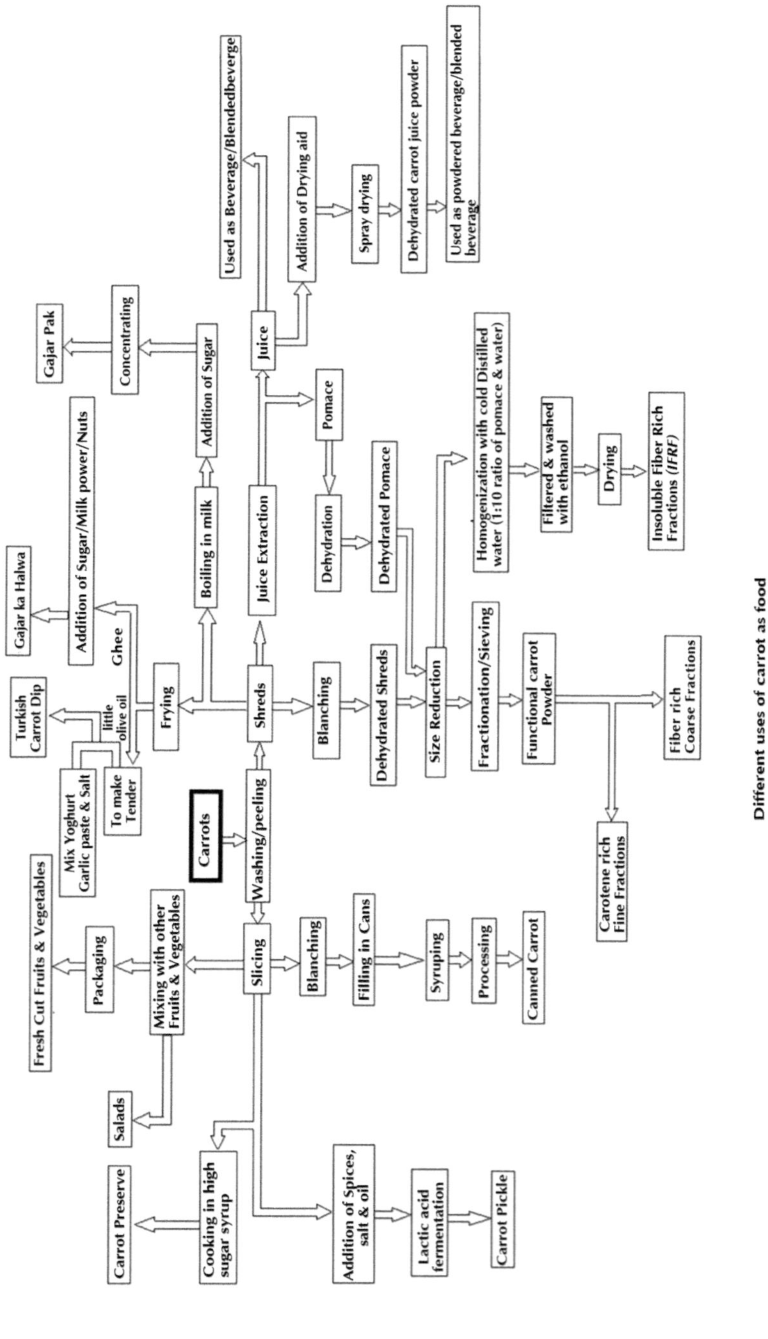

FIGURE 4.7 Uses of carrot in product formulation.

be popularized. Carrot pomace, the by-product of the juice industry, can be exploited in different processed products like extrusion, bakery, and beverages as a cheap source of the vitamin A that is one of the major micronutrient deficiency problems in developing countries.

KEYWORDS

- **carrot**
- ***Daucus carota***
- **phytochemical**
- **carotenoid**
- **secondary metabolite**
- **food product**
- **therapeutic application**

REFERENCES

Aalbersberg, B. Carotene Analysis. Proceedings of the Third OCEANIAFOODS Conference, 3–5 December 1991. Auckland, New Zealand, 1991, pp 91–97.

Abdulaali, N. A. Effect of Carrot Extracts on *Pseudomonas aeruginosa*. *Pak. J. Nutr.* **2009,** *8*(4), 373–376.

Acworth, I.; Plante, M.; Bailey, B.; Crafts, C.; Waraska, J. Simple and Direct Analysis of Falcarinol and Other Polyacetylenic Oxylipins in Carrots by Reversed-Phase HPLC and Charged Aerosol Detection. In *Plant Medica*; RUDIGERSTR 14, D-70469, Georg Thieme Verlag: Stuttgart, Germany, KG, 2011, vol. 77 (12), p 1263.

Adlercreutz, H.; Mazur, W. Phyto-Oestrogens and Western Diseases. *Ann. Med.* **1997,** *29*(2), 95–120.

Afify, A. M. R.; Romeilah, R. R. M.; Osfor M. H.; Elbahnasawy, A. S. M. Evaluation of Carrot Pomace (*Daucus carota* L.) as Hypocholesterolemic and Hypolipidemic Agent on Albino Rats. *Notulae Sci. Biol.* **2013,** *5*(1), 7–14.

Aguilo-Aguay, I.; Brunton, N.; Rai, D. K.; Balaguero, E.; Hossain, M. B.; Valverde J. Polyacetylene Levels in Carrot Juice, Effect of pH and Thermal Processing. *Food Chem.* **2014,** *152*, 370–377.

Ahmed, A. A.; Bishr, M. M.; El-Shanawany, M. A.; Attia, E. Z.; Ross, S. A.; Pare, P. W. Rare Trisubstituted Sesquiterpenes Daucines from the Wild *Daucus carota*. *Photochem. J.* **2005,** *66*, 1680–1684.

Ahuja, S. Overview. In *Comprehensive Analytical Chemistry*, vol. 47; Elsevier: Amsterdam, The Netherlands, 2006.

Ajila, C. M.; Brar, S. K.; Verma, M.; Tyagi, R. D.; Godbout, S.; Valero, J. R. Extraction and Analysis of Polyphenols: Recent Trends. *Crit. Rev. Biotechnol.* **2011,** *31*(3), 227–249.

Alasalvar, C.; Grigor, J. M.; Quantick, P. C. Method for the Static Headspace Analysis of Carrot Volatiles. *Food Chem.* **1999,** *65*, 391–397.

Alasalvar, C.; Grigor, J. M.; Zhang, D.; Quantick, P. C.; Shahidi, F. Comparison of Volatiles, Phenolics, Sugars, Antioxidant Vitamins, and Sensory Quality of Different Colored Carrot Varieties. *J. Agric. Food Chem.* **2001,** *49*, 1410–1416.

Algarra, M.; Fernandes, A.; Mateus, N.; de Freitas, V.; da Silva, J. C.; Casado, J. Anthocyanin Profile and Antioxidant Capacity of Black Carrots (*Daucus carota* L. ssp. *sativus* var. *atrorubens* Alef.) from Cuevas Bajas, Spain. *J. Food Compos. Anal.* **2014,** *33*, 71–76.

Alklint, C.; Wadso, L.; Sjoholm, I. Effects of Modified Atmosphere on Shelf-Life of Carrot Juice. *Food Control*, **2004,** *15*, 131–137.

Ames, B. N. Dietary Carcinogens and Anti-Carcinogens Oxygen Radicals and Degenerative Diseases. *Science*, **1983,** *221*, 1256–1262.

Ames, B. N.; Gold, L. S. Endogenous Mutagens and the Causes of Aging and Cancer. *Mutat. Res.* **1991,** *250,* 3–16.

Ames, B. N.; Shigenaga, M. K.; Gold, L. S. DNA Lesions, Inducible DNA Repair, and Cell Division: The Three Key Factors in Mutagenesis and Carcinogenesis. *Environ. Health Perspect.* **1993,** *101*(5), 35–44.

Andersen, O. M.; Jordheim, M. The Anthocyanins. In *Flavonoids: Chemistry, Biochemistry and Applications*. CRC Press: London, 2006, pp 471–552.

Anderson, J. W. Dietary Fibre in Diabetes. In *Medical Aspects of Dietary Fiber*; Spiller, G. A., Kay, R., Eds.; Plenum Press: New York, 1980.

Anderson, J. W.; Bridges, S. R. Dietary Fibre Content of Selected Foods. *Am. J. Clin. Nutr.* **1988,** *47*, 440–447.

Anderson, J. W.; Hanna, T. J. Impact of Nondigestible Carbohydrates on Serum Lipoproteins and Risk for Cardiovascular Disease. *J. Nutr.* **1999,** *129*, 1457S–1466S.

Anderson, J. W.; Smith, B. M.; Guftanson, N. S. Health Benefit and Practical Aspects of High Fiber Diets. *Am. J. Clin. Nutr.* **1994,** *595*, 1242–1247.

Anon. *Area and Production Statistics: Second Advance Estimates for the Horticulture Crops*. National Horticulture Board: New Delhi, 2012.

Anon. *The Wealth of India: Raw Materials.* Council of Scientific and Industrial Research: New Delhi, 1995.

Anon. World Cancer Research Fund International, & Washington American Institute for Cancer Research. *Food, Nutrition and the Prevention of Cancer: A Global Perspective.* World Cancer Research Fund, 1997.

AOAC International. *Official Methods of Analysis of AOAC International.* AOAC International, 1995.

Ascherio, A.; Rimm, E. B.; Giovannicci, E. L.; Colditz, G. A.; Rosner, B.; Willett, W. C.; Sacks, F.; Stampfer, M. J. A Prospective Study of Nutritional Factors and Hypertension among US Men. *Circulation* **1992,** *86*(5), 1475–1484.

Assous, M. T. M.; Abdel-Hady, M. M.; Medany, G. M. Evaluation of Red Pigment Extracted from Purple Carrots and Its Utilization as Antioxidant and Natural Food Colorants. *Ann. Agric. Sci.* **2014,** *59*, 1–7.

Awad, A. B.; Fink, C. S. Phytosterols as Anticancer Dietary Components: Evidence and Mechanism of Action. *J. Nutr.* **2000,** *130*, 2127–2130.
Babic, I.; Amiot, M. J.; Ngugen, C.; Aubert, S. Changes in Phenolic Content in Fresh, Ready-To-Use and Shredded Carrots during Storage. *J. Food Sci.* **1993,** *58*, 351–356.
Baker, D. C.; Dougall, D. K.; Glassgen, W. E.; Johanson, S. C.; Metzger, J. W.; Rose, A.; Seitz, H. U. Effects of Supplied Cinnamic Acid and Biosynthetic Intermediates on the Anthocyanins Accumulated by Wild Carrot Suspension Cultures. *Plant Cell, Tissue Organ. Cult.* **1994,** *39*, 79–91.
Baljeet, S. Y.; Ritika, B. Y.; Reena, K. Effect of Incorporation of Carrot Pomace Powder and Germinated Chickpea Flour on the Quality Characteristics of Biscuits. *Int. Food Res. J.* **2014,** *21*, 217–222.
Banga, O. Origin of the European Cultivated Carrot. *Euphytica* **1957b,** *6*, 54–63.
Banga, O. The Development of the Original European Carrot Material. *Euphytica* **1957a,** *6,* 64–76.
Bao, B.; Chang, K. C. Carrot Pulp Chemical Composition, Color and Water-Holding Capacity as Affected by Blanching. *J. Food Sci.* **1994,** *59*, 1159–1161.
Baranowitz, S. A.; Starrett, B.; Brookner, A. R. Carotene Deficiency in HIV Patients. *AIDS* **1996,** *10*, 115–118.
Barclay, G. Plant Anatomy. In *Encyclopedia of Life Sciences*. Nature Publishing Group, 2002.
Barth, M. M.; Zhou, C.; Kute, K. M.; Rosenthal, G. A. Determination of Optimum Conditions for Supercritical Fluid Extraction of Carotenoids from Carrot (*Daucus carota* L.) Tissue. *J. Agric. Food Chem.* **1995,** *43*, 2876–2878.
Basantpure, D.; Kumbhar, B. K.; Awashthi, P. Optimization of Levels of Ingredients and Drying Air Temperature in Development of Dehydrated Carrot Halwa using Response Surface Methodology. *J. Food Sci.* **2003,** *40*, 40–44.
Bauernfeind, J. C. Carotenoid Vitamin A Precursors and Analogs in Foods and Feeds. *J. Agric. Food Chem.* **1972,** *20*, 456–473.
Benakmoum, A.; Abbeddou, S.; Ammouche, A.; Kefalas, P.; Gerasopoulos, D. Valorisation of Low Quality Edible Oil with Tomato Peel Waste. *Food Chem.* **2008,** *110*, 684–690.
Bender, D. A. *Nutritional Biochemistry of the Vitamins*, 2nd ed. Cambridge University Press: Cambridge, 2003, pp 512.
Bendich, A. Carotenoids and the Immune Response. *J. Nutr.* **1989,** *119*, 112–115.
Bernart, M. W.; Cardellina, J. H.; Balaschak, M. S.; Alexander, M. R.; Shoemaker, R. H.; Boyd, M. R. Cytotoxic Falcarinol Oxylipins from *Dendropanax arboreus*. *J. Nat. Prod.* **1996,** *59*, 748–753.
Bijttebier, S.; D'Hondt, E.; Noten, B.; Hermans, N.; Apers, S.; Voorspoels, S. Ultra High Performance Liquid Chromatography versus High Performance Liquid Chromatography: Stationary Phase Selectivity for Generic Carotenoid Screening. *J. Chromatogr. A* **2014,** *1332*, 46–56.
Block, G. Nutrient Source of Pro-Vitamin A Carotenoids in American Diet. *Am. J. Epidemiol.* **1994,** *139*, 290–293.
Block, G.; Patterson, B.; Subar, A. Fruit, Vegetables and Cancer Prevention: A Review of the Epidemiological Evidence. *Nutr. Cancer* **1992,** *18*, 1–29.
Boik, J.; Jungi, W. F. Cancer and Natural Medicine. *Ann. Oncol.* **1996,** *7*(5), 432–432.

Borek, C. Antioxidant Health Effects of Aged Garlic Extract. *J. Nutr.* **2001,** *131*, 1010S–1015S.

Bradeen, J. M.; Simon, P. W. Carrot. In *Genome Mapping and Molecular breeding in Plants*, vol. 5; Kole, C., Ed.; Springer-Verlag: Berlin Heidelberg, New York, 2007.

Bravo, I. Polyphenols: Chemistry, Dietary Source, Metabolism, and Nutritional Significance. *Nutr. Rev.* **1998,** *56*, 317–333.

Brignole, A. E. Supercritical Fluid Extraction. *Fluid Phase Equilib.* **1986,** *29*, 133–144.

Britton, G. Structure and Properties of Carotenoids in Relation to Function. *FASEB J.* **1995,** *9*, 1551–1558.

Britton, G.; Liaaen-Jensen, S.; Pfander, H. *Carotenoids: Natural Functions*. Birkhauser Verlag: Basel, Boston, Berlin, 2008.

Brunner, G. Supercritical Fluids: Technology and Application to Food Processing. *J. Food Eng.* **2005,** *67*, 21–33.

Bunzel, M.; Seiler, A.; Steinhart, H. J. Characterization of Dietary Fiber Lignins from Fruits and Vegetables Using the DFRC Method. *Agric. Food Chem.* **2005,** *53*(24), 9553–9559.

Burton, G. W. Antioxidant Action of Carotenoids. *J. Nutr.* **1989,** *119*, 109–111.

Burton, G. W.; Ingold, K. U. Beta-carotene: An Unusual Type of Lipid Antioxidant. *Science*, **1984,** *224*, 569–573.

Burzo, I.; Voirica, V.; Luchian, V. *Fiziologia plantelor de cultura*, vol. V, *Fiziologia plantelor legumicole*, Editura Elisavaros: Bucuresti, 2005.

Bushway, R. J.; Wilson, A. M. Determination of (α and β-carotene in fruit and Vegetables by High Performance Liquid Chromatography. *Can. Inst. Food Sci. Technol. J.* **1982,** *15*, 165–169.

Butt, M. S.; Sultan, M. T. Coffee and Its Consumption: Benefits and Risks. *Crit. Rev. Food Sci. Nutr.* **2011,** *51*(4), 363–373.

Calixeto, F. S. Antioxidant Dietary Fiber Product: A New Concept and a Potential Food Ingredient. *J. Agric. Food Chem.* **1998,** *46*, 4303–4306.

Caragy, A. B. Cancer Preventive Foods and Ingredients. *Food Tech.* **1992,** *46*, 65–68.

Careri, M.; Elviri, L.; Mangia, A. Liquid Chromatography Electro-spray Mass Spectrometry of b-carotene and Xanthophylls: Validation of the Analytical Method. *J. Chromatogr. A* **1999,** *854*, 233–244.

Carr, A. C.; Frei, B. Toward a New Recommended Dietary Allowance for Vitamin C Based on Antioxidant and Health Effects in Humans. *Am. J. Clin. Nutr.* **1999,** *69*, 1086–1107.

Castaneda-Ovando, A.; Pacheco-Hernandez, M. L.; Paez-Hernandez, M. E.; Rodriguez, J. A.; Galan-Vidal, C. A. Chemical Studies of Anthocyanins: A Review. *Food Chem.* **2009,** *113*, 859–871.

Ceska, O.; Chaudhary, S. K.; Warrington, P. J.; Ashwood-Smith, M. J. Furocoumarins in the Cultivated Carrot (*Daucus carota*). *Phytochemistry* **1986,** *25*, 81–83.

Chandra, A.; Nair, M. G. Supercritical Fluid Carbon Dioxide Extraction of a- and b-Carotene from Carrot (*Daucus carota* L.). *Phytochem. Anal.* **1997,** *8*, 244–246.

Chandrika, U. G.; Jansz, E. R.; Wickramasinghe, S. M. D.; Warnasuriya, N. D. Carotenoids in Yellow- and Red-Flesh Papaya (*Carica papaya* L.). *J. Sci. Food Agric.* **2003,** *83*, 1279–1282.

Chau, C. F.; Chen, C. H.; Lee, M. H. Comparison of the Characteristics, Functional Properties, and In Vitro Hypoglycemic Effects of Various Carrot Insoluble Fiber-Rich Fractions. *Lebensmitt. Wissensch. Technol.* **2004a,** *37*, 155–160.

Chau, C. F.; Cheung, P. C. Effects of Physico-Chemical Properties of Legume Fibers on the Cholesterol Absorption in Hamsters. *Nutr. Res.* **1999,** *19*, 257–265.

Chau, C. F.; Huang, Y. L.; Lin, C. Y. Investigation of the Cholesterol-Lowering Action of Insoluble Fibre Derived from the Peel of *Citrus sinensis* L. cv. Liucheng. *Food Chem.* **2004b,** *87*, 361–366.

Chemat F; Abert-Vian, M.; Zill-e-Huma, Y. J. Microwave Assisted Separations: Green Chemistry in Action. In *Green Chemistry Research Trends*; Pearlman, J. T. Ed.; Nova Science Publishers: New York, 2009, pp 33–62.

Cheng, J. C.; Dai, F.; Zhou, B.; Yang, L.; Liu, Z. L. Antioxidant Activity of Hydroxycinnamic Acid Derivatives in Human Low Density Lipoprotein: Mechanism and Structure–Activity Relationship. *Food Chem.* **2007,** *104*(1), 132–139.

Cherng, J. M.; Chiang, W.; Chiang, L. C. Immunomodulatory Activities of Common Vegetables and Spices of *Umbelliferae* and its Related Coumarins and Flavonoids. *Food Chem.* **2008,** *106*(3), 944–950.

Chetreanu, D.; Atanasiu, N. Effect of Plant Density and Fertilizer on Crop Growth Root Yield and Quality of Carrot. *Lucrari stiintifice USAMV Bucuresti, Ser. B, Veg. Grow.* **2011,** *LV*, 48–51.

Chew, B. P.; Park, J. S. Carotenoid Action on the Immune Response. *J. Nutr.* **2004,** *134*, 257–261.

Cho, J. Y.; Kim, A. R.; Park, M. H. Lignans from the Rhizomes of *Coptis japonica* Differentially Act as Anti-Inflammatory Principles. *Planta Med.* **2001,** *67*, 312–316.

Cho, J. Y.; Park, J.; Yoo, E. S.; Baik, K. U.; Yoshikawa, K.; Lee, J.; Park, M. H. Inhibitory Effect of Lignans from the Rhizomes of *Coptis japonica* var. *dissecta* on Tumor Necrosis Factor-Alpha Production in Lipopolysaccharide-Stimulated RAW264.7 Cells. *Arch. Pharm. Res.* **1998,** *21*, 12–16.

Chobanian, V. A.; Dolecek, A. T.; Dustan, P. H. National Education Program Working Group. Report on the Management of Patients with Hypertension and High Blood Cholesterol. *Ann. Intern. Med.* **1991,** *114*, 3.

Choi, S. K.; Kim, J. H.; Park, Y. S.; Kim, Y. J.; Chang, H. I. An Efficient Method for the Extraction of Astaxanthin from the Red Yeast *Xanthophyllomyces dendrorhous*. *J. Microbiol. Biotechnol.* **2007,** *17*(5), 847–852.

Christensen, L. P.; Brandt, K. Bioactive Polyacetylenes in Food Plants of the Apiaceae Family: Occurrence, Bioactivity and Analysis. *J. Pharm. Biomed. Anal.* **2006,** *41,* 683–693.

Christensen, L. P.; Edelenbos, M.; Kreutzmann, S. Fruits and Vegetables of Moderate Climate. In *Flavours and Fragrances Chemistry, Bioprocessing and Sustainability*; Berger, R. G. Ed.; Springer: Berlin, 2007; pp 135–181.

Christensen, L. P.; Kreutzmann, S. Determination of Polyacetylenes in Carrot Roots (*Daucus carota* L.) by High-Performance Liquid Chromatography Coupled with Diode Array Detection. *J. Sep. Sci.* **2007,** *30*, 483–490.

Chu, Y. F.; Sun, J.; Wu, X.; Liu, R. H. Antioxidant and Antiproliferative Activities of Vegetables. *J. Agric. Food Chem.* **2002,** *50*, 6910–6916.

Chumnanpaison, N.; Niamnuy, C.; Devahastin, S. Mathematical Model for Continuous and Intermittent Microwave-Assisted Extraction of Bioactive Compound from Plant Material: Extraction of β-carotene from Carrot Peels. *Chem. Eng.* **2014,** *116*, 442–451.

Chun, J.; Lee, J.; Yea, L.; Exler, J.; Eitenmiller, R. R. Tocopherol and Tocotrienol Contents of Raw and Processed Fruits and Vegetables in the United States Diet. *J. Food Compos. Anal.* **2006,** *19*, 196–204.

Chung, K. T.; Wong, T. Y.; Wei, C. I.; Huang, Y. W.; Lin, Y. Tannins and Human Health: A Review. *Crit. Rev. Food Sci. Nutr.* **1998,** *38*(6), 421–464.

Clarke, I.; Lee, H. *Name that Flower: The Identification of Flowering Plants*. Melbourne University Press: Australia, 2003.

Clements, R. S.; Darnell, M. S. Myo-Inositol Content of Common Foods: Development of a High-Myo-Inositol Diet. *Am. J. Clin. Nutr.* **1980,** *33*, 1954–1967.

Cliff, M. A.; Fan, L.; Sanford, K.; Stanich, K.; Doucette, C.; Raymond N. Descriptive Analysis and Early-Stage Consumer Acceptance of Yogurts Fermented with Carrot Juice. *J. Dairy Sci.* **2013,** *96,* 4160–4172.

Clydesdale, F. M.; Ho, C.; Lee, C. Y.; Mondy, N. I.; Shewfelt, R. L. The Effects of Postharvest Treatment and Chemical Interactions on the Bioavailability of Ascorbic Acid, Thiamin, Vitamin A, Carotenoids and Minerals. *Crit. Rev. Food Sci. Nutr.* **1991,** *30*(6), 599–638.

Combs, G. F. *The Vitamins: Fundamental Aspects in Nutrition and Health*; Academic Press: San Diego, CA, 1998; pp 618.

Cordell, G. A.; Lemos, T. L. G.; Monte, F. J. Q.; de Mattos, M. C. Vegetables as Chemical Reagents. *J. Nat. Prod.* **2007,** *70*, 478–492.

da Costa, C. T.; Horton, D.; Margolis, S. A. Analysis of Anthocyanins in Foods by Liquid Chromatography, Liquid Chromatography–Mass Spectrometry and Capillary Electrophoresis. *J. Chromatogr. A* **2000,** *881*, 403–410.

Dahiya, M. S.; Yadav, A. C.; Singh, V. P.; Malik, Y. S. Effect of Time and Method of Sowing on Root Quality of Carrot cv. Hisar Gairic. *Haryana J. Hortic. Sci.* **2007,** *36*, 377–378.

Damon, M.; Zhang, N. Z.; Haytowitz, D. B.; Booth, S. L. Phylloquinone (Vitamin K1) Content of Vegetables. *J. Food Compos. Anal.* **2005,** *18*, 751–758.

Dar, A. H.; Sharma, K.; Kumar, N. Effect of Extrusion Temperature on the Microstructure, Textural and Functional Attributes of Carrot Pomace-Based Extrudates. *J. Food Process. Preserv.* **2012,** *38*(1), 212–222..

de Almeida, V. V.; Bonafe, E. G.; Muniz, E. C.; Matsushita, M.; de Souza, N. E.; Visentainer, J. V. Optimization of the Carrot Leaf Dehydration Aiming at the Preservation of Omega-3 Fatty Acids. *Quim. Nova* **2009,** *32*(5), 1334–1337.

de Pascual-Teresa, S.; Sanchez-Ballesta, M. T. Anthocyanins: From Plant to Health. Phytochem. Rev. **2008,** *7*(2), 281–299.

de Quiros, A. R.; Costa, H. S. Analysis of Carotenoids in Vegetable and Plasma Samples: A Review. *J. Food Compos. Anal.* **2006,** *19*, 97–111.

de Vries. J. W. On Defining Dietary Fiber. *Proc. Nutr. Soc.* **2003,** *62*(1), 37–43.

Decoteau, D. R. Root crops. In *Vegetable Crops*; Prentice Hall: Upper Saddle River, NJ, 2000; pp 290–323.

Delgado-Vargas, F.; Jimenez, A. R.; Paredes-Lopez, O. Natural Pigments: Carotenoids, Anthocyanins, and Betalains—Characteristics, Biosynthesis, Processing and Stability. *Crit. Rev. Food Sci. Nutr.* **2000,** *40*, 173–289.

Delgado-Vargas, F.; Paredes-Lopez, O. *Natural Colorants for Food and Nutraceutical Uses*. Library of Congress Cataloging-in-Publication Data, CRC Press: Boca Raton, FL, 2003.

Demmig-Adams, B.; Gilmore, A. M.; Adams, W. W. In Vivo Functions of Carotenoids in Higher Plants. *FASEB J.* **1996,** *10*, 403–412.
Deshpande, S. S.; Deshpande, U. S.; Salunkhe, D. K. Nutritional and Health Aspects of Food Antioxidants. In *Food Antioxidants—Technological, Toxicological and Health Perspectives*; Madhavi, D. L.; Deshpande, S. S, Salunkhe, D. K., Eds.; Marcel Dekker: New York, 1995. pp 361–382.
Dev, H. Effect of Root Size on Yield and Quality of Carrot Seed Crop cv. Early Nantes. *Haryana J. Hortic. Sci.* **2009,** *38*, 119–121.
Dixon, R. A.; Paiva, N. L. Stress Induced Phenol Propanoid Metabolism. *Plant Cell* **1995,** *7*, 1085–1097.
Donaldson, M. S. Nutrition and Cancer: A Review of the Evidence for an Anti-Cancer Diet. *Nutr. J.* **2004,** *3*, 19.
Dougall, D. K.; Baker B. C.; Gakh, E. G.; Redus, M. A.; Whittemore, N. A. Studies on Stability and Conformation of Monoacylated Anthocyanins. Part 2—Anthocyanins from Wild Carrot Suspension Cultures Acylated with Supplied Carboxylic Acids. *Carbohydr. Res.* **1998,** *310*, 177–189.
Downham, A.; Collins, P. Coloring Our Foods in the Last and Next Millennium. *Int. J. Food Sci. Technol.* **2000,** *35*, 5–22.
Dragsted, L. O.; Strube, M.; Larsen, J. C Cancer-Protective Factors in Fruits and Vegetables: Biochemical and Biological Background. *Pharmacol. Toxicol.* **1993,** *72*, 116–135.
Duan, H.; Takaishi, Y.; Momota, H.; Ohmoto, Y.; Taki, T. Immunosuppressive Constituents from *Saussurea medusa*. *Phytochemistry* **2002,** *59*(1), 85–90.
Duke, J. A. *Handbook of Phytochemical Constituents of Grass Herbs and Other Economic Plants,* CRC Press: Boca Raton, FL, 2006.
Duke, J. A. *The Green Pharmacy*. Rodale Press: Emmaus, PA, 1997.
Durante, M.; Lenucci, M. S.; Giovanni, M. Supercritical Carbon Dioxide Extraction of Carotenoids from Pumpkin (*Cucurbita* spp.): A Review. *Int. J. Mol. Sci.* **2014,** *15*, 6725–6740.
Durrani, A. M.; Srivastava, P. K.; Verma, S. Development and Quality Evaluation of Honey Based Carrot Candy. *Food Sci. Technol.* **2011,** *48*, 502–505.
Dutta, P. C. *Phytosterols as Functional Food Components and Nutraceuticals*. Marcel Dekker: New York, 2003, 450 pp.
Eastwood, M. A. Interaction of Dietary Antioxidants In Vivo: How Fruit and Vegetables Prevent Disease?. *QJM* **1999,** *92*, 527–530.
Eastwood, M. Dietary Fiber, How Did We Get Where We Are? *Annu. Rev. Nutr.* **2005,** *25*, 1–8.
Edge, R.; McGarvey, D. J.; Truscott, T. G. J. The Carotenoids as Antioxidants: A Review. *J. Photochem. Photobiol. B* **1997,** *41*(3), 189–200.
Eggers, R. Supercritical Fluid Extraction (SFE) of Oilseeds/Lipids in Natural Products. In *Supercritical Fluid Technology in Oil and Lipid Chemistry*; King, J. W., List, G. R., eds.; AOCS Press: Urbana, IL, 1996, pp 35–65.
El-Arab, A. E.; Khalil, F.; Hussein, L. Vitamin A Deficiency among Pre-school Children in a rural area of Egypt: The Results of Dietary Assessment and Biochemical Assay. *Int. J. Food Sci. Nutr.* **2002,** *53*, 465–474.

Elham, G.; Reza, H.; Jabbar, K.; Parisa, S.; Rashid, J. Isolation and Structure Characterization of Anthocyanin Pigments in Black Carrot (*Daucus carota* L). *Pak. J. Biol. Sci.* **2006,** *9*(15), 2905–2908.

Elkner, K. Jakosc i przydatnosc wa-rzyw do przetwórstwa. Konferencja naukowa "Uprawa warzyw do prze-twórstwa" I. Warz. Skierniewice, **2003,** 4–9. [in Polish].

Elson, C. E.; Yu, S. G. The Chemoprevention of Cancer by Mevalonate Derived Constituents of Fruits and Vegetables. *J. Nutr.* **1994,** *124*, 607–614.

Englberger, L.; Schierle, J.; Marks, G. C.; Fitzgerald, M. H. Micronesian Banana, Taro, and Other Foods: Newly Recognized Sources of Provitamin A and Other Carotenoids. *J. Food Compos. Anal.* **2003,** *16*, 3–19.

Erkkila, A. T.; Sarkkinen, E. S.; Lehto, S.; Pyorala, K.; Uusitupa, M. I. Dietary Associates of Serum Total, LDL and HDL Cholesterol and Triglycerides in Patients with Coronary Heart Disease. *Prevent. Med.* **1999,** 28, 558–565.

Eskilsson, C. S.; Bjorklund, E. Analytical-Scale Microwave-assisted Extraction. *J. Chromatogr. A* **2000,** *902*, 227–250.

Esterbauer, H.; Gebicki, J.; Puhl, H.; Jurgens, G. The Role of Lipid Peroxidation and Antioxidants in Oxidative Modification of LDL. *Free Radic. Biol. Med.* **1992,** *13*, 341–390.

FAOSTAT. *Production Year Book*, vol. 59. Food and Agriculture Organization of the United Nations: Rome, 2013.

Fatland, B. L.; Ke, J.; Anderson, M. D.; Mentzen, W. I.; Cui, L. W.; Allred, C. C.; Johnston, J. L.; Nikolau, B. J.; Wurtele, E. S. Molecular Characterization of a Heteromeric ATP-Citrate Lyase that Generates Cytosolic Acetyl-Coenzyme A in *Arabidopsis*. *Plant Physiol.* **2004,** 130, 740–756.

Faulks, R. M.; Southon, S. *Carotenoids, Metabolism and Disease, Handbook of Nutraceuticals and Functional Foods.* CRC Press, Boca Raton, FL, 2001.

Figuerola, F.; Hurtado, M. L.; Estevez, A. M.; Choffelle, I.; Asenjo, F. Fiber Concentrates from Apple Pomace and Citrus Peels as Potential Fiber Sources for Food Enrichment. *Food Chem.* **2005,** *91*, 395–401.

Fikselova, M.; Marecek, J.; Mellen, M. Carotenes Content in Carrot Roots (*Daucus carota* L.) as Affected by Cultivation and Storage. *Vegetable Crops Research Bull.* **2010,** *73*, 47–54.

Filotheou, A.; Nanou, K.; Roukas, T.; Kotzekidou, P.; Papaioannou, E.; Liakopoulou-Kyriakides, M. Application of Response Surface Methodology to Improve Carotene Production from Synthetic Medium by *Blakeslea trispora* in Submerged Fermentation. *Food Bioprocess. Technol.* **2010**. DOI:10.1007/s11947-010-0405-6.

Fisher, C.; Kocis, J. A. Separation of Paprika Pigment by HPLC. *J. Agric. Food Chem.* **1987,** *35*, 55–57.

Fossen, T.; Andersen, O. M.; Ovstedal, D. O.; Pedersen, A. T.; Raknes, A. Characteristic Anthocyanin Pattern from Onions and Other *Allium* spp. *J. Food Sci.* **1996,** *61*, 703–706.

Franceschi, S.; Parpinel, M.; Lavecchia, C.; Favero, A.; Talamini, R.; Negri, E. Role of Different Types of Vegetables and Fruit in the Prevention of Cancer of the Colon, Rectum, and Breast. *Epidemiology* **1998,** *9*, 338–341.

Francis, F. J.;. Markakis, P. C. Food Colorants: Anthocyanins. *Crit. Rev. Food Sci. Nutr.* **1989,** *28*(4), 273–314.

Fraser, P. D.; Bramley, P. M. The Biosynthesis and Nutritional Uses of Carotenoids. *Prog. Lipid Res.* **2004,** *43*, 228–265.

French, S. J and Read, N. W. Effect of Guar Gum on Hunger and Satiety after Meals of Differing Fat Content: Relationship with Gastric Emptying. *Am. J. Clin. Nutr.* **1994,** *59*, 87–91.

Frey, H. P.; Zieloff, K. *Qualitative und Quantitative Dünnschicht Chromatographie*. VCH-Verlagsgesellschaft: Berlin, 1993.

Fukuda, T.; Okazaki, K.; Shinano, T. Aroma Characteristic and Volatile Profiling of Carrot Varieties and Quantitative Role of Terpenoids Compounds for Carrot Sensory Attributes. *J. Food Sci.* **2013,** *78*, s1800–s1806.

Furr H. C. Analysis of Retinoids and Carotenoids: Problems Resolved and Unsolved. *J. Nutr.* **2004,** *134*, 281–285.

Ganzler, K.; Salgo, A.; Valko, K. Microwave Extraction: A Novel Sample Preparation Method for Chromatography. *J. Chromatogr.* **1986,** *371*, 299–306.

Garcia-Ayuso, L. E.; Luque De.; Castro, M. D. A Multivariate Study of the Performance of a Microwave-Assisted Soxhlet Extractor for Olive Seeds. *Anal. Chim. Acta* **1999,** *382*, 309–316.

Garcia-Ayuso, L. E.; Luque De.; Castro, M. D. Employing Focused Microwaves to Counteract Conventional Soxhlet Extraction Drawbacks. *Trends Anal. Chem.* **2001,** *20*, 28–34.

Gayas, B.; Shukla, R. N. Khan, B. M. Physico-Chemical and Sensory Characteristics of Carrot Pomace Powder Enriched Defatted Soyflour Fortified Biscuits. *Int. J. Sci. and Res. Public.* **2012,** *2*, 8.

Gazalli, H.; Malik, A. H.; Jalal, H.; Afshan, S.; Ambreen, M. Proximate Composition of Carrot Powder and Apple Pomace Powder. *Int. J. Food Nutr. Saf.* **2013,** *3*(1), 25–28.

Gazzani, G.; Papetti, A.; Massolini, G.; Daglia, M. Anti- and Pro-oxidant Activity of Water Soluble Components of Some Common Diet Vegetables and Effect of Thermal Treatment. *J. Agric. Food Chem.* **1998,** *46*, 4118–4122.

Gebhardt, S. E.; Matthews, R. H. *Nutritive value of Foods*. US Government Printing Office: Washington, DC, 1991.

Gill, H. S.; Kataria, A. S. Some Biochemical Studies in European and Asiatic Varieties of Carrot (*Daucus carota*). *Curr. Sci.* **1974,** *43*, 184–185.

Gill-Chavez, C. J.; Villa, J. A.; Ayala-zavala, J. F.; Heredia, J. B.; Sepulveda, D.; Yahia, E. M.; Gonzalez-Aguilar, G. A. Technologies for Extraction and Production of Bioactive Compounds to be used as Nutraceuticals and Food Ingredients: An Overview. *Compr. Rev. Food Sci. Food Saf.* **2013,** *12*. DOI:10.1111/1541-4337.12005.

Giovannucci, E.; Ascherio, A.; Rimm, E. B.; Stampfer, M. J.; Colditz, G. A.; Willett, W. C. Intake of Carotenoids and Retinol in Relation to Risk of Prostate Cancer. *J. Nat. Cancer Inst.* **1995,** *87*, 1767.

Giugliano, D. Dietary Antioxidants for Cardiovascular Prevention. *Nutr. Metab. Cardiovasc. Dis.* **2000,** *10*, 38–44.

Goncalves, E. M.; Pinheiro, J.; Abreu, M.; Brandão, T. R.; Silva, C. L. Carrot (*Daucus carota* L.) Peroxidase Inactivation, Phenolic Content and Physical Changes Kinetics due to Blanching. *J. Food Eng.* **2010,** *97*, 574–581.

Goodwin, T. W.; Mercer, E. I. *Introduction to Plant Biochemistry*, 2nd ed. Pergamon Press: New York, 1983, p 677.

Gopalan, C.; Ramasastri, B. V and Balasubramaniam, S. C. *Nutritive Value of Indian Foods*. NIN, Indian Council of Medical Research: Hyderabad, 1996.

Gorinstein, S.; Zachwieja, Z.; Folta, M.; Barton, H.; Piotrowicz, J.; Zember, M.; Trakhtenberg, S.; Martin-Belloso, O. Comparative Content in Persimmons and Apples. *J. Agric. Food Chem.* **2001,** *49*, 952–957.

Goto, M.; Sato, M.; Hirose, T. Supercritical Carbon Dioxide Extraction of Carotenoids from Carrots. In *Developments in Food Engineering—Proceedings of the Sixth International Congress on Engineering and Food*; Yano, T., Matsuno, R., Nakamura, K., Eds.; Springer US: New York, 1994; pp 835–837.

Granado, F.; Olmedilla, B.; Blanco, I.; Rojas-Hidalgo, E. Carotenoid Composition in Raw and Cooked Spanish Vegetables. *J. Agric. Food Chem.* **1992,** *40*, 2135–2140.

Grassmann, J.; Schnitzler, W. H.; Habegger, R. Evaluation of Different Colored Carrot Cultivars on Antioxidative Capacity Based on their Carotenoid and Phenolic Contents. *Int. J. Food Sci. Nutr.* **2007,** *58*(8), 603–611.

Grigelmo-Miguel, N.; Martın-Belloso, O. Comparison of Dietary Fiber from By-products of Processing Fruits and Greens and from Cereals. *Lebensmitt.-Wissensch. Technol.* **1999,** *32*, 503–508.

Grillini, P. M. Thin-layer Chromatography. In *Comprehensive Analytical Chemistry*; Ahuja, S., Jespersen, N., Eds., Elsevier: Amsterdam, The Netherlands, 2006.

Gross, J. *Pigments in Vegetables—Chlorophylls and Carotenoids*. Chapman & Hall: New York, 1991.

Gu, L.; Kelm, M. A.; Hammerstone, J. F.; Beecher, G.; Holden, J.; Haytowitz, D.; Gebhardt, S.; Prior, R. L. Concentrations of Proanthocyanidins in Common Foods and Estimations of Normal Consumption. *J. Nutr.* **2004,** *134*(3), 613–617.

Guo X; Zhang S; Shan X. Q. Adsorption of Metal Ions on Lignin. *J. Hazard. Mater.* **2008,** *151*(1), 134–142.

Hahn-Deinstrop, E. *Applied Thin-Layer Chromatography*. Wiley-VCH Verlag GmbH & Co. KGaA: Weinheim, 2007.

Hajslova, J.; Cajka, T. Gas Chromatography in Food Analysis. In *Handbook of Food Analysis Instruments*; Ötleş, S., Ed.; CRC Press, Taylor & Francis Group: Boca Raton, FL, 2008, pp 119–144.

Hajslova, J.; Cajka, T. Gas Chromatography–Mass Spectrometry (GC–MS). In *Food Toxicants Analysis*; Pico, Y. Ed.; Elsevier: Oxford, UK, 2006; pp 419–473.

Halliwell, B.; Murcia, M. A.; Chirico, S.; Auroma, O. I. Free Radicals and Antioxidants in Food and In Vivo: What They Do and How They Work?. *Crit. Rev. Food Sci. Nutr.* **1995,** *35*, 7–20.

Hallmann, E.; Sikora, M.; Rembialkowska, E.; Marszalek, K.; Lipowski, J. The Influence of Pasteurization Process on Nutritive Value of Carrot Juices from Organic and Conventional Production. *J. Res. Appl. Agric. Eng.* **2011,** *56*(3), 133–137.

Hanif, R.; Iqbal, Z.; Iqbal, M.; Hanif, S.; Rasheed, M. Use of Vegetables as Nutritional Food: Role in Human Health. *Agric. Biol. Sci.* **2006,** *1*(1), 18–22.

Hansen, S. L.; Purup, S.; Christensen, L. P. Bioactivity of Falcarinol and the Influence of Processing and Storage on Its Content in Carrots (*Daucus carota* L.). *J. Sci. Food Agric.* **2003,** *83*, 1010–1017.

Haq, Raees-ul.; Singh, Y.; Kumar, P.; Prasad, K. Development of Dehydrated Carrot Powder Fractions as Functional Ingredient. Proceedings of Sixth International Conference on

Fermented Foods, Health Status and Social Well-Being, Organized by Swedish South Asian Network on Fermented Foods and in Collaboration with Hildur on December 6–7, 2013 at Anand Agricultural University, Anand, 2013b, pp 94–95.

Haq, Raees-ul.; Singh, Y.; Kumar, P.; Prasad, K. Quality of Dehydrated Carrot Shreds as Affected by Partial Juice Extraction through Hydraulic Press. *Int. J. Agric. Food Sci. Technol.* **2013a,** *4*(4), 331–336.

Harborne, J. B. General Procedures and Measurement of Total Phenolics. *Methods Plant Biochem.* **1989,** *1*, 1–28.

Hartmann, M. A. Plant Sterols and the Membrane Environment. *Trends Plant Sci.* **1998,** *3*(5), 170–175.

Hartwell, J. L. Plants against Cancer. A Survey. *Lloydia* **1971,** *34*(2), 204–255.

Hassanin, N. I. Stability of Aflatoxin M1 during Manufacture and Storage of Yoghurt–Yoghurt Cheese. *J. Food Protect.* **1999,** *48*, 67–73.

Hatan, B. S.; Malhotra, T. Drying Kinetics of Osmotically Pretreated Carrot Shreds To Be Used for Preparation of Sweet Meat. *Agric. Eng. Int.: CIGR J.* **2012,** *14*(1), Manuscript No. 1973.

Hayashi, S.; Funatogawa, K.; Hirai, Y. Antibacterial Effects of Tannins in Children and Adults. In *Botanical Medicine in Clinical Practice*; Watson, R.; Preedy, V., Eds.; CABI Publishing: Wallingford, 2008, pp 141–151.

Heim, K. E.; Tagliaferro, A. R.; Bobilya, D. J. Flavonoid Antioxidants: Chemistry Metabolism and Structure–Activity Relationships. *J Nutr Biochem.* **2002,** *13*(10), 572–584.

Heinonen, M. I Carotenoids and Pro-vitamin A Activity of Carrot (*Daucus carota* L.) Cultivars. *J. Agric. Food Chem.* **1990,** *38*, 609–612.

Heinonen, M. I.; Ollilainen, V.; Linkola, E. K.; Varo, P. T.; Koivistonen, P. E. Carotenoids in Finnish Foods: Vegetables, Fruits, and Berries. *J. Agric. Food Chem.* **1989,** *37*, 655–659.

Hernandez-Ortega, M.; Kissangou, G.; Necoechea-Mondragon, H.; Sanchez-Pardo, M. E.; Ortiz-Moreno, A. Microwave Dried Carrot Pomace as a Source of Fiber and Carotenoids. *Food Nutr. Sci.* **2013,** *4*, 1037–1046.

Herrero-Martinez, J. M.; Eeltink, S.; Schoenmakers, P. J.; Kok, W. T.; Ramis-Ramos, G. Determination of Major Carotenoids in Vegetables by Capillary Electrochromatography. *J Sep Sci.* **2006,** *29*, 660–665.

Hertog, M. G. L.; Hollman, P. C. H.; Katan, M. B.; Kromhout, D. Intake of Potentially Anticarcinogenic Flavonoids and their Determinants in adults in The Netherlands. *Nutr. Cancer* **1993,** *20*, 21–29.

Heywood, V. H. Relationships and Evolution in the *Daucus carota* Complex. *Israel J. Bot.* **1983,** *32*, 51–65.

Hiranvarachat, B.; Devahastin, S. Enhancement of Microwave-assisted Extraction via Intermittent Radiation: Extraction of Carotenoids from Carrot Peels. *J. Food Eng.* **2014,** *126*, 17–26.

Hiranvarachat, B.; Devahastin, S.; Chiewchan, N.; Raghavan, G. S. V. Structural Modification by Different Pretreatment Methods to Enhance Microwave-assisted Extraction of β-Carotene from Carrots. *J. Food Eng.* **2013,** *115*, 190–197.

Hirsh, J.; Dalen, J.; Anderson, D. R.; Poller, L.; Bussey, H.; Ansell, J and Deykin, D. Oral Anticoagulants: Mechanism of Action, Clinical Effectiveness, and Optimal Therapeutic Range. *Chest* **2001,** *119*, 8S–21S.

Holland, B.; Unwin, J. D.; Buss, D. H. *Vegetables, Herbs and Spices: Fifth Supplement to McCance and Widdowson's*. Royal Society of Chemistry: London, 1991.

Hollman, P. C. H.; Arts, I. C. W. Flavonols, Flavones and Flavanols—Nature, Occurrence and Dietary Burden. *J. Sci. Food Agric.* **2000,** *80*, 1081–1093.

Hollman, P. C. H.; Katan, M. B. Dietary Flavonoids: Intake Health Effects and Bioavailability. *Food Chem. Toxicol.* **1999,** *37*, 937–942.

Hollman, P. C.; Hertog, M. G.; Katan, M. B. Analysis and Health Effects of Flavonoids. *Food Chem.* **1996,** *57*, 43–46.

Hong, V.; Wrolstad, R. E. Use of HPLC Separation/Photodiode Array Detection for Characterization of Anthocyanins. *J. Agric. Food Chem.* **1990,** *38*, 708–715.

Hoult, J. R.; Paya, M. Pharmacological and Biochemical Actions of Simple Coumarins: Natural Products with Therapeutic Potential. *Gen. Pharmacol.* **1996,** *27*(4), 713–722.

Howard, F. D.; MacGillivary, J. H.; Yamaguchi, M. *Nutrient Composition of Fresh California Grown Vegetables*. Bull Nr. 788, Calif. Agric. Expt. Stn., University of California: Berkeley, CA, 1962.

Howard, L. A.; Wong, A. D.; Perry, A. K.; Klein, B. P. Beta Carotene and Ascorbic Acid Retention in Fresh and Processed Vegetables. *J. Food Sci.* **1999,** *64*, 929–936.

Howard, L. R.; Braswell, D.; Heymann, H.; Lee, Y.; Pike, L. M.; Aselage, J. Sensory Attributes and Instrumental Analysis Relationships for Strained Processed Carrot Flavor. *J. Food Sci.* **1995,** *60*, 145–148.

Hsu, P. K.; Chien, P.; Chen, C. H and Chau, C. F. Carrot Insoluble Fiber-rich Fraction Lowers Lipid and Cholesterol Absorption in Hamsters. *LWT—Food Sci. Technol.* **2006,** *39*, 337–342.

Hui Y. H. Handbook of Food Science, Technology and Engineering. In *Vegetables: Types and Biology*; Tsao, J. S., Lo, H. F., Eds.; CRC Press, Taylor and Francis Group USA: Boca Raton, FL, 2006.

Jaga, L.; Duvvi, H. Risk Reduction for DDT Toxicity and Carcinogenesis through Dietary Modification. *J. R. Soc. Health.* **2001,** *121*, 107–113.

Jain, S.; Patni, D.; Bhati, D. Enhancing Nutritional Quality of Extruded Product by Incorporating an Indigenous Composite Powder. *Food Sci. Res. J.* **2012,** *3*, 232–235.

Jain, T.; Jain, V.; Pandey, R.; Vyas, A.; Shukla, S. S. Microwave Assisted Extraction for Phytoconstituents: An Overview. *Asian J. Res. Chem.* **2009,** *2*(1), 19–25.

Jang, S.; Dilger, R. N.; Johnson, R. W. Luteolin Inhibits Microglia and Alters Hippocampal-Dependent Spatial Working Memory in Aged Mice. *J. Nutr.* **2010,** *140*, 1892–1898.

Janve, B.; Prasad, K. K.; Prasad, K. Development of Fibre Rich Functional Mango Jam: Studies on its Formulation, Lambert Academic Publishing: Saarbrücken, Germany, 2014.

Jassie, L.; Revesz, R.; Kierstead, T.; Hasty, E.; Metz, S. Microwave-Assisted Solvent Extraction. In *Microwave-Enhanced Chemistry*; Kingston, H. M. S., Haswell, S. J., Eds.; American Chemical Society: Washington, DC, 1997; pp 569.

Jensen, C. D.; Haskel, W.; Shittam, J. H. Long-Term Effects of Water-Soluble Dietary Fiber in the Management of Hypercholesterolemia in Healthy Men and Women. *Am J Cardiol.* **1997,** *79*, 34–37.

Jiang, J.; Lin, L.; Lian, G.; Greiner, T. Vitamin A Deficiency and Child Feeding in Beijing and Guizhou, China. *World J. Pediatr.* **2008,** *4*, 20–25.

Johnson, E. J. The Role of Carotenoids in Human Health. *Nutr. Clin. Care.* **2002,** *5*(2), 56–65.

Jones, S. B.; Luchsinger, A. E. *Plant Systematics (2nd edition)*. McGraw-Hill: New York, 1987.

Jork, H.; Funk, W.; Fischer, W.; Wimmer, H. *Thin-layer Chromatography-Reagents and Detection Methods. Vol 1a: Physical and Chemical Detection Methods: Fundamentals, Reagents I*; VCH: Weinheim, 1990; pp 464.

Jork, H.; Funk, W.; Fischer, W.; Wimmer, H. Thin-layer Chromatography—Reagents and Detection Methods, Vol 1b. *Physical and Chemical Detection Methods: Activation Reactions, Reagent Sequences, Reagents II*; VCH: Weinheim, 1994; pp 496.

Ju, Y. H.; Allred, C. D.; Allred, K. F.; Karko, K. L.; Doerge, D. R.; Helferich, W. G. Physiological Concentrations of Dietary Genistein Dose-dependently Stimulate Growth of Estrogen-dependent Human Breast Cancer (MCF-7) Tumors Implanted in Athymic Nude Mice. *J. Nutr.* **2001,** *131*, 2957–2962.

Kahkonen, M. P.; Hopia, A. I.; Vuorela, H. J.; Rauha, J. P.; Pihlaja, K.; Kujala, T. S.; Heinonen, M. Antioxidant Activity of Plant Extracts Containing Phenolic Compounds. *J. Agric. Food Chem.* **1999,** *47*(10), 3954–3962.

Kaklamanos, G.; Aprea, E.; Theodoridis, G. Mass Spectrometry. In *Chemical Analysis of Food: Techniques and Applications*; Pico, Y. Ed.; Elsevier: USA, 2012; pp 249–284.

Kammerer, D. N.; Carle, R.; Schieber, A. Detection of Peonidin and Pelargonidin Glycosides in Black Carrots (*Daucus carota* ssp. *Sativus* var. *atrorubens* Alef.) by HPLC/electrospray Ionization Mass Spectrometry. *Rapid Commun. Mass Spectrom.* **2003,** *17*, 2407–2412.

Kammerer, D. N.; Carle, R.; Schieber, A. Quantification of Anthocyanins in Black Carrot (*Daucus carota* ssp. *Sativus* var. *atrorubens* Alef.) Extracts and Evaluation of their Color Properties. *Eur. Food Res. Technol.* **2004,** *219*, 479–486.

Kang, J. H.; Park, Y. H.; Choi, S. W.; Yang, E. K.; Lee, W. J. Resveratrol Derivatives Potently Induce Apoptosis in Human Promyelocytic Leukemia Cells. *Exp. Mol. Med.* **2003,** *35*(6), 467–474.

Kapasakalidis, P. G.; Rastall, R. A.; Gordon, M. H. Extraction of Polyphenols from Processed Black Currant (*Ribes nigrum* L.) Residues. *J. Agric. Food Chem.* **2006,** *54*(11), 4016–4021.

Karakaya, S.; Sedef, N. E. L. Quercetin, Luteolin, Apigenin and Kaempferol Contents of Some Foods. *Food Chem.* **1999,** *66*, 289–292.

Karangwa, E.; Khizar, H.; Rao, L.; Nishimiyimona, D. S.; Foh, M. B. K.; Li, L.; Xia, S. Q.; Zhang, X. M. Optimization of Processing Parameters for Clarification of Blended Carrot–Orange Juice and Improvement of Its Carotene Content. *Adv. J. Food Sci. Technol.* **2010,** *2*(5), 268–278.

Karnjanawipagul, P.; Nittayanuntawech, W.; Rojsanga Pand Suntornsuk L. Analysis of β-Carotene in Carrot by Spectrophotometry. *Mahid. Univ. J. Pharm. Sci.* **2010,** *37*, 8–16.

Kaufmann, B.; Christen, P. H. Recent Extraction Techniques for Natural Products: Microwave-assisted Extraction and Pressurised Solvent Extraction. *Phytochem. Anal.* **2002,** *13*, 105–113.

Kaur, C.; Kapoor H. C. Antioxidants in Fruits and Vegetables—the Millennium's Health. *Int. J. Food Sci. Technol.* **2001,** *36*, 703–725.

Kaur, K.; Shivhare, U. M.; Basu, S.; Raghavan, G. S. V. Kinetics of Extraction of β-Carotene from Tray Dried Carrots by Using Supercritical Fluid Extraction Technique. *Food Nutr. Sci.* **2012,** *3*, 591–595.

Kebede, B. T.; Grauwet, T.; Palmers, S.; Vervoort, L.; Carle, R.; Hendrickx, M.; Loey, A. V. Effect of High Pressure High Temperature Processing on the Volatile Fraction of Differently Colored Carrots. *Food Chem.* **2014,** *153*, 340–352.

Kelly, W. T.; MacDonald, G.; Phatak, S. C. *Commercial Production and Management of Carrots (B 1175).* UGA Extension, 2012. http://extension.uga.edu/publications/detail.cfm?number=B1175.

Khachik, F.; Beecher, G. R.; Goli, M. B. Separation, Identification and Quantification of Carotenoids in Fruits, Vegetables and Human Plasma by High Performance Liquid Chromatography. *Pure Appl. Chem.* **1991,** *63*, 71–80.

Khandare, V.; Walia, S.; Singh, M.; Kaur, C. Black Carrot (*Daucus carota* ssp. *sativus*) Juice: Processing Effects on Antioxidant Composition and Color. *Food Bioprod. Process.* **2011,** *89*, 482–486.

Khanum, F. M.; Swamy, M. S.; Krishna, K. R.; Santhanam, K.; Visvanathan, K. R. Dietary Fiber Content of Commonly Fresh and Cooked Vegetables Consumed in India. *Plant Foods Hum. Nutr.* **2000,** *55*, 207–218.

Kikuzaki, H.; Kayano, S.; Fukutsuka, N.; Aoki, A.; Kasamatsu. K.; Yamasaki, Y. Y.; Mitani, T.; Nakatani, N. *J. Agric. Food Chem.* **2004,** *52*, 344.

Kingsley, R. S.; Bidlack, J. E.; Shelly, H. J. *Introductory Plant Biology*. McGraw-Hill: New York, 2008.

Kingston, H. M.; Jassie, L. B. *Introduction to Microwave Sample Preparation: Theory and Practice*. Vol. 61. American Chemical Society: Washington, DC, 1988.

Kırca, A.; Ozkan, M.; Cemerog, B. Effects of Temperature, Solid Content and pH on the Stability of Black Carrot Anthocyanins. *Food Chem.* **2007,** *101*, 212–218.

Kjeldsen, F.; Christensen, L. P.; Edelenbos, M. Changes in Volatile Compounds of Carrots (*Daucus carota* L.) during Refrigerated and Frozen Storage. *J. Agric. Food Chem.* **2003,** *51*(18), 5400–5407.

Knauf, V. C.; Facciotti, D. Genetic Engineering of Foods to Reduce the Risk of Heart Disease and Cancer. *Adv. Exp. Med. Biol.* **1995,** *369*, 221–228.

Knekt, P.; Kumpulainen, J.; Järvinen, R.; Rissanen, H.; Heliövaara, M.; Reunanen, A.; Hakulinen, T.; Aromaa. A. Flavonoid Intake and Risk of Chronic Diseases. *J. Clin. Nutr.* **2002,** *76*, 560–568.

Knopp, R. H.; Superko, H. S.; Davidson, M.; Insull, W.; Dujovne, C. A.; Kwiterovich, P. O.; Zavoral, J. H.; Graham, K.; O'Connor, R. R.; Edelman, D. A. Long-term Blood Cholesterol Lowering Effects of a Dietary Fiber Supplement. *Am. J. Prevent. Med.* **1999,** *17*, 18–23.

Koca, N.; Burdurlu, H. S.; Karadeniz, F. Kinetics of Color Changes in Dehydrated Carrots. *J. Food Eng.* **2007,** *78*, 449–455.

Kohlmeier, L.; Hastings, S. B. Epidemiological Evidence of a Role of Carotenoids in CVD Prevention. *Am. J. Clin. Nutr.* **1995,** *62*, 1370–1393.

Koidis, A.; Rawson, A.; Tuohy, M.; Brunton, N. Influence of Unit Operations on the Levels of Polyacetylenes in Minimally Processed Carrots and Parsnips: An Industrial Trial. *Food Chem.* **2012,** *132*, 1406–1412.

Koley, T. K.; Singh, S.; Khemariya, P.; Sarkar, A.; Kaur, C.; Chaurasia, S. N. S.; Naik, P. S. Evaluation of Bioactive Properties of Indian Carrot (*Daucus carota* L.): A Chemometric Approach. *Food Res. Int.* **2014,** *60*, 76–85.

Konczak, I.; Zhang, W. Anthocyanins—More than Nature's Colors. *J. Biomed. Biotechnol.* **2004,** *5*, 239–240.

Kotecha, P. M.; Desai, B. B.; Madhavi, D. L. Carrot. In *Handbook of Vegetable Science and Technology: Production, Composition, Storage and Processing*; Salunke, D. K.; Kadam, S. S., Eds.; Marcel Dekker: New York, 1998; pp 119–140.

Kreutzmann, S.; Christensen, L. P.; Edelenbos, M. Investigation of Bitterness in Carrots (*Daucus carota* L.) Based on Quantitative Chemical and Sensory Analyses. *LWT—Food Sci. Technol.* **2008a,** *41*, 193–205.

Kreutzmann, S.; Thybo, A. K.; Edelenbos, M.; Christensen, L. P. The Role of Volatile Compounds on Aroma and Flavor Perception in Colored Raw Carrot Genotypes. *Int. J. Food Sci. Technol.* **2008b,** *43*(9), 1619–1627.

Krinsky, N. I. Overview of Lycopene, Carotenoids, and Disease Prevention. *Proc. Soc. Exp. Biol. Med.* **1998,** *218*, 95–97.

Krinsky, N. I.; Johnson, E. J. Carotenoid Actions and their Relation to Health and Disease. *Mol. Aspects Med.* **2005,** *26*, 459–516.

Krinsky, N.; Mayne, S.; Sies, H. *Carotenoids in Health and Disease*. CRC Press: Boca Raton, FL, 2004.

Kumar, J. Y.; Sivan, Y. S.; Arumughan, C.; Sundaresan, A.; Balachandran, C.; Job, J. Consumption Profile of Preschool Children Supplemented with β-Carotene Through Red Palmoil in a Rural Community in Tamil Nadu. *Indian J. Nutr. Dietician* **2001,** *38*, 199–208.

Kumar, N.; Kumar, K. Development of Carrot Pomace and Wheat Flour Based Cookies. *J. Pure Appl. Sci. Technol.* **2011,** *1*, 4–10.

Kurzer, M. S.; Xu, X. Dietary Phytoestrogens. *Annu. Rev. Nutr.* **1997,** *17*, 353–381.

Lacker, T.; Strohschein, S.; Albert, K. Separation and Identification of Various Carotenoids by C30 Reversed-phase High Performance Liquid Chromatography Coupled to UV and Atmospheric Pressure Chemical Ionization Mass Spectrometric Detection. *J. Chromatogr. A* **1999,** 854, 37–44.

Lain, C. S. Research on *Daucus* L. (*Umbelliferae*). *Anal. Jardin Bot. Madrid.* **1981,** *37*(2), 481–533.

Lal, G.; Sidappa, G. S.; Tandon, G. L. *Preservation of fruits and vegetables.* ICAR Publication: New Delhi, India, 1986.

Lambert, L. A.; Koch, W. H.; Wamer, W. G.; Kornhauser, A. Antitumor Activity in Skin of SKH and Sencar Mice by Two Dietary Betacarotene Formulations. *Nutr. Cancer* **1990,** *13*, 213–221.

Lang, Q.; Wai, M. C. Supercritical Fluid Extraction in Herbal and Natural Product Studies—A Practical Review. *Talanta* **2001,** *53*(4–5), 771–782.

Laura, B. Polyphenols: Chemistry, Dietary Sources, Metabolism, and Nutritional Significance. *Nutr. Rev.* **1998,** *56*, 317–333.

Lawrence, G. H. M. *Taxonomy of Vascular Plants*. Macmillan Company: New York, 1951.

Lee, G. T, Shallenberger R. S.; Vittum, M. T. Free Sugars in Fruits and Vegetables. *N. Y. Food Life Sci. Bull.* **1970,** *1*, 1–12.

Lee, H. J.; Park, Y. K.; Kang, M. H. The Effect of Carrot Juice, β-Carotene Supplementation on Lymphocyte DNA Damage, Erythrocyte Antioxidant Enzymes and Plasma Lipid Profiles in Korean Smoker. *Nutr. Res. Pract. (Nutr. Res. Pract.)* **2011,** *5*(6), 540–547.

Lee, J.; Finn, C. E.; Wrolstad, R. E. Comparison of Anthocyanin Pigment and Other Phenolic Compounds of *Vaccinium membranaceum* and *Vaccinium ovatum* Native to the Pacific Northwest of North America. *J. Agric. Food Chem.* **2004,** *52*, 7039–7044.

Leite, C. W.; Boroski, M.; Boeing, J. S.; Aguiar, A. C.; Franca, P. B.; de Souza, N. E.; Visentainer, J. V. Chemical Characterization of Leaves of Organically Grown Carrot (*Dacus carota* L.) in Various Stages of Development for Use as food. *Cienc. Tecnol. Aliment, Campinas*. **2011,** *31*(3), 735–738.

Leja, M.; Kamińska, I.; Kramer, M.; Maksylewicz-Kaul, A.; Kammerer, D.; Carle, R.; Baranski, R. Origin and Root Color. *Plant Foods Hum. Nutr.* **2013,** *68*, 163–170.

Lenucci, M. S.; Caccioppola, A.; Durante, M.; Serrone, L.; Rescio, L.; Piro, G.; Dalessandro, G. Optimisation of Biological and Physical Parameters for Lycopene Supercritical CO_2 Extraction from Ordinary and High-Pigment Tomato Cultivars. *J. Sci. Food Agric.* **2010,** *90*, 1709–1718.

Letellier, M.; Budzinski, H.; Garrigues, P. Focused Microwave Assisted Extraction of Polycyclic Aromatic Hydrocarbons. *LC–GC Int.* **1999,** *12*, 222–225.

Li, H.; Deng, Z.; Zhu, H.; Hu, C.; Liu, R.; Young, J. C.; Tsao, R. Highly Pigmented Vegetables: Anthocyanin Compositions and Their Role in Antioxidant Activities. *Food Res. Int.* **2012,** *46*, 250–259.

Li, Y.; Skouroumounis, G. K.; Elsey, G. M.; Taylor, D. K. Microwave-Assistance Provides Very Rapid and Efficient Extraction of Grape Seed Polyphenols. *Food Chem.* **2011,** *129*(2), 570–576.

Ligor, M.; Kovacova, J.; Gadzala-Kopciuch, R. M.; Studzinska, S.; Bocian, S.; Lehotay, J.; Buszewski, B. Study of RP HPLC Retention Behaviours in Analysis of Carotenoids. *Chromatographia* **2013**. DOI:10.1007/s10337-014-2657-1.

Lin, M. L.; Durance, T. D.; Scaman, H. S. Characterization of Vacuum Microwave Air, and Freeze Dried Carrot Slices. *Food Res. Int.* **1998,** *31*, 111–117.

Lineback, D. R. *The Chemistry of Complex Carbohydrates*. Marcel Dekker: New York, 1999, pp 1–17.

Liu, C.; Zhong, M.; Shen, Z. Extraction of Wheat Germ Oil by Supercritical Carbon Dioxide. *Food Sci. Sin.* **1994,** *3*, 14–17.

Liu, R. H Health Benefits of Fruits and Vegetables are from Additive and Synergistic Combination of Phytochemicals. *Am. J. Clin. Nutr.* **2003,** *78*, 517S–520S.

Liu, R. H. Potential Synergy of Phytochemicals in Cancer Prevention: Mechanism of Action. *J. Nutr.* **2004,** *134*, 3479–3485.

Liu, R. H.; Hotchkiss, J. H. Potential Genotoxicity of Chronically Elevated Nitric Oxide: A Review. *Mutat. Res.* **1995,** *339*, 73–89.

Longo, L.; Scardino, A. and Vasapollo, G. Identification and Quantification of Anthocyanins in the Berries of *Pistacia lentiscus* L., *Phillyrea latifolia* L. and *Rubia peregrina* L. *Innov. Food Sci. Emerg. Technol.* **2007,** *8*, 360–364.

Longo, L.; Vasapollo, G. Extraction and Identification of Anthocyanins from *Smilax aspera* L. berries. *Food Chem.* **2006,** *94*, 226–231.

Lund, E. D.; Bruemmer, J. H. Sesquiterpene Hydrocarbons in Processed Stored Carrot Sticks. *Food Chem.* **1992,** *43*, 331–335.

Lund, E. D.; White, J. M. Polyacetylenes in Normal and Water-stressed "Orlando Gold" Carrots (*Daucus carota*). *J. Sci. Food Agric.* **1990,** *51*, 507–516.

Luo, Y.; Suslow, T.; Cantwell, M. Carrot. In *Agriculture Handbook Number 66: The Commercial Storage of Fruits, Vegetables, and Florist and Nursery Stocks*; Gross, K. C., Wang, C. Y.; Saltveit, M., Eds.; USDA, Agricultural Research Service, Beltsville Agricultural Research Center: Ithaca, NY, 2014.

Mackevic, V. I. The Carrot of Afghanistan. *Bull. Appl. Bot. Genet. Plant Breed.* **1929,** *20,* 517–562.

Madan, S.; Dhawan, S. S. Development of Value Added Product 'CANDY' from Carrots. *Process Food Ind.* **2005,** *8,* 26–29.

Madukwe E. U.; Eme, P. E.; Okpara, C. E. Nutrient Content and Microbial Quality of Soymilk-Carrot Powder Blend. *Pak. J. Nutr.* **2013,** *12*(2), 158–161.

Makris, M.; Watson, H. G. The Management of Coumarin induced Over-anticoagulation. *Br. J. Haematol.* **2001,** *114,* 271–280.

Mandal, V.; Yogesh, M. Hemalatha, S. Microwave Assisted Extraction—An Innovative and Promising Extraction Tool for Medicinal Plant Research. *Pharmacogn. Rev.* **2007,** *1*(1), 7–18.

Mangels, A. R.; Holden, J. M.; Beecher, G. R.; Forman, M.; Lanza, E Carotenoid Content of Fruits and Vegetables: An Evaluation of Analytical Data. *J. Am. Diet. Assoc.* **1993,** *93,* 284–296.

Manjunatha, S. S.; Kumar, B. L.; Mohan, G.; Das, D. K. Development and Evaluation of Carrot Kheer Mix. *J. Food Sci. Technol.* **2003,** *40,* 310–312.

Manosa, N. A.; Engelbrecht, G. M.; Allemann, J.; Adipala, E.; Tusiime, G.; Majaliwa, J. G. M. Influence of Temperature on Yield of Carrots. In Second RUFORUM Biennial Regional Conference on "Building capacity for food security in Africa", Entebbe, Uganda, 20–24 September 2010, RUFORUM, 2010, pp 205–208.

Mansoor, G. Z.; Khursheed, K.; Jairajpuri, D. S. Preparation, Processing and Packaging of Pre-mix for the Production of Carrot Dessert. *IOSR J. Environ. Sci., Toxicol. Food Technol.* **2013,** *3*(6), 38–42.

Marinova, D.; Ribarova, F.; Atanassova, M. Total Phenolics and Total Flavonoids in Bulgarian Fruits and Vegetables. *J. Univ. Chem. Technol. Metall.* **2005,** *40,* 255–260.

Marinova, E.; Toneva, A.; Yanishlieva, N. Synergistic Antioxidant Effect of α-Tocopherol and Myricetin on the Autoxidation of Triacylglycerols of Sunflower Oil. *Food Chem.* **2008,** *106*(2), 628–633.

Markham, K. R. *Techniques in Flavonoid Identification,* vol. 31. London Academic Press: London, 1982.

Marsili, R.; Callahan, D. Comparison of a Liquid Solvent Extraction Technique and Supercritical Fluid Extraction for the Determination of Alpha- and Beta-carotene in Vegetables. *J. Chromatogr. Sci.* **1993,** *31,* 422–428.

Marta, G. T.; Măniutiu, D. N.; Andreica, I.; Balcau, S.; Lazar, V.; Bogdan, I. Sugar Content of Carrot Roots Influenced by the Sowing Period. *J. Hortic., Forestry Biotechnol.* **2013,** *17,* 66–69.

Martinez, J. L. *Supercritical Fluid Extraction of Nutracueticals and Bioactive Compounds*; Taylor and Francis Group CRC Press: Boca Raton, FL, 2008; pp 85–91.

Mathiasson, L.; Turner, C.; Berg, H.; Dahlberg, L.; Theobald, A.; Anklam, E.; Ginn, R.; Sharman, M.; Ulberth, F.; Gabernig, R. Development of Methods for the Determination of Vitamins A, E and Beta-carotene in Processed Foods Based on Supercritical Fluid Extraction: A Collaborative Study. *Food Addit. Contam.* **2002,** *19,* 632–646.

Mattila, P.; Hellstro, M J. Phenolic Acids in Potatoes, Vegetables, and Some of Their Products. *J Food Comp. Anal.* **2007,** *20*, 152–160.

Mayne, S. T. Beta-carotene, Carotenoids, and Disease Prevention in Humans. *FASEB J.* **1996,** *10*, 690–701.

Mazur, W. Phytoestrogen Content in Foods. *Baillieres Clin. Endocrinol. Metab.* **1998,** *12*, 729–742.

Mazza, G.; Miniati, E. *Anthocyanins in Fruits, Vegetables and Grains*. CRC Press: London, 1993.

Mercadante, A. Z. Chromatographic Separation of Carotenoids. *Arch. Latino-Am. Nutr.* **1999,** *49*, 52–57.

Mercier, J. J.; Arul, J.; Julien, C. Effect of Food Preparation on the Isocoumarin 6-Methoxymellein Content of UV-treated Carrots. *Food Res. Int.* **1994,** *27*, 401–404.

Mikanagi, Y.; Saito, N.; Yokoi, M.; Tatsuzawa, F. Anthocyanins in Flowers of Genus *Rosa*, Sections *Cinnamomeae (=Rosa)*, *Chinenses, Gallicanae* and Some Modern Garden Roses. *Biochem. Syst. Ecol.* **2000,** *28*(9), 887–902.

Milder, I. E. J.; Arts, I. C. W.; van de Putte, B.; Venema, D. P.; Hollman, P. C. H. Lignan Contents of Dutch Plant Foods: A Database Including Lariciresinol, Pinoresinol, Secoisolariciresinol and Matairesinol. *Br. J. Nutr.* **2005,** *93*(3), 393–402.

Mishra, H.; Das, C. A review on Biological Control and Metabolism of Aflatoxins. *Crit. Rev. Food Sci. Nut.* **2003,** *43*, 245–264.

Miura, S. Watanabe, J.; Sano, M.; Tomita, T.; Osawa, T.; Hara, Y.; Tomita, I Effects of Various Natural Antioxidants on the Cu^{+} Mediated Oxidative Modification of Low Density Lipoproteins. *Biol. Pharm. Bull.* **1995,** *18*, 1–4.

Mohammad, A.; Rehman, A. M. Fifty Years of Indian Agriculture—Production and Self-sufficiency. In *Fifty years of vegetable research in India: A brief background*; Arora, S. K, Mangal, J. L., Eds.; Concept Publishing Company: New Delhi, 116, 2007.

Molldrem, K. L.; Li, J.; Simon, P. W.; Tanumihardjo, S. A. Lutein and β-Carotene from Lutein-Containing Yellow Carrots are Bioavailable in Humans. *Am. J. Clin. Nutr.* **2004,** *80*, 131–136.

Moniruzzaman, M.; Akand, M. H.; Hossain, M. I.; Sarkar, M. D.; Ullah, A. Effect of Nitrogen on the Growth and Yield of Carrot (*Daucus carota* L.). *Agriculturists* **2013,** *11*, 76–81.

Montilla, E. C.; Arzaba, M. R.; Hillebrand, S.; Winterhalter, P. Anthocyanin Composition of Black Carrot (*Daucus carota* ssp. *sativus* var. *atrorubens* Alef.) Cultivars Antonina, Beta Sweet, Deep Purple, and Purple Haze. *J. Agric. Food Chem.* **2011,** *59*, 3385–3393.

Moreira, M. M.; Morais, S.; Barros, A. A.; Delerue-Matos, C.; Guido, L. F. A Novel Application of Microwave-assisted Extraction of Polyphenols from Brewer's Spent Grain with HPLC–DAD–MS Analysis. *Anal. Bioanal. Chem.* **2012,** *403*(4), 1019–1029.

Mosig, A.; Schramm, S. G. Der Arzneipflanzen und Drogenschatz Chinas und die Bedeutung des Pentsao Kang-mu als: Standardwerk der Chinensischen Materia Medica. *Beih. Pharm. Heft* **1955,** *4*, 1–71.

Muller, H. Determination of the Carotenoid Content in Selected Vegetables and Fruit by HPLC and Photodiode Array Detection. *Zeitschr. Lebensmitt.-Untersuch.-Forsch. A* **1997,** *204*(2), 88–94.

Murakoshi, M.; Nishino, H.; Satomi, Y.; Takayasu, J.; Hasegawa, T.; Tokuda, H.; Iwashima, A.; Okuzumi, J.; Okabe, H.; Kitano, H.; Iwasaki, R. Potent Preventive Action of α-carotene against Carcinogenesis: Spontaneous Liver Carcinogenesis and Promoting Stage of Lung and Skin Carcinogenesis in Mice are Suppressed More Effectively by α-carotene Than by β-carotene. *Cancer Res.* **1992,** *52*, 6583–6587.

Mustapha, Y.; Babura, S. Determination of Carbohydrate and β-carotene Content of Some Vegetables Consumed in Kano Metropolis, Nigeria. *Bayero J. Pure Appl. Sci.* **2009,** *2*(1), 119–121.

Nagai, T.; Reiji, I.; Hachiro, I.; Nobutaka, S. Preparation and Antioxidant Properties of Water Extract of Propolis. *Food Chem.* **2003,** *80*, 29–33.

Nandal, J. K.; Tehlan, S. K.; Gupta, V.; Partap, P. S. Effect of Time and Method of Sowing on Root Quality of Carrot. *Haryana J. Hortic. Sci.* **2008,** *37*, 295–296.

Narayan, M. S.; Venkataraman, L. V. Characterisation of Anthocyanins Derived from Carrot (*Daucus carota*) Cell Culture. *Food Chem.* **2000,** *70*, 361–363.

Narisawa, T.; Fukaura, Y.; Hasebe, M.; Ito, M.; Aizawa, R.; Murakoshi, M.; Uemura, S.; Khachik, F.; Nishino, H. Inhibitory Effects of Natural Carotenoids, α-carotene, β-carotene, Lycopene and Lutein, on Colonic Aberrant Crypt Foci Formation in Rats. *Cancer Lett.* **1996,** *107*, 137–142.

NAS-NCR (National Academy of Science/National Council Research). *Recommended Dietary Allowances*; 9th ed. NAS-NCR: Washington, DC, 1980; pp 51–71.

Nawirska, A.; Kwasniewska, M. Dietary Fiber Fractions from Fruit and Vegetable Processing Waste. *Food Chem.* **2005,** *91*, 221–225.

Negi, P. S.; Roy, S. K. The Effect of Blanching on Quality Attributes of Dehydrated Carrots During Long Term Storage. *Eur. Food Res. Technol.* **2001,** *212*, 445–448.

Nicolle, C.; Simon, G.; Rock, E.; Amouroux, P.; Remesy, C. Genetic Variability Influences Carotenoid, Vitamin, Phenolic, and Mineral Content in White, Yellow, Purple, Orange, and Dark-Orange Carrot Cultivars. *J. Amer. Soc. Hortic. Sci.* **2004,** *129*, 523–529.

Nonnecke, I. L. *Vegetable Production*. Van Nostrand Reinhold New York, 1989, ISBN 0-442-26721-5.

Normen, L.; Johnsson, M.; Andersson, H.; van Gameren, Y.; Dutta, P. Plant Sterols in Vegetables and Fruits Commonly Consumed in Sweden. *Eur. J. Nutr.* **1999,** *38*(2), 84–89.

Northolt, M.; Burgt, G. J.; Buisman, T.; Bogaerde, A. V. *Parameters for Carrot Quality and the Development of the Inner Quality Concept.* Louis Bolk Institute Driebergen: The Netherlands, 2004.

O'Keefe, S. J. Nutrition and Gastrointestinal Disease. *Scan J. Gastroenterol.* **1996,** *220*, 52–59.

Okuda, K. Hepatocellular Carcinoma: Recent Progress. *Hepatology* **1992,** *15*(5), 948–963.

Oliver J.; Palou, A.; Pons, A. Semi-quantification of Carotenoids by HPLC: Saponification Induced Losses in Fatty Foods. *J. Chromatogr. A* **1998,** *1051*, 517–226.

Oliver, J.; Palou, A. Chromatographic Determination of Carotenoids in Foods. *J. Chromatogr. A* **2000,** *881*, 543–555.

Ollanketo, M.; Hartonen, K.; Riekkola, M. L.; Holm, Y.; Hiltunen, R. Supercritical Carbon Dioxide Extraction of Lycopene in Tomato Skins. *Eur. Food Res. Technol.* **2001,** *212*, 561–565.

Ostlund, R. E. Phytosterols in Human Nutrition. *Annu. Rev. Nutr.* **2002,** *22*, 533–549.

Ozcan, M. M.; Chalchat, J. C. Chemical Composition of Carrot Seeds (*Daucus carota* L.) Cultivated in Turkey: Characterization of the Seed Oil and Essential Oil. *Gras. Y Aceit.* **2007,** *58*(4), 359–365.

Paiva, S. A. R.; Russell, R. M. Beta-carotene and Other Carotenoids as Antioxidants. *J. Am. College Nutr.* **1999,** *18*(5), 426–433.

Pakin, C.; Bergaentzle, M.; Hubscher, V.; Aoude-Werner, D.; Hasselmann C. Fluorimetric Determination of Pantothenic Acid in Foods by Liquid Chromatography with Post-Column Derivatization. *J. Chromatogr. A* **2004,** *1035*(1), 87–95.

Panalaks, T.; Murray, Y. K. Effect of Processing on the Content of Carotene Isomers in Vegetables and Peaches. *J. Can. Inst. Food Technol.* **1970,** *3*, 145–151.

Pandey, D.; Rajput, A.; Kulshrestha, K. Carotene Rich Food Supplements Based on Carrot Powder. *J. Dairy. Foods Homesci.* **2003,** *22*, 239–241.

Park, Y. W. Effect of Freezing, Thawing, Drying and Cooking on Carotene Retention in Carrots, Broccoli and Spinach. *J. Food Sci.* **1987,** *52*, 1022–1025.

Parveen, N.; Akhtar, M. S.; Abbas, N.; Abid, A. R. Effects of Carrot Residue Fiber on Body Weight Gain and Serum Lipid Fractions. *Int. J. Agric. Biol.* **2000,** *2*, 1–2.

Pasquet, V.; Cherouvrier, J. R.; Farhat, F.; Thiery, V.; Piot, J. M.; Berard, J. B.; Kaas, R.; Serive, B.; Patrice, T.; Cadoret, J. P. Study on the Microalgal Pigments Extraction Process: Performance of Microwave Assisted Extraction. *Process. Biochem.* **2011,** *46*(1), 59–67.

Pearce, C. E.; Parker, R. A.; Deason, M. E.; Quershi, A. A.; Wright, J. J. Hypocholesterolemic Activity of Synthetic and Natural Tocotrienols. *J. Med. Chem.* **1992,** *35*, 3595–3606.

Pennington, J. A. T.; Douglass, J. S. *Food Values of Portions Commonly Used,* 18th ed. Lippincott Williams and Wilkins: Baltimore, MD, 2005.

Pferschy-Wenzig, E. V.; Getzinger, V.; Kunert, O.; Woelkart, K.; Zahrl, J.; Bauer, R. Determination of Falcarinol in Carrot (*Daucus carota* L.) Genotypes Using Liquid Chromatography/Mass Spectrometry. *Food Chem.* **2009,** *114*, 1083–1090.

Pimenov, M. G.; Leonov, M. V. *The Genera of the Umbelliferae: A Nomenclature.* Royal Botanical Gardens: Kew, UK, 1993.

Pinheiro-Santana, H. M.; Stringheta, P. C.; Brandao, S. C. C.; Paez, H. H.; de Queiroz, V. M. V. Evaluation of Total Carotenoids, α- and β-Carotene in Carrots (*Daucus carota* L.) during Home Processing. *Food Sci. Technol. (Campinas)* **1998.** On-line version, ISSN 1678-457X.

Prasad, K.; Ghubade, V.; Prasad, K. K. Development and Characterization of Carrot Based Beverage Base: Studies on its Formulation, Lambert Academic Publishing: Germany, 2014.

Premachandra, B. R. A Simple TLC Method for the Determination of Pro-Vitamin A Content of Fruits and Vegetables. *Int. J. Vitam. Nutr. Res.* **1985,** *55*, 139–147.

Prior, R. L.; Cao, G. Antioxidant Phytochemicals in Fruits and Vegetables: Diet and Health Implications. *Hortic. Sci.* **2000,** *35*, 588–592.

Profir, A. G.; Vizireanu, C. Evolution of Antioxidant Capacity of Blend Juice Made from Beetroot, Carrot and Celery During Refrigerated Storage. *Ann. Univ. Dun. Jos Galati Fascicle VI—Food Technol.* **2013,** *37*(2), 93–99.

Radovich, T. J. K. Biology and classification of Vegetables. In *Handbook of Vegetables and Vegetable Processing*; Sinha, N. K., Hui, Y. H., Eds.; Wiley-Blackwell Publishing: USA, 2011.

Rakcejeva, T.; Augspole, I.; Dukalska, L.; Dimins, F. Chemical Composition of Variety 'Nante' Hybrid Carrots Cultivated in Latvia. *World Acad. Sci., Eng. Technol.* **2012,** *64*, 1120–1126..

Ranalli, A.; Contento, S.; Lucera, L.; Pavone, G.; Di Giacomo, G.; Aloisio, L.; Di Gregorio, C.; Mucci, A.; Kourtikakis, I. Characterization of Carrot Root Oil Arising from Supercritical Fluid Carbon Dioxide Extraction. *J. agric. food chem.* **2004,** *52*(15), 4795–4801.

Rathore, D. S. *Indian Horticulture*, 2001, pp 8–12.

Rawson, A.; Koidis, A.; Patras, A.; Tuohy, M. G.; Brunton, N. P. Modelling the Effect of Water Immersion Thermal Processing on Polyacetylene Levels and Instrumental Colour of Carrot Discs. *Food Chem.* **2010,** *121*, 62–68.

Rawson, A.; Tiwari, B. K.; Tuohy, M. G.; O'Donnell, C. P.; Brunton, N. Effect of Ultrasound and Blanching Pretreatments on Polyacetylene and Carotenoid Content of Hot Air and Freeze Dried Carrot Discs. *Ultrason. Sonochem.* **2011,** *18*, 1172–1179.

Rein, M. *Co-pigmentation Reactions and Color Stability of Berry Anthocyanins*; University of Helsinki: Helsinki, 2005; pp 10–14.

Reverchon, E.; de Marco, I. Supercritical Fluid Extraction and Fractionation of Natural Matter. *J. Supercrit. Fluids* **2006,** *38*, 146–166.

Reverchon, E.; Donsi, G.; Osseo, L. S. Modeling of Supercritical Fluid Extraction from Herbaceous Matrices. *Ind. Eng. Chem. Res.* **1993,** *32*, 2721–2726.

Rice-Evans, C. A.; Miller, N. J.; Paganga G. Antioxidant Properties of Phenolic Compounds. *Trends Plant Sci.* **1997,** *2*(4), 1360–1385.

Rice-Evans, C. A.; Packer, L. *Flavonoids in Health and Disease*; Marcel Dekker: New York, 2003; p 504.

Rice-Evans, C.; Miller, N. J.; Paganga, G. Structure Antioxidant Activity Relationships of Flavonoids and Phenolic Acids. *Free Rad. Biol. Med.* **1996,** *20*, 933–956.

Rivera, S. M.; Canela-Garayoa. Analytical Tools for the Analysis of Carotenoids in Diverse Materials. *J. Chromatogr. A* **2012,** *1224*, 1–10.

Rodgureiz-Amaya, D. B.; Kimura, M. *HarvestPlus Handbook for Carotenoid Analysis*. HarvestPlus: Washington DC, 2004.

Rodriguez-Amaya, D. B. *A guide to Carotenoid Analysis in Foods*. ILSI Press: Washington, DC, 2001.

Rodriguez-Amaya, D. B. Critical Review of Provitamin A Determination in Plant Foods. *J. Micronutr. Anal.* **1989,** *5*, 191–225.

Rodriguez-Amaya, D. B.; Kimura, M.; Godoy, H. T.; Arima, H. K. Assessment of Provitamin A Determination by Open Column Chromatography/Visible Absorption Spectrophotometry. *J. Chromatogr. Sci.* **1988,** *26*, 624–629.

Rojas-Garbanzo, C.; Perez, A. M.; Bustos-Carmona, J.; Vaillant, F. Identification and Quantification of Carotenoids by HPLC–DAD during the Process of Peach Palm (*Bactris gasipaes* H.B.K.) Flour. *Food Res. Int.* **2011,** *44*, 2377–2384.

Rosamond, W. D. Dietary Fiber and Prevention of Cardiovascular Disease. *J. Am. Coll. Cardiol.* **2002,** *39*, 57–59.

Rosenfeld H. J.; Samuelsen R. T.; Lea, P. The Effect of Temperature on Sensory Quality, Chemical Composition and Growth of Carrots (*Daucus carota* L.). I. Constant Diurnal Temperature. *J. Hort. Sci. Biotechnol.* **1998,** *73*(2), 275–886.

Rosenfeld, H. J.; Aaby, K.; Lea, P. Influence of Temperature and Plant Density on Sensory Quality and Volatile Terpenoids of Carrot (*Daucus carota* L.) Root. *J. Sci. Food Agric.* **2002,** *82*, 1384–1390.

Rosenfeld, H. J.; Vogt, G.; Aaby, K.; Olsen, E. Interaction of Terpenes with Sweet Taste in Carrots (*Daucus carota* L.). *Acta Hortic.* **2004,** *637*, 377–386.

Ross, J. A.; Kasum, C. M. Dietary Flavonoids: Bioavailability, Metabolic Effects, and Safety. *Annu. Rev. Nutr.* **2002,** *22*, 19–34.

Routray, W.; Orsat, V. Microwave-Assisted Extraction of Flavonoids: A Review. *Food Bioprocess Technol.* **2011,** *5*(2), 1–16.

Roy, U.; Batish, V.; Grover, S.; Neelakantan, S. Production of Antifungal Substances by *Lactococcus lactis* subspecies *lactis*. *Int. J. Food Microbial.* **1996,** *32*, 27–34.

Rozzi, N. L.; Singh, R. K.; Vierling, R. A.; Watkins, B. A. Supercritical Fluid Extraction of Lycopene from Tomato Processing by Products. *J. Agric. Food Chem.* **2002,** *50*, 2638–2643.

Rubatzky, V. E.; Quiros, C. F.; Simon, P. W. *Carrots and Related Vegetable Umbelliferae.* CABI: New York, 1999.

Rude, R. K. Magnesium. In *Biochemical and Physiological Aspects of Human Nutrition*; Stipanuk, M. H, Ed.; W. B. Saunders Company: Philadelphia, PA, 2000; pp 671–685.

Ruzic, I.; Skerget, M.; Knez, Z. Potential of Phenolic Antioxidants. *Acta Chim. Slov*, **2010,** *57*, 263–271.

Sahena, F.; Zaidul, I. S. M.; Jinap, S.; Karim, A. A.; Abbas, K. A.; Norulaini, N. A. N.; Omar, A. K. M. Application of Supercritical CO_2 in Lipid Extraction—A Review. *J. Food Eng.* **2009,** *95*, 240–253.

Sakara, A.; Kalisz, A.; Cebula, S.; Grabowska, A. The Quality and Processing Usefulness of Chosen Polish Carrot Cultivars. *Acta Sci. Pol., Hort. Cultus* **2012,** *11*(5), 101–112.

Saleem, M.; Kim, H. J.; Ali, M. S.; Lee Y. S. An Update on Bioactive Plant Lignans. *R. Soc. Chem.* **2005,** *22*, 696–716.

Salwa, A. A.; Galal, E. A.; Neimat, A. E. Carrot Yoghurt: Sensory, Chemical and Microbiological Properties and Consumer Acceptance. *Pak. J. Nutr.* **2004,** *3*(6), 322–330.

Sampathu, S. R.; Chakraberty, S.; Kamal, P.; Bisht, H. C.; Agrawal, N. D.; Saha, N. K. Standardization and Preservation of Carrot Halwa—An Indian Sweet. *Indian Food Packer,* **1981,** *35*, 60–67.

Sanchez-Moreno, C.; Larrauri, J. A.; Saura-Calixto, F. A Procedure to Measure the Antiradical Efficiency of Polyphenols. *J. Sci. Food Agric.* **1998,** *76*, 270–276.

Sancho, L. E. G.; Yahia, E. M.; Gonzalez-Aguilar, G. A. Identification and Quantification of Phenols, Carotenoids, and Vitamin C from Papaya (*Carica papaya* L., cv. Maradol) Fruit Determined by HPLC–DAD–MS/MS–ESI. *Food Res. Int.* **2011,** *44*, 1284–1291.

Sander, L. C.; Sharpless, K. E.; Pursh, M. C. C_{30} Stationary Phases for the Analysis of Food by Liquid Chromatography. *J. Chromatogr. A* **2000,** *880*, 189–202.

Sandra, G.; Dusan, M.; Irena, Z.; Marko, S.; Ruzica, A.; Dejan, S.; Gogu, G.; Ursula, S.; Constantin, T. Carrot Fruit Essential Oil and Supercritical Fluid Extract—The Chemical Composition and Antimicrobial Activity. Fourth Conference on Medicinal and Aromatic Plants of South-East European Countries. Ninth National Symposium 'Medicinal Plants-Present and Perspectives'. Third National Conference of Phytotherapy, Proceedings. Iaşi, Romania, 28–31 May, 2006. Association for Medicinal and Aromatic Plants of Southeast European Countries (AMAPSEEC), pp 107–112.

Schneeman, B. O. Carbohydrates: Significance for Energy Balance and Gastrointestinal Function. *J. Nutr.* **1994,** *124*, 1747S–1753S.

Schwarz, M.; Wray, V.; Winterhalter, P. Isolation and Identification of Novel Pyranoanthocyanins from Black Carrot (*Daucus carota* L.) Juice. *J. Agric. Food Chem.* **2004,** *52*, 5095–5101.

Seo, A.; Yu, M. Toxigenic Fungi and Mycotoxins. In *Handbook of Industrial Mycology*; Andrea, Z., Ed.; Academic: London. 2003; pp 233–246.

Serrano, J.; Puupponen-Pimi, A. R.; Dauer, A.; Aura, A. M.; Saura-Calixto, F. Tannins: Current Knowledge of Food Sources, Intake, Bioavailability and Biological Effects. *Mol. Nutr. Food Res.* **2009,** *53*(2), S310–S329.

Sethi, V.; Anand, J. C. Studies on the Preparation, Quality and Storage of Intermediate Moisture Vegetables. *J. Food Sci. Technol.* **1982,** *19*, 168–170.

Sevimli-Gur, C.; Cetin, B.; Akay, S.; Guice-Iz, S.; Yasil-Celiktas, O. Extracts from Black Carrot Tissue Culture as Potent Anticancer Agents. *Plant Foods Hum. Nutr.* **2013,** *68*, 293–298.

Sharma, O. P. *Plant Taxonomy*. Tata McGraw-Hill: New Delhi, 2009.

Sharma, S. L.; Caralli, S. *A Dictionary of Food and Nutrition*. CBS: New Delhi, 1998.

Sheng, H. P. Sodium, Chloride, and Potassium. In *Biochemical and Physiological Aspects of Human Nutrition*; Stipanuk, M. H., Ed.; W. B. Saunders Company: Philadelphia, PA, 2000; pp 686–710.

Shi, J.; Dai, Y.; Kakuda, Y.; Mittal, G.; Xue, J. Effect of Heating and Light Irradiation on the Stability of Lycopene in Tomato Puree. *Food Control* **2008,** *19*, 514–520.

Shi, J.; Khatri, M.; Xue, S. J.; Mittal, G. S.; Ma, Y.; Li, D. Solubility of Lycopene in Supercritical CO_2 Fluid as Affected by Temperature and Pressure. *Sep. Purif. Technol.* **2009,** *66*, 322–328.

Shi, J.; Mittal, G.; Kim, E.; Xue, J. Solubility of Carotenoids in Supercritical CO_2. *Food Rev. Int.* **2007,** *23*, 341–371.

Shi, X.; Wu, H.; Shi, J.; Xue, S. J.; Wang, D.; Wang, W.; Cheng, A.; Gong, Z.; Chen, X.; Wang, C. Effect of Modifier on the Composition and Antioxidant Activity of Carotenoid Extracts from Pumpkin (*Cucurbita maxima*) by Supercritical CO_2. *LWT—Food Sci. Technol.* **2013,** *51*, 433–440.

Shilpi, A.; Shivhare, U. S.; Basu, S. Supercritical CO_2 Extraction of Compounds with Antioxidant Activity from Fruits and Vegetables Waste—A Review. *Focus. Modern Food Ind. (FMFI),* **2013,** *2*(1), 43–62.

Shymala, B. N.; Jamuna, P. Nutritional Content and Antioxidant Properties of Pulp Waste from *Daucus carota* and *Beta vulgaris*. *Malaysian J. Nutr.* **2010,** *16*, 397–408.

Sies, H.; Stahl, W. Vitamins E and C, β-Carotene, and other Carotenoids as Antioxidants. *Am. J. Clin. Nutr.* **1995,** *62*, 1315–1321.

Sihvonen, M.; Jarvenpaa, E.; Hietaniemi, V.; Huopalahti, R. Advances in Supercritical Carbon Dioxide Technologies. *Trends Food Sci. Technol.* **1999,** *10*(6–7), 217–222.

Simon, P. W.; Freeman, R. E.; Vieira, J. V.; Boiteux, L. S.; Briard, M.; Nothnagel, T.; Michalik, M.; Kwon, S. K. Carrot. In *Handbook of Plant Breeding—Vegetables II*, vol. II; Prohens, J., Nuez, F.; Springer Science Business Media: New York, 2008; pp 324–357.

Simon, P. W.; Peterson, C. E.; Lindsay, R. C. Correlations between Sensory and Objective Parameters of Carrot Flavor. *J. Agric. Food Chem.* **1980,** *28*, 559–562.

Simon, P. W.; Rubatzky, V. E.; Bassett, M. J.; Strandberg, J. O.; White, J. M. B7262, Purple Carrot Inbred. *HortScience* **1997,** *32*, 146–147.

Simon, P. W.; Wolff, X. Y. Carotene in Typical and Dark Orange Carrots. *J. Agric. Food Chem.* **1987,** *35,* 1017–1022.

Simopoulos, A. P. The Importance of Ratio of Omega-6/Omega-3 Essential Fatty Acids. *Biomed. Pharmacother.* **2002,** *56*, 365–379.

Simpson, K. L. Relative Value of Carotenoids as Precursors of Vitamin A. *Am. Proc. Nutr. Soc.* **1983,** *42*, 7–17.

Simsek, M.; Sumnu, G.; Sahin, S. Microwave-Assisted Extraction of Phenolic Compounds from Sour Cherry Pomace. *Separ Sci. Technol.* **2012,** *47*(8), 1248–1254.

Singh, A.; Singh, N.; Chandy, A.; Manigauha, A. In Vivo Antioxidant and Hepato-Protective Activity of Methanolic Extracts of *Daucus carota* Seeds in Experimental Animals. *Asian Pac. J. Trop Biomed.* **2012,** *2*(5), 385–388.

Singh, B.; Panesar, P. S.; Nanda, V. Utilization of Carrot Pomace for the Preparation of a Value Added Product. *World J. Dairy Food Sci.* **2006,** *1*, 22–27.

Singh, D. P.; Beloy, J.; Mclnerney, J. K.; Day, L. Impact of Boron, Calcium and Genetic Factors on Vitamin C, Carotenoids, Phenolic Acids, Anthocyanins and Antioxidant Capacity of Carrots (*Daucus carota*). *Food Chem.* **2011,** *132*, 1161–1170.

Singh, G.; Kawtra, A.; Sehgal, S. Nutritional Composition of Selected Green Leafy Vegetables, Herbs and Carrots. *Plant Foods Human Nutr.* **2001,** *56*, 359–364.

Singh, K. The Potential and Problems for Horticultural Crops in Haryana. In *Lecture Compendium of Agro techniques of Horticultural Crops*. Directorate of Human Resource Management, CCS HAU: Hisar, Haryana, 1998.

Singh, P.; Kulshrestha, K.; Kumar, S. Effect of Storage on β-carotene Content and Microbial Quality of Dehydrated Carrot Products. *Food Biosci.* **2013,** *2*, 39–45.

Singh, S.; Shivhare, U. S.; Ahmed, J.; Raghavan, G. S. Osmotic Concentration Kinetics and Quality of Carrot Preserve. *Food Res. Int.* **1999,** *32*, 509–514.

Singh, V.; Katiyar, P. Nutritional Evaluation of a Mixture Powder (Fruits, Vegetables and Pulses) for Children as Supplementary Food. *Food Sci. Res. J.* **2014,** *5*, 72–73.

Sinha, N. K.; Hui, Y. H. Handbook of Vegetables and Vegetable Processing. Wiley-Blackwell: USA, 2011.

Siong, T.; Lam, L. Analysis of Carotenoids in Vegetables by HPLC. *ASEAN Food J.* **1992,** *7*(2), 91–99.

Small, E.; Catling, P. M. *Canadian Medicinal Crops*. NRC Research Press: Ottawa, ON, 1999.

Smith, J. G.; Yokoyama, W. H.; German, G. B. Butyric Acid from Diet: Actions at the Levels of Gene Expression. *Crit. Rev. Food Sci.* **1998,** *38*, 259–267.

Sontag, S. J. Defining GERD. *Yale J. Biol. Med.* **1999,** *72*, 69–80.

Soria, A. C.; Sanz, J.; Villamiel, M. Analysis of Volatiles in Dehydrated Carrot Samples by Solid-Phase Micro-Extraction Followed by GC–MS. *J. Separ. Sci.* **2008,** *31*, 3548–3555.

Southon, S. Increased Fruit and Vegetable Consumption within the EU: Potential Health Benefits. *Food Res. Int.* **2000,** *33*, 211–217.

Sowani, H. M.; Thorat, P. Antimicrobial Activity Studies of Bacteriocin Produced by *Lactobacilli* Isolates from Carrot Kanji. *Online J. Biol. Sci.* **2012,** *12*(1), 6–10.

Spiller, R. C. Pharmacology of Dietary Fibre. *Pharmacol. Ther.* **1994,** *62*, 407–427.

Sroka, Z.; Cisowski, W. Hydrogen Peroxide Scavenging, Antioxidant and Anti-Radical Activity of Some Phenolic Acids. *Food Chem. Toxicol.* **2003,** *41*(6), 753–758.

Staggs, C. G.; Sealey, W. M.; McCabe, B. J.; Teague, A. M.; Mock, D. M. Determination of the Biotin Content of Select Foods Using Accurate and Sensitive HPLC/Avidin Binding. *J. Food Compos.* **2004,** *17*(6), 767–776.

Stahl, E. Thin-Layer Chromatography: Methods, Influencing Factors and an Example of Its Use. *Pharmazie* **1956,** *11*(10), 633–637.

Stahl, W.; Sies, H. Lycopene: A Biologically Important Carotenoid for Humans. *Arch. Biochem. Biophys.* **1996,** *336*, 1–9.

Stalikas, C. D. Extraction, Separation, and Detection Methods for Phenolic Acids and Flavonoids. *J. Sep. Sci.* **2007,** *30*, 3268–3295.

Steinmetz, K. A.; Potter, J. D. Vegetables, Fruit and Cancer Prevention—A Review. *J. Am. Diet. Assoc.* **1996,** *96*, 1027–1039.

Stolarczyk, J.; Janick, J. Carrot: History and Iconography. *Chron. Hortic.* **2011,** *51*, 12–18.

Suman, M.; Kumari, K. A Study on Sensory Evaluation, Beta Carotene Retention and Shelf-Life of Dehydrated Carrot Products. *J. Food Sci. Technol.* **2002,** *39*, 677–681.

Sun, J.; Chu, Y. F.; Wu, X.; Liu, R. H. Antioxidant and Anti-Proliferative Activities of Fruits. *J. Agric. Food Chem.* **2002,** *50*, 7449–7454.

Sun, M. S.; Mihyang, K.; Song, J. B. Cytotoxicity and Quinine Reductase Induced Effects of *Daucus carrot* Leaf Extracts on Human Cells. *Korean Food Sci.* **2001,** *30*, 86–91.

Sun, M.; Temelli, F. Supercritical Carbon Dioxide Extraction of Carotenoids from Carrot Using Canola Oil as a Continuous Co-Solvent. *J. Supercrit. Fluids* **2006,** *37*, 397–408.

Sun, T.; Simon, P. W.; Tanumihardjo, S. A. Antioxidant Phytochemicals and Antioxidant Capacity of Biofortified Carrots (*Daucus carota* L.) of Various Colors. *J. Agric. Food Chem.* **2009,** *57*, 4142–4147.

Surles, R. L.; Weng, N.; Simon, P. W.; Tanumihardjo, S. A. Carotenoid Profiles and Consumer Sensory Evaluation of Specialty Carrots (*Daucus carota* L.) of Various Colors. *J. Agric. Food Chem.* **2004,** *52*(11), 3417–3421.

Swartz, M. E. UPLC: An Introduction and Review. *J. Liq. Chromatogr. Relat. Technol.* **2005,** *28*, 1253–1263.

Sweeney, F. P.; Marsh, A. C. Effect of Processing on Provitamin A in Vegetables. *J. Am. Diet. Assoc.* **1971,** *59*, 238–243.

Takaichi, S. Characterization of Carotenes in a Combination of a C_{18} HPLC Column with Isocratic Elution and Absorption Spectra with a Photodiode-Array Detector. *Photosynth. Res.* **2000,** *65*, 93–99.

Tanaka, T.; Shnimizu, M.; Moriwaki, H. Cancer Chemoprevention by Carotenoids. *Molecules,* **2012,** *17*, 3202–3242.

Temelli, F.; Saldana, M. D. A.; Comin, L. Application of Supercritical Fluid Extraction in Food Processing. In *Comprehensive Sampling and Sample Preparation*, vol. 4; Janusz, P. Ed.; Elsevier: Amsterdam, 2012; pp 415–440.

Terao, J.; Piskuli, M.; Yao, Q. Protective Effect of Epicatechin, Epicatechin Gallate and Quercetin on Lipid Peroxidation in Phospholipid Bilayers, *Arch. Biochem. Biophys.* **1994,** *308*, 278–284.

Thomas, S. C. L. *Vegetables and Fruits: Nutritional and Therapeutic Values*. Taylor and Francis Group, CRC Press: Boca Raton, FL, 2008.

Toor, R. K.; Savage, G. P.; Lister, C. E. Seasonal Variations in the Antioxidant Composition of Greenhouse Grown Tomatoes. *J. Food Compos. Anal.* **2006,** *19*, 1–10.

Torronen, R.; Lehmusaho, M.; Hakkinen, S.; Hanninen, O.; Mykkanen, H. Serum β-Carotene Response to Supplementation with Raw Carrots, Carrot Juice or Purified β-Carotene in Healthy Non-smoking Women. *Nutr. Res.* **1996,** *16*, 565–575.

Touchstone, J. C.; Dobbins, M. F. *Practice of Thin Layer Chromatography*, Wiley: New York, 1978, pp 103–109.

Trombino, S.; Serini, S.; Di Nicuolo, F.; Celleno, L.; Ando, S.; Picci, N.; Calviello, G.; Palozza, P. Antioxidant Effect of Ferulic Acid in Isolated Membranes and Intact Cells: Synergistic Interactions with Alpha-Tocopherol, Beta-Carotene, and Ascorbic Acid. *J. Agric. Food Chem.* **2004,** *52*(8), 2411–2420.

Tsao, R. Chemistry and Biochemistry of Dietary Polyphenols. *Nutrients* **2010,** *2*, 1231–1246.

Tsao, R.; McCallum, J. Chemistry of Flavonoids. In *Fruit and Vegetable Phytochemicals: Chemistry, Nutritional Value and Stability*; de la Rosa, L. A., Alvarez-Parrilla, E., Gonzalez-Aguilar, G. Eds.; Blackwell Publishing: Ames, IA, 2009; pp 131–153.

Uddin, A. S.; Hoque, A. K. Effect of Nutrients on the Yield of Carrot. *Pak. J. Biol. Sci.* **2004,** *7*(8), 1407–1409.

Unlukara, A.; Cemek, B.; Kesmez, D.; Ozturk, A. Carrot (*Daucus carota* L.): Yield and Quality under Saline Conditions. *Anadolu J. Agr. Sci.* **2011,** *26*(1), 51–56.

USDA. *Agriculture Handbook Number 8-11: Composition of Foods: Vegetables and Vegetables Products, Raw, Processed, Prepared.* U.S. Government Printing Office: Washington, DC, 2012.

Valko, M.; Rhodes, C. J.; Moncol, J.; Izakovic, M.; Mazur, M. Free Radicals, Metals and Antioxidants in Oxidative Stress-Induced Cancer. *Chem. Biol. Interact.* **2006,** *160*(1), 1–40.

van den Berg, H.; Faulks, R.; Granado, H. F.; Hirschberg, J.; Olmedilla, B.; Sandmann, G.; Southon, S.; Stahl, W. The Potential for the Improvement of Carotenoid Levels in Foods and the Likely Systemic Effects. *J. Sci. Food Agric.* **2000,** *80*, 880–912.

van Poppel, G.; Goldbohm, R. A. Epidemiological Evidence for Beta Carotene and Cancer Prevention. *Am. J. Clin Nutr.* **1995,** *62*, 1393–1402.

Vasapollo, G.; Longo, L.; Rescio, L.; Ciurlia, L. Innovative SC-CO_2 Extraction of Lycopene from Tomato in Presence of Vegetable Oil as Co-solvent. *J. Supercrit. Fluids* **2004,** *29*, 87–96.

Vasudevan, M.; Gunnam, K. K.; Parle, M. Antinociceptive and Anti-Inflammatory Properties of *Daucus carota* Seed Extract. *J. Health Sci.* **2006,** *52*, 598–606.

Vavilov, N. I. The Origin, Variation, Immunity and Breeding of Cultivated Plants. *Chron. Bot.* **1951,** *13*, 1–366.

Vega, P. J.; Balaban, M. O.; Sims, C. A.; O'Keefe, S. F.; Cornell, J. A. Supercritical Carbon Dioxide Extraction Efficiency for Carotenes from Carrots by RSM. *J. Food Sci.* **1996,** *61*(4), 757–759.

Velazques, D. A. J. *Bioactive Compounds in Carrot Can Be Used as Antiviral for Treatment against Flu.* FEMSA Center of Biotechnology at Technologic of Monterrey, 2013. http://www.news-medical.net/news/20131005/Bioactive-compounds-in-carrot-can-be-used-as-antiviral-for-treatment-against-flu.aspx.

Vickie, A. V.; Elizabeth, W. C. *Essentials of Food Science*. 3rd ed., Springer Science Business Media, LLC: New York, USA, 2007.

Villanueva-Suarez, M. J.; Redonda-Cuenca, A.; Rodriguez-Sevilla, M. D.; Heras, M. Characterization of Non-starch Polysaccharides Content from Different Edible Organs of Some Vegetables, Determined by GC and HPLC: Comparative Study. *J. Agric. Food Chem.* **2003,** *51*, 5950–5955.

Vukasin, B.; Djordje, M.; Slavica, J.; Damir, B. Effect of Cultivar on Carotene and Vitamin C Content in Carrot Root. Fifth Conference on Medicinal and Aromatic Plants of Southeast European Countries, Brno, Czech Republic, 2008, Proceedings CD 04-16:1-4.

Waladkhani, A. R.; Clemens, M. R. Effect of Dietary Phytochemicals on Cancer Development. *Int. J. Mol. Med.* **1998,** *1*, 747–753.

Wang, L.; Weller, C. L. Recent Advances in Extraction of Nutraceuticals from Plants. *Trends Food Sci. Technol.* **2006,** *17*, 300–312.

Wang, Y. M.; van Eys, J. Nutritional Significance of Fructose and Sugar Alcohols. *Annu. Rev. Nutr.*, 1, 437–475.

Webb, A. L.; McCullough, M. L. Dietary Lignans: Potential Role in Cancer Prevention. *Nutr. Cancer* **2005,** *51*(2), 117–131.

Weisburger, J. H. Approaches for Chronic Disease Prevention Based on Current Understanding of Underlying Mechanisms and Discussion. *Am. J. Clin. Nutr.* **2000,** *71*, 1710S–1719S.

Winston, J. C. Phytochemicals: Guardians of Our Health. *J. Am. Diet. Assoc.* **1996,** *5*(3), 6–8.

Woese, K.; Lange, D.; Boess, C.; Bogl, K. W. A Comparison of Organically and Conventionally Grown Foods—Results of a Review of the Relevant Literature. *J. Sci. Food Agric.* **1997,** *74*, 281–293.

Wolf, G. Vitamin A. In *Human Nutrition—A Comprehensive treatise. Nutrition and the adult: micro-nutrients*, vol. 3B; Alfin-Slater, R. B., Kritchevsky, D., Eds.; Plenum Press: New York, 1980; pp 97–203.

Wolterbeek, A. P. M.; Schoevers, E. J.; Bruyntjes, J. P.; Rutten, A. A. J.; Feron, V. J. Benzo[a]Pyrene-Induced Respiratory Tract Cancer in Hamsters Fed a Diet Rich in Beta-Carotene. A Histomorphological Study. *J. Environ. Pathol. Toxicol. Oncology* **1995,** *14*, 35–43.

Wrzodak, A.; Szwejda-Grzybowska, J.; Elkner, K. Comparison of the Nutritional Value and Storage Life of Carrot Roots from Organic and Conventional Cultivation. *Vegetable Crops Res. Bull.* **2012,** *76*, 137–150.

Yates, S. G.; England, R. E.; Kwolex, W. F. Analysis of Carrot Constituents: Myristicin, Falcarinol, and Falcarindiol [Determined by a Sequence of Dichloromethane Extraction, Column Chromatographic Purification, and Gas–Liquid Chromatographic Analysis]. *ACS Symp. Ser. Am. Chem. Soc.* **1983,** 333–344..

Yen, C. C.; Chen, H. H.; Duh, P. D. Extraction and Identification of an Anti-Oxidative Component from Jue Ming Zi (*Cassia tora* L.). *J. Agric. Food Chem.* **1998,** *46*, 820–824.

Yoon, K. Y.; Cha, M.; Shin, S. R.; Kim, K. S. Enzymatic Production of a Soluble Fiber Hydrolysate from Carrot Pomace and its Sugar Composition. *Food Chem.* **2005,** *92*, 151–157.

You, Q.; Wang, B.; Chen, F.; Huang, Z.; Wang, X.; Luo, P. G. Comparison of Anthocyanins and Phenolics in Organically and Conventionally Grown Blueberries in Selected Cultivars. *Food Chem.* **2011,** *125*, 201–208.

Young, I. S.; Woodside, J. V. Antioxidants in Health and Disease. *J. Clin. Pathol.* **2001,** *54*, 176–186.

Young, J. C. Microwave Assisted Extraction of the Fungal Metabolite Ergosterol and Total Fatty Acids. *J. Agric. Food Chem.* **1995,** *43*, 2904–2910.

Young, J. F.; Nielsen, S. E.; Haraldsdóttir, J.; Daneshvar, B.; Lauridsen, S. T.; Knuthsen, P.; Crozier, A. Effect of Fruit Juice Intake on Urinary Quercetin Excretion and Biomarkers of Antioxidative Status. *Am. J. Clin. Nutr.* **1999,** *69*, 87–94.

Yun, T. K. Update from Asia. Asian Studies on Cancer Chemoprevention. *Ann. N.Y. Acad. Sci.* **1999,** *889*, 157–192.

Zaini, R. G.; Brandt, K.; Clench, M. R.; Le Maitre, C. L. Effects of Bioactive Compounds from Carrots (*Daucus carota* L.), Polyacetylenes, Beta-carotene and Lutein on Human Lymphoid Leukaemia Cells. *Anticancer Agents Med. Chem.* **2012,** *12*, 640–652.

Zeisel, S. H.; Mar, M. H.; Howe, J. C.; Holden, J. M. Concentrations of Choline-Containing Compounds and Betaine in Common Foods. *J. Nutr.* **2003,** *133*, 1302–1307.

Zhang, D.; Hamauzee, Y. Phenolic Compounds and Their Antioxidant Properties in Different Tissues of Carrots (*Daucus carota* L.). *Food Agric. Environ.* **2004,** *2*, 95–101.

Zhang, H. F.; Yang, X. H.; Wang, Y. Microwave Assisted Extraction of Secondary Metabolites from Plants: Current Status and Future Directions. *Trends Food Sci. Technol.* **2011,** *22*(12), 672–688.

Zhao, L.; Zhao, G.; Chen, F.; Wang, Z.; Wu, J.; Hu, X. Different Effects of Microwave and Ultrasound on the Stability of Astaxanthin. *J. Agric. Food Chem.* **2006,** *54*(21), 8346–8351.

Zidorn, C.; Jöhrer, K.; Ganzera, B.; Schubert, B.; Sigmund, E. M.; Mader, J.; Greil, R.; Ellmerer, E. P.; Stuppner, H. Polyacetylenes from the Apiaceae Vegetables Carrot, Celery, Fennel, Parsley and Parsnip and their Cytotoxic Activities. *J. Agric. Food Chem.* **2005,** *53*(7), 2518–2523.

Zobel, A. M. Coumarins in Fruit and Vegetables. In *Phytochemistry of Fruit and Vegetables*; Tom, A. S., Barber, A. F. A., Robins, R. J. Eds.; Oxford Science: Oxford, 1997; pp 173–204.

Zweig, F.; Sharma, J. Analytical Methods for Pesticides and Plant Growth Regulators. *Acad. Press Inc.* **1984,** *13*, 126–130.

CHAPTER 5

APPLICATIONS OF PLANT SECONDARY METABOLITES IN FOOD SYSTEMS

JULIO CESAR LOPEZ-ROMERO[1], ROBERTA ANSORENA[2,3], GUSTAVO A. GONZALEZ-AGUILAR[1], HUMBERTO GONZALEZ-RIOS[1], JESUS FERNANDO AYALA-ZAVALA[1*], and MOHAMMED WASIM SIDDIQUI[4]

[1]*Centro de Investigación en Alimentación y Desarrollo, AC, Carretera a la Victoria km. 0.6. Apartado Postal 1735, Hermosillo 83000, Sonora, México*

[2]*Consejo Nacional de Investigaciones Científicas y Técnicas, Argentina*

[3]*Grupo de Investigación en Ingeniería en Alimentos, Facultad de Ingeniería, Universidad Nacional de Mar del Plata. J. B. Justo 4302, Mar del Plata, Buenos Aires, Argentina*

[4]*Department of Food Science and Postharvest Technology, Bihar Agricultural University, Sabour, Bhagalpur 813210, Bihar, India*

**Corresponding author*

CONTENTS

ABSTRACT

Global food production is affected by microbiological and oxidative factors, resulting in raising food waste. In order to reduce the incidence of these factors, food industry use synthetic additives, but this trend is declining due to their association with different diseases. Thus, the food industry is searching for safe and innovative alternatives for consumers, taking into account the consumer preference toward healthy and natural additives. This issue gives a real prospection for using natural additives in food such as plants additives that contain a variety of bioactive secondary metabolites. Also, their activity of prolonging shelf life of foods has been demonstrated by several authors, which reduces the impact of microbiological and oxidative factors. Likewise, these compounds showed other functional and interesting properties such as nutrient supplement, colorant, flavor, sweeteners, and texturizer.

5.1 INTRODUCTION

Food processing is essential for development and maintenance of human life; however, several factors such as processing deficiency, oxidative stress, and microbiological growth reduce the product shelf life (Fung, 2010). In this regard, the Food and Agriculture Organization (FAO) estimated that approximately between 15% and 20% of worldwide food production is wasted, with fruits, vegetables, and tubers the most affected (approximately 50%), followed by seafood (35%), meat (20%) and dairy (20%) products (FAO, 2014).

The main agents that affect the stability and quality of foods are microbial and oxidative factors that may occur in any or all steps of the food chain such as growth, harvest, processing, storage, and distribution (Shahidi & Zhong, 2010; Ye et al., 2013). About the microbiological factor, the food is highly susceptible to microbial contamination, representing an ideal substrate for bacteria and fungi growth, resulting in highly risk food for human consumption. In this way, it is estimated that foodborne diseases cause 128,000 hospitalizations and 3000 deaths every year in the United States (CDC, 2015a). The high incidence of pathogens in foods have been demonstrated in the 24 foodborne outbreaks presented in the United States during 2103–2014, where vegetables, fruits, meat, fish, dairy, and fish products were the most affected (CDC, 2015b).

Food oxidation results from several autocatalytic reactions which carry rancidity development. This process causes functional and nutritional compound alterations as fatty acids damage and oxidize polymer production, which can cause food safety problems (Medina-Meza et al., 2014). Based on the above, the food industry is facing a challenge to provide safe foods to consumers. Accordingly, food industry traditionally used synthetic preservatives trying to reduce microbial and oxidative process in food; however, the use of these preservatives is decreasing due to carcinogenic and teratogenic aspects. With this in mind, the food industry is searching for new safe alternatives for food consumers (Kim et al., 2013).

The increasing demand and preference for natural products by consumers make the plants sources an excellent option for its high secondary metabolite content (phenolic compounds, terpenes and steroids, and alkaloids). These bioactive compounds are organic molecules that do not play an important role in the growth and development of plants, but have important functions as defense mechanisms with metabolic functions over the stress physiology (Vasconsuelo & Boland, 2007; Shan et al., 2009). Therefore, plants can be used to develop extracts and purify compounds. On the other hand, these compounds could provide various biological activities, being used in different areas such as pharmaceutical, agriculture, cosmetic, chemistry, and food (Lopez-Romero et al., 2015). Studies related to the application of these substances in foods demonstrated that secondary metabolites from plants prolonged shelf life in vegetable and animal food retarding microbial spoilage and oxidation (Burt, 2004). Additionally, the incorporation of secondary metabolites in different foods can provide technological benefits for the food industry as coloring, flavoring, sweeteners, or texturizing agents (Gök et al., 2013). Also, secondary metabolites could be considered as nutraceuticals for their bioactive behavior after product technological processing, providing some nutritional and pharmaceutical properties in degenerative illnesses reduction.

5.2 SECONDARY METABOLITES APPLIED IN PLANT FOOD

Plant-derived compounds are mostly secondary metabolites, most of which are phenols or their oxygen-substituted derivatives that can be found in leaves, seeds, flowers, bulbs, and roots.

5.2.1 ANTIMICROBIAL COMPOUNDS

Plant-derived secondary metabolites possess various benefits including antimicrobial properties against pathogenic and spoilage microbes (Hayek et al., 2013). Plant products have also been used since ancient times flavoring food and beverages, and for medicinal purposes. It is estimated that there are 250,000–500,000 plant species on the earth (McChesney et al., 2007) and only one-tenth of these have been explored till date. In the last few years, a number of studies have been conducted in different countries to prove their efficacy (Bhatt & Negi, 2012; Butkhup et al., 2010; Zeng et al., 2012), and thousands of compounds have been isolated from plants, which are claimed to have antimicrobial or medicinal properties. The use of plant extracts with known antimicrobial properties can be of great significance in food preservation. The value of plants lies in some chemical substances that produce a definite action on the microbiological, chemical, and sensory quality of foods, and these phytochemicals have been grouped in several categories including phenolics, phenolic acids, quinones, saponins, flavonoids, tannins, coumarins, terpenoids, and alkaloids (Lai & Roy, 2004). Natural products, such as plant extracts, either as pure compounds or as standardized extracts, provide unlimited opportunities for control of microbial growth owing to their chemical diversity. The structural diversity of compounds is immense, and the impact of antimicrobial action they produce against microorganisms depends on their structural configuration.

The mechanisms of action of active compounds in plant extracts are not completely understood yet. However, there are three aspects on which most authors agree in attributing their inhibitory function (da Cruz Cabral et al., 2013): (1) the presence of OH groups able to form hydrogen bonds that have effects on enzymes, modifying a variety of intracellular functions; (2) action on bacterial and mold morphology due to interactions with membrane enzymes, resulting in the loss of rigidity and integrity of the cell wall; and (3) changes in permeability of cell membranes, granulation of the cytoplasm, and cytoplasmic membrane rupture. All three aspects are intimately interrelated. The hydrophobicity of these compounds leads them to cross cell membrane and interact with cell compounds, so they could affect both membrane and intracellular enzymes. It has been suggested that some hydrophobic compounds present in plant extracts could change the permeability of the microbial membranes for cations such as H^+ and K^+, and so could cause a change in the flow of protons, modifying cell pH and affecting chemical composition of the cells and their activity. The ability of hydrophobic compounds to partition or dissolve in the lipid phase of the cytoplasmic membrane is the key for their activity, but higher solubility does not always mean greater antimicrobial action. The loss of differential permeability of the cytoplasmic membrane is generally considered the cause of cell death because this may result in an imbalance in intracellular osmotic pressure, subsequent disruption of intracellular organelles, leakage of cytoplasmic contents, and finally cell death.

Phenolic compounds probably exert their toxic effects at the level of the membrane, as high correlation between the toxicity and hydrophobicity of different phenolic compounds was observed (Borges et al., 2013). Phenol changes membrane functioning and influences protein–lipid ratios in the membrane and induces efflux of potassium ions (Devi et al., 2010; Oliveira et al., 2015). The catechins have shown to disrupt membrane integrity, as they cause leakage from liposomes (He et al., 2014). Catechins and epigallocatechin gallate interact in the outer polar zone of lipid bilayers in liposomes and cause membrane disruption (Sirk et al., 2008). Terpenes accumulate in the membrane and cause a loss of membrane integrity and dissipation of the proton motive force as well as disruption of the lipid structures (Ultee et al., 2002; Gill & Holley, 2006). Bacterial cell wall lysis also has been reported after treatment with phenolic compounds (Di Pasqua et al., 2007).

The antimicrobials are added to food for two purposes: (1) to control natural spoilage of food and/or (2) to avoid/control contamination by microorganisms. The major targets for such antimicrobials are food-poisoning microorganisms (infective agents and toxin producers) and spoilage microorganisms whose metabolic end products or enzymes cause off-odors, off-flavors, texture problems, and discoloration (Tajkarimi et al., 2010).

The efficacy of an antimicrobial compound depends on the type, genus, species, and strain of the target microorganism, besides the environmental factors such as pH, water activity, temperature, atmospheric composition, and initial microbial load of the food substrate (Negi, 2012). The antimicrobial nature of phytochemical is determined by its chemical properties, such as p*K*a value, hydrophobicity/lipophilic ratios, solubility, and volatility (Stratford & Eklund, 2003). The pH and polarity are the most prominent factors influencing the effectiveness of a food antimicrobial. Therefore, it is very important to know the specific characteristics of the food system that needs to be preserved since a high proportion of lipids could limit the effectiveness of some antimicrobial agents (Owen & Palombo, 2007). Further, hydrophobic properties of some antimicrobial substances can make their dissolution difficult in water limiting their use in foods. The concentration thresholds required for inhibition or inactivation of microorganisms will depend on the specific targets of the antimicrobial substance, including cell wall, cell membrane, metabolic enzymes, protein synthesis, and genetic systems.

Dietary herbs and spices have been traditionally used as food additives throughout the world not only to improve the sensory characteristics of foods but also to extend their shelf life by reducing or eliminating survival of pathogenic bacteria. Many herbs and spice extracts possess antimicrobial activity against a range of bacteria, yeast, and molds (Friedman et al., 2004; Raybaudi-Massilia et al., 2009; Tajkarimi et al., 2010, González Aguilar et al., 2013). A wide variety of phenolic substances derived from herbs and spices possess potent biological activities, which contribute to their preservative potential (Xia et al., 2010). The antimicrobial activity of polyphenols found in fruit, vegetables, and medicinal plants has been extensively investigated against a wide range of microorganisms. The most effective polyphenolic compounds are the flava-3-ols (e.g., catechin), flavonols (e.g., quercetin), and tannins, given their broad spectrum of action and their high antimicrobial power. Some classes of polyphenols have been proposed as alternatives for the development of new food preservatives because of their effectiveness as antimicrobial. In fact, some studies have tested plant extracts rich in polyphenols to ensure safety, extend shelf life, and reduce spoilage of fresh and processed fruit and vegetables, meat and fish products, mayonnaise and juices, as well as factors influencing its effectiveness (Viacava et al., 2013; Gyawali & Ibrahim, 2012; Hayek et al., 2013; Tajkarimi et al., 2010, Alvarez et al., 2013; Ayala-Zavala et al., 2010; Cruz-Valenzuela et al., 2013; Jiménez et al., 2005; Kim et al., 2011; Martín Diana et al., 2008; Vega-Vega et al., 2013).

Isothiocyanates consist of aliphatic and aromatic compounds resulting from the reaction between glucosinolates and the endogenous enzyme myrosinase in cruciferous vegetables (cauliflower, broccoli, and cabbage) that occurs when tissues are damaged (Wilson et al., 2013). It has been reported that these compounds exert inhibitory activity against several fungi and pathogenic and food spoilage bacteria (Davidson et al., 2013; Mari et al., 2008). There are many components that give fruits and vegetables their characteristic flavors. Some of these volatile compounds are aldehydes produced through the lipoxygenase pathway to exert a key role in the defense system of plants against pathogens attack such as hexanal, 2-(*E*)-hexenal, *trans*-2-hexenal, and hexyl acetate. These components have demonstrated antimicrobial activity against spoilage and pathogenic species (Lanciotti et al., 1999; Abanda-Nkpwatt et al., 2006).

Essential oils (EOs) are aromatic oily liquids that can be obtained by various methods from plant material (flowers, leaves, seeds, buds, twigs, bark, herbs, wood, fruits, and roots) and are usually mixtures of several components (Burt, 2004). Their inherent antimicrobial activity is commonly related to the chemical structure of their components, the concentration in which they are present, and their interactions, which can affect their bioactive properties. Volatile oils are a complex mixture of compounds, mainly monoterpenes, sesquiterpenes, and their oxygenated derivatives (alcohols, aldehydes, esters, ethers, ketones, phenols, and oxides). Other volatile compounds include phenylpropenes and specific sulfur- or nitrogen-containing substances. Because antimicrobial activity depends not only on chemical composition but also on lipophilic properties, the potency of functional groups or aqueous solubility and the mixture of compounds with different biochemical properties may increase the efficacy of EOs (Gutierrez et al., 2009). This antimicrobial activity could also vary depending on the type of microorganisms, extraction method, culture medium, size of inoculum, and method of determination (Tajkarimi et al., 2010). Of these, oregano oil, thyme oil, clove and cinnamon oil are some of the most important.

Although *in vitro* screening of plant extracts and EOs is an important first step in identifying potential plants for this purpose, *in vivo* confirmation of activity is essential because food matrices may interact with the bioactive compounds, decreasing their efficacy. A lot of research has been done to test the antimicrobial activity of EOs against a wide range of pathogenic microorganisms *in vitro* and *in vivo* assays during recent years. There is a vast number of foodstuffs where EOs have been applied, namely meat, fish, dairy products, vegetables, rice, and fruits (Burt, 2004; Gyawali & Ibrahim, 2014; Ponce et al., 2004, 2011; Moreira et al., 2005, 2007; Ayala-Zavala et al., 2008, 2009). In general, to obtain the same effect in food products as those observed in *in vitro* assays, higher concentrations of EOs or plant extracts must be utilized. A possible explanation for this is that when they are in contact with the food surface, highly hydrophobic and volatile active substances are bound by food components (carbohydrates, fat, and proteins), while other components are partitioned through the product according to their affinity with water. If this is so, undesirable flavor and sensory changes may occur (Gyawali & Ibrahim, 2014).

The major limitation of antimicrobial compounds derived from plants is the strong flavor they may impart, thus restricting their applicability only to products with compatible flavor. Researchers have proposed different ideas, such as using the plant extract not only as a preservative but also as a flavor component. Alternatively, if the product in which the extracts are incorporated already has strong flavor, it may mask that of the natural antifungal. Several authors have reported that better results can be obtained by incorporating volatile components of EOs in films or edible coatings. The application of antimicrobial coatings presents several advantages over the direct application of natural antimicrobials on food. These coatings can be developed to reduce the diffusion rate from the coated surface inward. In this way, the activity of the antimicrobial agent is maintained at the surface of the food. Thus, smaller amounts of the antimicrobial come into contact with the food compared to other application methods such as immersion or spray (Hygreeva et al., 2014). This results in a smaller impact on the sensory attributes of the food product when they are treated with antimicrobial coatings. In this way, Ayala-Zavala et al. (2013) demonstrated the antimicrobial activity of an edible film formulated with cinnamon leaf oil that can be useful in preserving the quality of fresh-cut peaches. Similarly, Raybaudi-Massilia et al. (2008) reported the reduction of *E. coli* O157:H7 population on fresh-cut Fuji apples with cinnamon, clove, and lemongrass oils and their active compounds, cinnamaldehyde, eugenol, and citral, incorporated into alginate films with no compromise of sensory attributes. In another study, the addition of grapefruit seed extract to the rape seed protein–gelatin film inhibited the growth of *E. coli* O157:H7 and *L. monocytogenes* in strawberries (Jang et al., 2011).

Recently, microencapsulation technology for EOs and plant extracts has received special attention (Donsi et al., 2011; Ayala-Zavala et al., 2014). The use of encapsulation as a delivery system is suggested as an alternative to the direct applications of EOs on food products (Ayala-Zavala et al., 2010; Ansorena, 2015).

5.2.2 ANTIOXIDANT

Lipid oxidation is a deleterious chemical reaction that occurs in foods that renders them inedible. This is a major cause of food quality deterioration and product rejection and can lead to the formation of undesirable off-flavors and off-odors as well as harmful compounds (Decker et al., 2010). In addition to product quality loss due to development of rancid flavor, changes in color and texture and consumer acceptance, there is also nutritive quality losses due to degradation of essential fatty acids and vitamins. Also, there are health risks associated with lipid or oil oxidation due to the formation of toxic compounds when fats and oil undergo oxidative degradation. These oxidation products can cause damage in living organisms as well as mutagenesis and carcinogenesis (e.g., lipid peroxide, malondialdehyde or MDA). Antioxidants are mainly used in food to prevent off-flavors by oxidation of fats, therefore halting their peroxidation in the initiation or propagation phases. There are five types of antioxidants; radical scavengers or chain-breaking antioxidants; chelators, that bind to metals and prevent them from initiating radical formation; quenchers, which deactivate high-energy oxidant species; oxygen scavengers, that remove oxygen from systems, avoiding their destabilization; and finally the antioxidant regenerators, that regenerate other antioxidants when these become radicalized (Karre et al., 2013).

Plant secondary metabolites with antioxidant activity can be obtained from different sources such as fruits, vegetables, herbs, and spices. Spices and herbs are rich sources of phytochemicals with antioxidant capacity (Shan et al., 2005; Srinivasan, 2014). The major active components/phytochemicals responsible for the antioxidant activity of plant derivatives are polyphenols, flavonoids, phenolic diterpenes, carotenoids, and tannins (Zhang et al., 2010).

Phenolics render antioxidant activity mainly due to their role as reducing agents, hydrogen donors, and singlet oxygen quenchers. Some phenolics also have the ability to chelate metal ions which act as catalysts in oxidation reactions. Flavonoids are natural polyhydroxylated aromatic compounds that are widely distributed in plants and have the ability to scavenge free radicals, including hydroxyl, peroxyl, and superoxide radicals and can form complexes with catalytic metal ions rendering them inactive. It has also been found that flavonoids can inhibit lipoxygenase and cyclooxygenase enzymes, the enzymes responsible for development of oxidative rancidity in foods.

The main foods where antioxidant compounds derived from plant extracts are used are meats, oils, fried foods, dressings, dairy products, baked goods, and extruded snacks (Hygreeva et al., 2014; Baines & Seal, 2012). Polyphenols are some of the most interesting groups of natural compounds and due to their strong antioxidant capacity they display interesting effects toward human health, namely, against cancer, osteoporosis, cataracts, cardiovascular dysfunctions, brain diseases, and immunological conditions (Embuscado, 2015). Due to their high efficacy in preserving food and their wide acceptance from the general public, it is desirable to add them to food. In all the classes of polyphenols (phenolic acids—hydroxybenzoic or hydroxycinnamic acids, flavonoids including anthocyanins, tannins, lignans, stilbenes, and coumarins), some stand out with higher potential than others. Plant extracts rich in polyphenols such as grape seed, cocoa leaves, broccoli, and green tea have shown antioxidant activity in meat and meat products (Zuo et al., 2002). They can be added as plant extracts, taking advantage of the synergistic effects between compounds, or by further purification to individual molecules, adding the most bioactives to the food system. Polyphenolic extracts like rosemary and other extracts from plants have been used to act as antioxidants in food, in terms of rosemary, it has been identified as a food additive by the Council Regulation (EC) 1129/2011, with the number E392. The role of herbs and spice extracts, including rosemary, oregano, clove, thyme, and citrus fruits have been studied for their antioxidant potential in cooked, fermented, and irradiated meat products (Hygreeva et al., 2014; Rodríguez Vaquero et al., 2010). Although the synergistic effects between the compounds are important for the extract's antioxidant activity, some industries seek specific molecules to carry out these effects.

Carnosic acid, a hydroxybenzoic acid derivative, is a known constituent of rosemary extract and is believed to have the most important antioxidant effect in it. It is used in oils, animal fats, sauces, bakery wares, meat patties (between 22.5 and 130 ppm for the meat patties) and fish, among others (Naveena et al., 2013). Ferulic acid, a hydroxycinnamic acid, is also used in the food industry as an antioxidant and a precursor of other preservatives, as well as in food gels and edible films (Kumar & Pruthi, 2014; Ou & Kwon, 2004). Catechin, a widely known flavon-3-ol, is also known for its antioxidant activity. It can be directly added to food, joined with other natural substances and even encapsulated to promote and extend its effects (Kaewprachu et al., 2015). Regarding other compounds with potential antioxidant activity, we could note that ascorbic acid, also known as vitamin C, is a high oxygen scavenger used in various foodstuffs. It regenerates phenolic oxidants and tocopherols that have suffered oxidation, due to its high oxidation potential. Ascorbic acid is particularly important to stabilize lipids and oils, but can be used in other matrices. In 2015, the European Food Safety Authority (EFSA) gathered a scientific opinion regarding ascorbic acid and determined that there was no risk in its consumption. Carotenoids are also known for their antioxidant potential as food additives, although their use is always limited by being very susceptible to oxidation by light exposure. Lycopene (E-160d) is the most abundant carotenoid, found mainly in tomatoes, although it is not widely used as a food antioxidant. On the other hand, β-carotene is used in baked goods, eggs, and dairy products, among others, as a singlet oxygen quencher (Smith & Hong-Shum, 2011). In many foodstuffs that use carotenes, ascorbic acid, or vitamin E (tocopherols) are used to benefit from synergies. Carotene mixes and β-carotene have been reviewed by the EFSA's scientific panel that ruled out any toxicity arising from its consumption (EFSA, 2012). Tocopherols, which are the building blocks of vitamin E, are also known as very strong antioxidants. Their main antioxidant function is by terminating free radicals in autoxidation reactions (Smith & Hong-Shum, 2011). In some cases, tocopherols are used in films and coatings (Barbosa-Pereira et al., 2013; Lin & Pascall, 2014), although they can be used as an additive as well (E- 306 to E-309). These compounds have been used in bacon (300 mg/kg), meats, dairy products, and oils, among others (Smith & Hong-Shum, 2011; Wang et al., 2015).

5.2.3 COLORANTS

Color is one of the most important quality attributes for the food industry. The colorants are used to enhance existing colors that can be lost either during the manufacture or over the shelf life, or even to attribute new ones to it. Consequently, there is a technological need for coloring food in order to restore its initial appearance, compensate for fluctuating quality of raw material and batch-to-batch variations, reinforce color to meet consumer expectations, protect photolabile vitamins and aromas and finally color food which otherwise would be colorless for reasons of visual appeal (Stintzing & Carle, 2004). While synthetic pigments are increasingly rejected by the consumer and are supposed to be unwholesome, proven or not, the acceptance of natural or nature-derived alternatives is promoted by their psychological comprehension of being healthy and of good quality (Stintzing & Carle, 2004).

Ever since, natural colors from spices and herbs, fruits and vegetables have been part of the everyday diet of humans. The most common plant pigments are carotenoids, chlorophylls, anthocyanins and, betalains. Whereas the first are located in specialized plastids, the later are deposited in the vacuole. Annatto is a permitted natural food colorant, extracted from the *Bixa orellana* L. tree, with the E number E160b. The main constituents of the annatto mixture are the carotenoids bixin and norbixin, which display a yellow to orange coloration. There are many foodstuffs where annatto is used, cakes (from 250 to 1000 mg/kg of dough), biscuits, rice, dairy products, flour, fish, soft drinks, snacks, and meat products (Rao et al., 2005; Scotter, 2009). Paprika is another mixture of two carotenoids, capsanthin and capsorubin, which is also approved in the EU (E160c) and displays an orange to red color (Schrader et al., 2013). The main applications of carotenoids in food are related to sauces, marinades, spice blends, coatings, beverages, milk, among others (Baker & Günther, 2004; Baines & Seal, 2012).

Anthocyanins (E163) are pigments in nature, namely, red, purple, violet, and blue and can be transposed to food when they are used as colorants. The main anthocyanins in nature are cyanidin, delphinidin, malvinidin, pelargonidin, peonidin, and petunidin, and have their main applications in soft drinks, confectionary products, and fruit preparations (Wu et al., 2006; Baines & Seal, 2012). Anthocyanins in black carrot (*Daucus carota* L.) extracts are accompanied by a large number of phenolic acids, especially the derivatives of hydroxycinnamic acids (Day et al., 2009). Most of them are monoacylquinic acids, composed of caffeic, *p*-coumaric or ferulic acids, and a molecule of quinic acid. Chlorogenic acid (5-caffeoylquinic acid, 5-CQA), an ester of caffeic and quinic acid, was found to be the predominant compound (Kammerer et al., 2004). The presence of a large number of phenolic acids contributes significantly to the stability of black carrot polyphenolic complex through intermolecular or intramolecular co-pigmentation effect (Rein & Heinonen, 2004). Concentrates obtained from black carrot enjoy increasing popularity as a natural food colorant in soft drinks and dried pasta (Day et al., 2009). The anthocyanin complex in black carrot consists of five major cyanidin-based (about 97%

of total anthocyanin) and minor peonidin- and pelargonidin-based (about 3% of total anthocyanin) pigments, as well as a rich mixture of at least 46 cinnamic acids (Kammerer et al., 2004). Black carrot polyphenolics provide an intense and relatively stable red color to food products due to the presence of anthocyanins with acetylated substituted molecular structure.

Other very similar compounds to anthocyanins are betalains, which display colors ranging from red-violet (betacyanins) to yellow-orange (betaxanthins). They are not exhaustively studied like anthocyanins, but still have some applicability in the food industry as natural colors due to having three times more coloring strength than anthocyanins (Stintzing & Carle, 2004; Moreno et al., 2008). The only betalain legislated for use is derived from beetroot (E 162-betanin), and has application in dairy products, meat products, and many others (Delgado-Vargas et al., 2000). Chlorophylls (E 140) are vegetable pigments that occur naturally in plants and confer color. Among the five different chlorophylls that exist, only two (*a* and *b*) are used in the food industry as colorants. Their complex structure is difficult to stabilize; this is the main drawback of their use in the industry, which has studied mechanisms of retaining or replacing the magnesium ion within the structure. The used commercial colorants of chlorophylls are extracted from alfalfa, and have been employed in dairy products, soups, drinks, and sugar confections (MacDougall, 2002). Curcumin (E 100), a pigment purified from turmeric which is extracted from the dried rhizomes of the plant *Curcuma longa* L., is another widespread used food colorant. It confers orange color to food, and is used in mustard, yoghurt, baked goods, dairy industry, ice creams, and salad dressings (MacDougall, 2002).

5.2.4 SWEETENERS AND FLAVORINGS AGENTS

Sweeteners have been used for centuries to make foodstuffs more appetizing and appealing to consumers. Natural sweeteners can be divided into two groups: bulk sweeteners and high-potency sweeteners. Sweeteners that belong to the first group have a potency of one or less sucrose molecule (sucrose is the international standard for sweetness), while the latter have higher sweetness than that of one sucrose molecule. For natural sweeteners, to be considered viable to be introduced in the markets and widely used, they need to have a good taste, be safe, and have a high solubility, a high stability, and an acceptable cost-on-use (Baines & Seal, 2012). Regarding the bulk sweeteners, the two main compounds of this group are erythritol and tagatose. Erythritol (E 968), a sugar alcohol (polyol), which occurs naturally in some fruits and vegetables, is used in baked goods, coatings, frostings, fermented milk, chocolate, low-calorie beverages, candy, chewing gums, among others (O'Brien-Nabors, 2001; Baines & Seal, 2012). Tagatose is a ketohexose, an enantiomer of fructose, and is also considered a prebiotic and a flavor enhancer. It occurs in very small quantities in fruits and heat-treated dairy products. The applications of tagatose in the food industry encompass cereals, beverages, yoghurts, frostings, chewing gum, chocolate, fudge, caramel, fondant, and ice cream (Dobbs & Bell, 2010; Baines & Seal, 2012).

Steviol glycosides (E 960) are an example of natural compounds with a high dissemination around the world. These glycosides, mainly steviosides and rebaudiosides are also known as stevia, stevioside, or steviol, and are purified from the plant *Stevia rebaudiana* Bertoni. Owing to various compounds in its formula, steviol glycosides have different potency, with the lowest ones being 30 times sweeter than sucrose (dulcoside A, rebaudioside C) and others which are about 300 times more potent (rebaudioside A). Their production relies on the harvest from the herb, which fostered a high production of these plants, especially in China. Steviol glycosides have been approved as a sweetener in many countries, including the EU and the USA, with great results regarding toxicity, cariogenicity, carcinogenicity, and allergic reactions. Among the food industry, steviol glycosides are used in beverages, dairy products, ice cream, frozen desserts, sugar-free confectionary, mints, dried sea-foods, and sauces.

Another high potency sweetener is glycyrrhizin (E 958) (Barclay et al., 2014), a triterpene glycoside extracted from *Glycyrrhiza glabra* L., the liquorice plant. This compound, also known as glycyrrhizic acid can act as a sweetener with a potency of 50 times sweeter than sucrose, and also as a foaming agent and flavor enhancer. This compound is legally used in the USA and the EU under the form of monoammonium glycyrrhizinate and ammoniated glycyrrhizin. It is manly used in liquorice, baked goods, frozen dairy products, beverages, confectionery, and chewing gum (Spillane, 2006; Baines & Seal, 2012). Thaumatin, a mixture of five proteins (taumatin I, I, III, a and b), is also used as a sweetener in many countries. Thaumatin is extracted from the fruit of *Thaumatococcus daniellii* Benth, a plant native to Africa. Owing to its liquorice cool aftertaste, thaumatin is not used at high quantities, although it is very useful to mix with other sweeteners to confer umami taste and to reduce bitterness in foodstuffs. The main foods where it is employed as either a sweetener or flavor enhancer are sauces, soups, fruit juices, poultry, egg products, chewing gum, processed vegetables, among others (O'Brien-Nabors, 2001; Baines & Seal, 2012). There are some other natural sweeteners that could be used in the future, but do not have any applications in foodstuffs today because of their scarcity and poor yields when isolated from plant matrices. Examples of these compounds are monatin and brazzein.

5.2.5 NUTRIENT SUPPLEMENT

The incorporation of bioactive plant extracts to provide bioactive functionality to food has aroused a great interest. Nevertheless, since these compounds have low stability under typical storage and end-use conditions, introducing plant extracts is still a challenge. Some research has been done to incorporate bioactive phytochemicals and plant extracts into foods to promote health. In this regard, El-Said et al. (2014) studied yoghurt fortified with pomegranate peel extracts and Chouchouli et al. (2013) studied the fortification of full-fat and non-fat yoghurts with grape seed extracts from two grape varieties (Moschofilero and Agiorgitiko). Grape seed extracts were successfully employed for the production of polyphenol-fortified yoghurts. Yoghurt with added grape seed polyphenols may be a convenient food format to satisfy consumer interest in original yoghurt nutrients, beneficiary effects of starter cultures, and health benefits of added polyphenols.

5.3 APPLICATION OF SECONDARY METABOLITES IN ANIMAL ORIGIN FOODS

5.3.1 PRESERVATIVE

Animal origin foods contain naturally high proteins, lipids, and minerals, as well as high water activity values and near neutral pH. In general, these conditions make the foods highly susceptible to oxidative process and contamination by pathogenic microorganisms. In this sense, natural additives (secondary metabolites) as preservatives are used for the research area to try to reduce the impact of these factors in the animal origin food quality and safety.

Regarding the incorporation of secondary metabolites as preservatives in meat and meat products, Ali et al. (2010), Khare et al. (2014), Mariem et al. (2014), and Bazargani-Gilani et al. (2015) analyzed the effect of propolis (flavonoids and phenolic acids), eugenol, *Nitraria retusa* fruit extracts (phenolic compounds, flavonoids, anthocyanins), and pomegranate juice (anthocyanins, phenolic acids, and flavanols) in oriental sausage, chicken noodles, ground beef patties, and chicken breast, respectively. Generally, in the microbial (total viable count, psychrophilic bacteria, coliform bacteria, and yeasts-molds) and oxidative evaluations (thiobarbituric acid, peroxide values, volatile bases nitrogen and carbonyl content) a significant reduction in both parameters was observed by the incorporation of natural preservatives compared with control treatments during the shelf life. In the same regard, Amalaradjou et al. (2010), Bukvički et al. (2014), and Vijayakumar and Wolf-Hall (2002) evaluated the antimicrobial effect of *trans*-cinnamaldehyde, *Satureja horvatii* (*p*-cymene, thymol, and thymol methyl ether), and green tea (polyphenols) added in ground beef patties, ground pork, and ground turkey inoculated with *L. monocytogenes*, *E. coli* O157:H7, and *L. monocytogenes*, respectively. The findings of these researches evidenced the antimicrobial effect of evaluated additives, because the bacterial growth of foodborne pathogens inoculated in different meats was affected by the incorporation of natural preservatives, having a parallel behavior doses-reduction, increasing antimicrobial effect by the natural compound concentration; additionally, in some cases a complete inhibition of growth was observed.

Mahmoud et al. (2006), Ojagh et al. (2010), and Zarei et al. (2015) evaluated the antioxidant and antimicrobial effect of carvacrol and thymol, orange, and cinnamon EOs (cinnamaldehyde, β-caryophyllene, linalool, and other terpenes), and pomegranate encapsulated in chitosan, and subsequently incorporated in carp fillets and rainbow trout fillets to be stored at different time and temperature conditions. All evaluated preservative treatments showed higher antioxidant effect compared to the control, having lower oxidation values in the different determinations, such as volatile bases nitrogen, peroxide value, thiobarbituric acid, which are the principal indicators of meat deterioration. On the other hand, the microbial growth is the principal factor of food spoilage; in this way, it was observed that preservatives analyzed showed significant reduction in the total microbial viable, mesophilic, and psychotropic count compared with the control solution after the evaluation condition treatment. On the same regard, other natural preservatives such as carvacrol, citral, geraniol, cinnamaldehyde, and thyme oil (thymol) showed antimicrobial activity causing reduction growth of *Salmonella typhimurium* and *Pseudomonas putida* inoculated in cooked shrimps and red grouper fillets, respectively, at different condition (Ouattara et al., 2001; Kim et al., 1995). Moreover, Luther et al. (2007) evaluated the oxidative stability index and fatty acid stability of fish oil added with Chardonnay and black raspberry extracts (phenolic compounds). The antioxidant treatments retarded lipid oxidation and rancidity of fish oils. Also, antioxidant treatments decreased the deterioration of polyunsaturated fatty acids (PUFA) in fish oil under oxidative conditions.

On the other hand, Tseng and Zhao (2013) observed that yoghurt added with pomace powder (phenolic compounds and condensed tannins) stored for 3 weeks exhibited less peroxide values during the shelf life, and it was observed that antioxidant activity increased by increased extract concentrations (7, 3, 2.6, and 2.4 milliequivalent peroxide/1000 g sample to 1%, 2%, and 3% of extract, respectively). Similarly, Asensio et al. (2015) reported that Compacto, Cordobes, Criollo, and Mendocino EOs (*trans*-Sabinene hydrate, thymol and Sabinene, majority compounds), and thymol added in organic cottage cheese presented lower hydroperoxides and conjugated dienes values, also, showed lower saturated/unsaturated fatty acids ratios compared to the control. In the same way, Shan et al. (2011) analyzed the antioxidant and antibacterial potential of different extracts such as cinnamon (cinnamaldehyde), oregano (thymol and carvacrol), clove (eugenol), pomegranate, and grape (phenolic compounds) incorporated in cheese. The findings showed antioxidant potential of extracts, because the preservative treatments presented a protective effect against lipid oxidation. Moreover, the preservatives were effective against growth of *L. monocytogenes, Staphylococcus aureus,* and *S. enterica* inoculated in cheese. On the other hand, it was observed that carvacrol and mint oil presented antimicrobial effect against *L. monocytogenes* and *S. enteritidis* inoculated in milk and tzatziki, achieving to reduce the population of inoculated pathogens (Tassou et al., 1995; Karatzas et al., 2001).

The obtained results in these research evidenced the antioxidant and antimicrobial effect of secondary metabolites present in naturals extracts.

Secondary metabolites decreased the deterioration process (oxidative and microbiological) and extended the shelf life of different animal origin food matrices stored at different conditions. Thus, the secondary metabolites presented an excellent option as natural food preservatives. Regarding the antimicrobial effects of secondary metabolites from plants it has been observed that bioactive compounds induce changes in the hydrophobicity causing disruption of cell wall and cytoplasmic membrane, promoting lysis and leakage of intracellular compounds (Borges et al., 2013; Lopez-Romero et al., 2015). Otherwise, the antioxidant mechanism is based on their ability to donate an electron or proton from functional groups and the functional groups position in the molecule (Perron & Brumaghim, 2009).

5.3.2 NUTRIENT SUPPLEMENT

The animal origin foods are essential in the human diet; however, some consumers perceive these foods as unhealthy because they contain high animal fat, cholesterol, and synthetic additives, which are associated with chronic degenerative diseases (Serrano et al., 2007; Hygreeva et al., 2014). Accordingly, the new consumer tendency is the preference for healthier food, such as, incorporation of bioactive or functional compounds, low fat level, and substitution or addition of fatty acids, which may provide healthy beneficial effects to consumers. In this regard, the food technologist is evaluating the incorporation of natural extracts in different food matrices.

Ribas-Agustí et al. (2014) evaluated the phenolic compounds stability of cocoa and grape extracts incorporated in dry fermented salchichon and fuet pork. The result showed that phenolic compounds presented a reduction at the final of aging process; however, the bioactive compounds values did not decreased significantly. Generally, the phenolic compounds stability was not affected for the sausages type. Also, the phenolic compounds concentration was at least 50% of their initial content at the end of shelf life. On the other hand, gallic acid, galloylated flavan-3-ols, oligomeric flavan-3-ols, glycosylated flavonols were the majority compounds in the evaluation process. Similarly, Fernández-López et al. (2007) tested the incorporation of orange by-products at different concentrations in dry-cured sausages. The main compound present in the orange extracts was hesperidin, and it was observed that this flavonoid presented stability after the incorporation in the dry-cured sausages over time.

Vural and Javidipour (2002) and Javidipour et al. (2005) studied the replacement of beef fat in frankfurters and salami by inter-esterified vegetable oils (palm, cotton, and olive). The incorporation of vegetable oils cause significant changes in the fat content and fatty acid profile, saturated-to-insaturated (SFA/UFA) and polyunsaturated-to-saturated (PUFA/SFA) ratio of beef frankfurters. For example, the vegetable oils decreased the fat content and inversely increased the oleic and linoleic acid content, also decreased the ratio of SFA/UFA and heighten the PUFA/SFA ratio, compared to the control (incorporated only with beef fat). In the same way, Gök et al. (2011) obtained that meat burgers incorporated with poppy seeds at different concentrations showed higher protein and lower fat content compared to the control. Also, it was observed that PUFA (linoleic and linolenic) content increased with the incorporation of poppy seeds. Added to this, the poppy treatment reduced between 25% and 89% cholesterol content in the analyzed burgers. Moreover, Valencia et al. (2008) evaluated the incorporation of linseed oil, green tea catechin, and green coffee in cooked pork sausages. The findings showed that evaluated treatments decreased the SFA, MUFA, and TRANS compositions, otherwise, the PUFA values increased specifically the n-3 fatty acids such as α-linoleic and eicosapentaenoic. Also the PUFA/SFA and MUFA + PUFA/SFA ratio increased in the treatments incorporated with natural additives.

Regarding the dairy products, Han et al. (2011a) developed a functional cheese product incorporating polyphenolic compounds. The polyphenolic compounds added in cheese as functional ingredients presented high retention coefficient values, where cranberry powder, tannic acid, hesperetin, and flavone presented higher stability. Added to this, the incorporated compounds are present in the final product, indicating that cheese elaboration process does not affect the phenolic compounds content. Similarly, Helal et al. (2015) elaborated enriched cheese with polyphenols. Initially the phenolic compounds showed high retention coefficient in curd, where catechin and tannic acid showed superior concentrations. Moreover, the phenolic compounds presented stability after the gastric and pancreatic digestion simulation. For example, after the gastric digestion, catechin and chlorogenic acid showed increased concentration, and after the pancreatic digestion, tannic, ferulic, 3,4-dihydroxyphenylacetic, and vanillic acids were the most stable. Added to this, the polyphenols increased the antioxidant activity (ABTS assay) after the digestion process. In a study by Rashidinejad et al. (2015), the addition of catechin in cheese increased the total phenolic compounds concentration in *in vitro* digestion during 90 days, also, catechin treatments presented higher antioxidant activity (FRAP, ORAC) through time and this activity increased as the concentration of catechin increased. On the other hand, Reddy et al. (2005) evaluated the *in vivo* effect (human) of consumption of green tea (catechins) added with milk. The results showed that catechins were present in plasma after the consumption and this reduced the plasma and urinary TBARS values. The findings suggest that beverage evaluated could prevent oxidative damage *in vivo*.

Karaaslan et al. (2011) and Tseng and Zhao (2013) tested the incorporation of grape extracts, grape pomace, and oleuropein as functional ingredients in different yoghurts stored at 4°C for 14, 21, and 35 days, respectively. All natural additives showed stability during the shelf life. For example, in yoghurts incorporated with grape extracts at 14 days presented total anthocyanin concentration between 2.2 and 10 mg malvidin-3-glucoside equivalents/kg, and the main bioactive compounds present were catechin (29.5 mg/kg), epicatechin (15.1 mg/kg), gallic acid (6.7 mg/kg), and coumaric acid (2.5 mg/kg). Yoghurts added with grape pomace showed at 21 storage days the total phenolic compounds concentration between 600 and 1000 mg gallic acid equivalents/kg, depending on the grape pomace treatment.

Generally evident is that bioactive compounds from plants could be incorporated as functional ingredients in the elaboration of different animal origin foods to offer novel functional foods to consumer. Added to this, the secondary metabolites can show beneficial effects to the consumer and prevent oxidative damage in humans.

5.3.3 COLORANT

Color is one of the principal attributes in the food product acceptability; for this reason the food industry utilized synthetic colorants to improve the appearance of foods; however, the use of these colorants is controversial, decreasing their application (Pazmiño-Duran et al., 2001). In this way, the worldwide tendency in the food industry and consumers toward the consumption of natural products opens the possibility of applying secondary metabolites from plants as natural colorants on meat products.

Calvo et al. (2008) evaluated the color properties of dry-fermented sausages incorporated with lycopene. The lycopene treatments cause changes in the color parameters. Generally, the lycopene treatments present highest values of redness (a^*), yellowness (b^*), hue angle, and saturation index in comparison to the control that presents higher lightness (L^*) values. In a study performed by Zhou et al. (2012), it was observed that *Amaranthus* pigments (0.1, 0.2, and 0.3%) added in pork sausages caused color changes. The control and nitrite treatments (control) presented higher L^* and b^* values and lower a^* values for the presence of anthocyanins and betalains in *Amaranthus* pigments. In the sensory evaluation, the pigment at 0.3% showed higher acceptation than nitrite treatments. Generally, the *Amaranthus* pigment treatments present greater acceptability than control. On the other hand, Gómez et al. (2008) reported that addition of paprika and rosemary in chorizo promotes higher redness degree values based in the red-green component, a^*, and spectrophotometric color. Added to this, the redness values are maintained until the end of shelf life.

The substitution of nitrate for natural colorants (curcumin, β-carotene, paprika, betanin) in frankfurters was analyzed by Bloukas et al. (1999). The natural colorants presented effect in the color attributes, for example, the addition of colorants showed lower L^* values than control, except curcumin treatment. Inversely, the colorant treatments showed higher a^* and b^* values compared to the control, where β-carotene, paprika, betanin presented the superior values after 4 weeks of storage. At the same time, the natural colorants showed higher stability after exposure to artificial light during 2 and 4 h, demonstrating that natural colorant are stable to the exposure light. Regarding the temperature (70 and 117°C) stability of natural colorants, it was observed that thermic process affected the color parameters causing an increase in the L^* and b^* values and decreased a^* values, where betanin and β-carotene were the most affected. Moreover, in consumer preference according to the color, the frankfurters added with betanin were the most acceptable followed by paprika treatments. Similarly, the incorporation of paprika and tomato paste in nitrite reduced pork meat batters analyzed. Generally, the natural colorant causes diminution in the L^* parameter and inversely increases a^*, b^*, Chroma, and hue angle values, specifically the paprika treatments. Additionally in the sensory color evaluation, the paprika treatments obtained higher acceptability in the color acceptability evaluation. This behavior is attributed to secondary metabolites present in the colorants such as β-carotene, zeaxanthin, capsanthin, and capsobirum in paprika, and β-carotene and lycopene in tomato paste (Bázan-Lugo et al., 2012).

Regarding fish products, Lauro (2000) and Park (2008) mentioned that surimi industry actually applied natural colorants for the consumer demand, for this reason the surimi industry utilized vegetable colorants such as paprika, lycopene, annatto, grape, and beet juice extracts to obtain dark-red (blood-red) and dark brown-red (orange-red) colors in the final products. On the other hand, Al-Bulushi et al. (2013) observed that fish sausages incorporated with natural vegetables additives such as garlic, onion, ginger, cumin, cinnamon, white pepper caused a diminution in a^* values and higher L^* and b^* values compared to the control, and this behavior was observed during the 12 storage weeks. In the same way, Al-Bulushi et al. (2005) reported similar color characteristics in fish hamburgers elaborated with similar ingredients to those mentioned above.

Regarding dairy products, Wallace and Giusti (2008) and Karaaslan et al. (2011) evaluated the color stability of yoghurt added with Peruvian berry and callus extracts as colorants. The addition of 20 mg of Peruvian berry powder and 100 mL of callus extract showed similar color values ($L^* = 65.5$, $a^* = 9.7$, $b^* = -3.6$, Chroma = 11.5, hue angle = 329.4 and $L^* = 88.4$, $a^* = -0.1$, $b^* = 8.9$, Chroma = 8.9, hue angle = −1.5 to the Peruvian berry and callus yoghurts, respectively) to the control yoghurts ($L^* = 65.4$, $a^* = 10$, $b^* = -3.47$, Chroma = 10.6 and hue angle = 341; $L^* = 88.41$, $a^* = -0.88$, $b^* = 9.61$, Chroma = 9.6 and hue angle = −1.4 to the Peruvian berry and callus, respectively). Additionally anthocyanins present (nine anthocyanins) in the Peruvian berry powder provided stability during the shelf life (60 days) and the major anthocyanins were malvidin-3-glucoside, petunidin-3-glucoside, delphinidin-3-glucoside, and cyaniding-3-glucoside. Regarding the callus extract treatment, the color sensory evaluation presented similar behavior to the control, and the major phenolic compounds in yoghurt were catechin, epicatechin, gallic acid, and coumaric acid.

5.3.4 FLAVORS

Flavor is a major sensory attribute that influences the perception of foods by consumers and food processing industry, and this attribute is related with the antioxidant effect of preservatives, as they prevent the formation of oxidation products which affect the sensory attributes and quality of foods, and it is considered that flavor parameter determines the consumer acceptation or rejection (Mohamed & Mansour, 2012).

Mohamed and Mansour (2012) and Bazargani-Gilani et al. (2015) evaluated the impact on the flavor of beef patties and chicken meat incorporated with rosemary, marjoram, and pomegranate. The beef patties added with rosemary and marjoram did not show statistical differences with the control during 3 storage months. Added to this, the phytochemical treatments presented greater acceptance by the panelist compared to the control. Similarly, the chicken meat incorporated with pomegranate presented higher acceptability by the panelist. In the same way, different flavor parameters of salchichon and fuet incorporated with cocoa EOs were evaluated. The flavor parameters such as sweetness, saltiness, acid taste, bitterness, piquantness, and ripeness were not altered by the incorporation of cocoa as the added treatments exhibited similar behavior to the control (Ribas-Agustí et al., 2014).

Maqsood et al. (2012) observed that tannic acid incorporated in fish emulsion sausages showed similar behavior to the control in the flavor evaluation. Yasin and Abou-Taleb (2007) analyzed the flavor changes during the shelf life of mullet fillets added with thyme and marjoram (carvacrol majority compounds in both additives). The finding showed that concentration of 0.5% of additives maintained acceptable fillets during 12 days, while control treatments presented unacceptable values after 8 evaluation days.

Gad et al. (2010) tested date fruit extract in flavor characteristic of yoghurt. The results showed that concentration of 2–10% trended to show superior values, however, were identical to the control. Similarly, Hala et al. (2010) found that higher (2–5%) concentration of rosemary extracts incorporated in soft cheese increased the flavor acceptability values in the sensory evaluation during the shelf life (30 days), and presented greater acceptability than control treatment.

5.3.4 TEXTURIZER

Texture parameter is a fundamental quality characteristic that determines food acceptance and preference by consumers (Piqueras-Fiszman & Spence, 2012). Accordingly, the food industry focuses attention on this parameter analyzing the incorporation of natural sources that do not affect the food acceptability and also provide a healthy alternative for consumers.

Viuda-Martos et al. (2010) observed that incorporation of oregano EO and orange fiber in bologna under air condition did not affect the textural properties during the shelf life (24 days), for example, the hardness, gumminess, and chewiness values were similar to the control. Similarly, de Oliveira et al. (2015) tested the effect of eugenol in textural parameters of turkey breast ham reduced in NaCl. The hardness, cohesiveness, adhesiveness, springiness, and chewiness values were statistically similar in eugenol and control treatments in the evaluation days (60 days). In the same way, Mercadante et al. (2010) evaluated the addition of lycopene and β-carotene in raw and cooked sausages hardness after 45 storage days. The sausages did not present texture differences before and after cooking, and treatments added with lycopene and β-carotene are statistically similar to the control in hardness evaluation.

Regarding fish products, addition of oregano EOs in raw sea bream under modified atmosphere packaging showed high acceptability values in the sensory evaluation. For example, treatment with 0.8% of oregano EOs presented acceptability values after 33 evaluations days, contrary to the control that after 24 days showed unacceptable values (Goulas & Kontominas, 2007). In the same sense, Maqsood et al. (2012) observed that tannin acid and kiam wood extract (tannic acid majority compound) added in fish emulsion sausages presented similar values to the control in the hardness, springiness, cohesiveness, gumminess, and chewiness parameters in the shelf life. Studies performed by Karaaslan et al. (2011) and El-Said et al. (2014) evaluated the sensory properties of yoghurt incorporated with callus extracts and pomegranate peel extracts. The texture evaluation showed that different concentrations of pomegranate (5, 10, 20, and 30%) and callus (100 mL) extracts were similar to the control. Han et al. (2011b) observed that addition of different phenolic compounds and extracts such as catechin, epicatechin, tannin acid, homovanillic acid, hesperetin, flavone, grape, green tea, and cranberry extracts to cheese did not affect the physical texture properties of final cheese.

The addition of secondary metabolites present in plant extracts represents an excellent alternative for the animal food industry because these compounds do no affect the textural properties after the incorporation and demonstrate stability during shelf life This suggests that secondary metabolites provide stability protein as protein alterations are related to textural loss in food (Mercadante et al., 2010). Therefore, these additives represent a healthy option for the food industry and consumers.

5.3.5 REGULATION

During the last few years, a number of EOs and their major components have been studied for their safety limit profile by using different toxicity testing methods (Burt, 2004). Some of the EOs and their major components thus are kept in "GRAS" category by US Code of Federal Regulations. Based on the code of Federal Regulation, EOs of cinnamon, clove, lemon grass, oregano, thyme, nutmeg, basil, and others, are generally recognized as safe (GRAS) in the United States (US Code of Federal Regulations, 2013). Carvacrol, carvone, cinnamaldehyde, citral, *p*-cymene, eugenol, limonene, menthol, linalool, vanillin, citral, and thymol are the constituent of EOs which have been registered by the European Commission for use as flavorings in foodstuffs in view of its nontoxic effect (Hyldgaard et al., 2012; Tajkarimi et al., 2010). The flavorings registered are considered to present no risk to the health of the consumer and include, among others, carvacrol, carvone, cinnamaldehyde, citral, *p*-cymene, eugenol, limonene, menthol, and thymol. Estragole and methyl eugenol were deleted from the list in 2001 due to their being genotoxic. New flavorings may only be evaluated for registration after toxicological and metabolic studies have been carried out, which could entail a considerable financial outlay.

The EU-registered flavorings also appear on the "Everything Added to Food in the US" (EAFUS), which means that the United States Food and Drug Administration (FDA) has classified the substances as generally recognized as safe (GRAS) or as approved food additives. Estragole, specifically prohibited as flavoring in the EU, is on the EAFUS list. In other countries, and if added to food for a purpose other than flavoring, these compounds may be treated as new food additives. Approval as a food additive would probably involve expensive safety and metabolic studies, the cost of which may be prohibitive. From a legislative point of view, it would be in those countries economically more feasible to use a whole spice or herb or a whole EO as an ingredient than to use individual EO components.

5.5 TOXICITY EVALUATION OF PLANT EXTRACTS

Most of the plant extracts, which might find application in foods, have been consumed by humans for thousands of years; however, typical toxicological information such as acceptable daily intake (ADI) or no observed adverse effect level (NOEL) is not available for them. Although international guidelines exist for the safety evaluation of food additives, due to problems in standardization of extracts or dried preparation owing to their batch-wise compositional variability, it becomes difficult to assign ADI or NOEL for the plant extracts. The marker compounds in extracts are affected by variety of plant, geographical origin, plant part used, age and growth condition of plants, method of extraction or drying, preparation, packaging and storage. According to Dietary Supplement Health and Education Act (DSHEA), 1994, botanicals are exempted from food additive category, and GRAS submission of safety evidence is not required as long as that ingredient was in market before October 1994. Recently, there have been an increase in the number of botanical products as food ingredients or supplements and these are a commercially important part of the health food market. Botanicals may be derived from conventional primary food sources (soy extracts, tomato extracts) or from secondary sources such as herbs and spices (garlic oil, rosemary extracts, green tea extracts). Some botanicals may have no significant history of use as food ingredients but may be derived from sources that have been used in herbal medicinal products in various regions of the world and considered for food use (e.g., *Ginkgo biloba*, Ginseng extract). Further, materials with no history of human use (phytostanols derived as a by-product from wood, shikimic acid isolated from water-soluble extract of pine needles of *C. deodara*) may be considered for use in foods. Therefore, a simple checklist of tests that will be appropriate for establishing the safety of phytochemicals to be added to foods is not yet available.

The International Life Sciences Institute Europe has developed a comprehensive document on the use of plant materials in food products (Schilter et al., 2003), which stresses that the ingredient for use in food products must be well identified and characterized. The starting material must be accurately identified in order to ensure that the plant materials for food use are consistent with respect to quality and quantity of active ingredient and the method of preparation must meet good manufacturing practices. Risk assessment of natural products may require adequate specification of identity and composition as it may be the whole plant, extracts thereof or purified components, and the variability among plant source and the process used to obtain the constituents will be a limiting factor in adopting a generic approach to their risk assessment. The nature of the compound, prior knowledge of human consumption, likely exposure, and nutritional impact will determine the approach for toxicological testing of such compounds. Generally, for herbs or complex extracts, it is not possible to make a risk assessment on the basis of a single active component as more than one component may be of toxicological significance and food matrix may affect their bioavailability. A decision tree has been suggested as an aid to the safety evaluation process for plant material intended for food use (Walker, 2004), and general framework for safety assessment of botanicals has been described (Speijers et al., 2010; van den Berg et al., 2011).

5.6 CONCLUSIONS

Consumer preference and current trends toward natural additives make the secondary metabolites from plants a viable option for the food industry. This is because they have proven to be effective preservatives retarding oxidative and microbiological processes. Also, secondary metabolites have shown functional characteristics in foods, such as, nutrient supplement, colorant, flavoring, sweetener, and texturizing. Added to this, these bioactive compounds have an important role as food ingredients increasing their incorporation in food products being influenced by changes in lifestyles and healthy trends.

KEYWORDS

- **food processing**
- **oxidation**
- **plant-derived compounds**
- **colorants**
- **secondary metabolites**

REFERENCE

Abanda-Nkpwatt, D.; Krimm, U.; Schreiber, L.; Schwab, W. Dual Antagonism of Aldehydes and Epiphytic Bacteria from Strawberry Leaf Surfaces against the Pathogenic Fungus *Botrytis cinerea* in vitro. *BioControl* **2006,** *51*(3), 279–291.

Al-Bulushi, I. M.; Kasapis, S.; Al-oufi, H.; Al-mamari, S. Evaluating the Quality and Storage Stability of Fish Burgers during Frozen Storage. *Fisheries Sci.* **2005,** *71*(3), 648–654.

Al-Bulushi, I. M.; Kasapis, S.; Dykes, G. A.; Al-Waili, H.; Guizani, N.; Al-Oufi, H. Effect of Frozen Storage on the Characteristics of a Developed and Commercial Fish Sausages. *J. Food Technol.* **2013,** *50*(6), 1158–1164.

Ali, F. H.; Kassem, G. M.; Atta-Alla, O. A. Propolis as a Natural Decontaminant and Antioxidant in Fresh Oriental Sausage. *Vet. Ital.* **2010,** *46*(2), 167–172.

Alvarez, M. V.; Ponce, A. G.; Moreira, M. D. R. Antimicrobial Efficiency of Chitosan Coating Enriched with Bioactive Compounds to Improve the Safety of Fresh Cut Broccoli. *LWT—Food Sci. Technol.* **2013,** *20*, 78–87.

Amalaradjou, M. A. R.; Baskaran, S. A.; Ramanathan, R.; Johny, A. K.; Charles, A. S.; Valipe, S. R.; Venkitanarayanan, K. Enhancing the Thermal Destruction of *Escherichia coli* O157: H7 in Ground Beef Patties by *Trans*-cinnamaldehyde. *Food Microbiol.* **2010,** *27*(6), 841–844.

Ansorena, M. R. Antimicrobial Protection and Antioxidant Enhancement of Minimally Processed Broccoli Treated with Thyme Essentials Oil Encapsulated in Beta-Cyclodextrin. International Conference of Innovations in Food Packaging, Shelf Life and Food Safety, Munich, Germany, 2015.

Arts, M. J. T. J.; Haenen, G. R. M. M.; Wilms, L. C.; Beetstra, S. A. J. N.; Heijnen, C. G. M.; Voss, H.; et al. Interactions between Flavonoids and Proteins: Effect on the Total Antioxidant Capacity. *J. Agric. Food Chem.* **2002,** *50*, 1184–1187.

Asensio, C. M.; Grosso, N. R.; Juliani, H. R. Quality Preservation of Organic Cottage Cheese Using Oregano Essential Oils. *LWT—Food Sci. Technol.* **2015,** *60*(2), 664–671.

Ayala-Zavala, J. F.; González-Aguilar, G. A. Optimizing the use of Garlic Oil as Antimicrobial Agent on Fresh-Cut Tomato through a Controlled Release System. *J. Food Sci.* **2010,** *75*, 398–405.

Ayala-Zavala, J. F.; González-Aguilar, G. A.; Ansorena, M. R.; Alvarez-Párrilla, E.; de la Rosa, L. Nanotechnology Tools to Achieve Food Safety. In *Practical Food Safety: Contemporary Issues and Future Directions*; Bhat, R., Gomez-Lopez, V., Eds.; Wiley-Blackwell: New Jersey, 2014, Chapter 17, vol. 632; pp 341–353.

Ayala-Zavala, J. F.; González-Aguilar, G. A.; Del Toro-Sánchez, L. Enhancing Safety and Aroma Appealing of Fresh-Cut Fruits and Vegetables Using the Antimicrobial and Aromatic 306 Handbook of Natural Antimicrobials for Food Safety and Quality power of essential oils. *J. Food Sci.* **2009,** *74*, 84–91.

Ayala-Zavala, J. F.; Oms-Oliu, G.; Odriozola-Serrano, I.; González-Aguilar, G. A.; ÁlvarezParrilla, E.; Martín-Belloso, O. Bio-Preservation of Fresh-Cut Tomatoes Using Natural Antimicrobials. *Eur. Food Res. Technol.* **2008,** *226*, 1047–1055.

Ayala-Zavala, J. F.; Rosas-Domínguez, C.; Vega-Vega, V.; González-Aguilar, G. A. Antioxidant Enrichment and Antimicrobial Protection of Fresh-Cut Fruits Using their Own by Products: Looking for Integral Exploitation. *J. Food Sci.* **2010,** *75*, 175–181.

Ayala-Zavala, J. F.; Silva-Espinoza, B.; Cruz-Valenzuela, M.; Leyva, J.; OrtegaRamírez, L.; Carrazco-Lugo, D.; et al. Pectine cinnamon Leaf Oil Coatings Add Antioxidant and Antibacterial Properties to Fresh-Cut Peach. *Flavour Frag. J.* **2013,** *28*(1), 39–45.

Baines, D.; Seal, R. Natural Food Additives, Ingredients and Flavourings. Woodhead Publishing: Cambridge, UK, 2012.

Baker, R.; Günther, C. The Role of Carotenoids in Consumer Choice and the Likely Benefits from their Inclusion into Products for Human Consumption. *Trends Food Sci. Technol.* **2004,** *15*, 484–488.

Barbosa-Pereira, L.; Cruz, J. M.; Sendón, R.; Quirós, A. R. B.; Ares, A.; Castro-López, M.; Abad, M.; Maroto, J.; Paseiro-Losada, P. Development of Antioxidant Active Films Containing Tocopherols to Extend the Shelf Life of Fish. *Food Control* **2013,** *31*, 236–243.

Barclay, A.; Sandall, P.; Shwide-Slavin, C. The Ultimate Guide to Sugars and Sweeteners: Discover the Taste, Use, Nutrition, Science, and Lore of Everything from Agave Nectar to Xylitol. The Experiment, LLC: New York, USA, 2014.

Bázan-Lugo, E.; García-Martínez, I.; Alfaro-Rodríguez, R. H.; Totosaus, A. Color Compensation in Nitrite-Reduced Meat Batters Incorporating Paprika or Tomato Paste. *J. Sci. Food Agric.* **2012,** *92*(8), 1627–1632.

Bazargani-Gilani, B.; Aliakbarlu, J.; Tajik, H. Effect of Pomegranate Juice Dipping and Chitosan Coating Enriched with *Zataria multiflora* Boiss Essential Oil on the Shelf-Life of Chicken Meat During Refrigerated Storage. *Innov. Food Sci. Emerg.* **2015,** *29*, 280–287.

Beltrame, P.; Beltrame, P. L.; Carniti, P.; Guardione, D.; Lanzetta, C. Inhibiting Action of Chlorophenols on Biodegradation of Phenol and Its Correlation with Structural Properties of Inhibitors. *Biotechnol. Bioeng.* **1998,** *31*, 821–828.

Bhatt, P.; Negi, P. S. Antioxidant and Antibacterial Activities of Indian Borage (*Plectranthus amboinicus*) Leaf Extracts. *Food Nutr. Sci.* **2012,** *3*, 146–152.

Bloukas, J. G.; Arvanitoyannis, I. S.; Siopi, A. A. Effect of Natural Colourants and Nitrites on Colour Attributes of Frankfurters. *Meat Sci.* **1999,** *52*(3), 257–265.

Borges, A.; Ferreira, C.; Saavedra, M. J.; Simões, M. Antibacterial Activity and Mode of Action of Ferulic and Gallic Acids against Pathogenic Bacteria. *Microb. Drug Resist.* **2013,** *19*(4), 256–265.

Borris, R. P. Natural Product Research: Perspectives from a Major Pharmaceutical Company. *J. Ethanopharmacol.* **1996,** *51*, 29–38.

Bukvički, D.; Stojković, D.; Soković, M.; Vannini, L.; Montanari, C.; Pejin, B.; Marin, P. D. *Satureja horvatii* Essential Oil: In vitro Antimicrobial and Antiradical Properties and In Situ Control of *Listeria monocytogenes* in Pork Meat. *Meat Sci.* **2014,** *96*(3), 1355–1360.

Burt, S. Essential Oils: Their Antibacterial Properties and Potential Applications in Foods—A Review. *Int. J. Food Microbiol.* **2004,** *94*(3), 223–253.

Butkhup, L.; Chowtivannakul, S.; Gaensakoo, R.; Prathepha, P.; Samappito, S. Study of the Phenolic Composition of Shiraz Red Grape Cultivar (*Vitis vinifera* L.) Cultivated in North-Eastern Thailand and Its Antioxidant and Antimicrobial Activity. *S. Afr. J. Enol. Vitic.* **2010,** *31*, 89–98.

Calvo, M. M.; García, M. L.; Selgas, M. D. Dry Fermented Sausages Enriched with Lycopene from Tomato Peel. *Meat Sci.* **2008,** *80*(2), 167–172.

Carocho, M.; Barreiro, M. F.; Morales, P.; Ferreira, I. C. F. R. Adding Molecules to Food, Pros and Cons: A Review of Synthetic and Natural Food Additives. *Compr. Rev. Food Sci. F.* **2014,** *13*, 377–399.

CDC (Centers for Disease Control and Prevention). *Foodborne Diseases Center for Outbreak Response Enhancement*, 2015a. http://www.cdc.gov/foodcore/ (accessed Jun 10, 2015).

CDC (Centers for Disease Control and Prevention). *List of Selected Multistate Foodborne Outbreak Investigation*, 2015b. http://www.cdc.gov/foodsafety/outbreaks/multistate-outbreaks/outbreaks-list.html (accessed Jun 10, 2015).

Chouchouli V.; Kalogeropoulos, N.; Konteles S. J.; Karvela, E.; Makris, D. P.; Karathanos, V. T. Fortification of Yoghurts with Grape (*Vitis vinifera*) Seed Extracts. *LWT—Food Sci. Technol.* **2013,** *53*, 522–529.

Cruz-Valenzuela, M. R.; Carrazco-Lugo, D. K.; Vega-Vega, V.; Gonzalez-Aguilar, G. A.; Ayala-Zavala, J. F. Fresh-cut Orange Treated with Its Own Seed By-products Presented Higher Antioxidant Capacity and Lower Microbial Growth. *Int. J. Postharvest Technol. Innov.* **2013,** *3*(1), 13–27.

Da Cruz Cabral, L.; Fernández Pinto, V.; Patriarca, A. Application of Plant Derived Compounds to Control Fungal Spoilage and Mycotoxin Production in Foods. *Int. J. Food Microbiol.* **2013,** *166*, 1–14.

Davidson, P. M.; Critzer, F. J.; Taylor, T. M. Naturally Occurring Antimicrobials for Minimally Processed Foods. *Annu. Rev. Food Sci. Technol.* **2013,** *4*, 163–190.

Day, L.; Seymour R. B.; Pitts K. F.; Konczak I.; Lundin L. Incorporation of Functional Ingredients into Foods. *Trends Food Sci. Technol.* **2009,** *20*, 388–395.

de Oliveira, T. L. C.; Junior, B. R. D. C. L.; Ramos, A. L.; Ramos, E. M.; Piccoli, R. H.; Cristianini, M. Phenolic Carvacrol as a Natural Additive to Improve the Preservative Effects of High Pressure Processing of Low-Sodium Sliced Vacuum-Packed Turkey Breast Ham. *LWT—Food Sci. Technol.* **2015,** *64*(2), 1297–1308.

Decker, E. A.; Elias, R. J.; McClements, D. J. Oxidation in Foods and Beverages and Antioxidant Applications. Woodhead Publishing: Oxford, 2010.

Decker, E. A.; Park, Y. Healthier Meat Products as Functional Foods. *Meat Sci.* **2010,** *86*(1), 49–55.

Delgado-Vargas, F.; Jiménez, A. R.; Paredes-López, O. Natural Pigments: Carotenoids, Anthocyanins, and Betalains—Characteristics, Biosynthesis, Processing, and Stability. *Crit. Rev. Food Sci. Nutr.* **2000,** *40*, 173–289.

Devi, K. P.; Nisha, S. A.; Sakthivel, R.; Pandian, S. K. Eugenol (An Essential Oil of Clove) Acts as an Antibacterial Agent Against *Salmonella typhi* by Disrupting the Cellular Membrane. *J. Ethnopharmacol.* **2010,** *130*(1), 107–115.

Di Pasqua, R.; Betts, G.; Hoskins, N.; Edwards, M.; Ercolini, D.; Mauriello, G. Membrane Toxicity of Antimicrobial Compounds from Essential Oils. *J. Agric. Food Chem.* **2007,** *55*(12), 4863–4870.

Dobbs, C. M.; Bell, L. N. Storage Stability of Tagatose in Buffer Solutions of Various Compositions. *Food Res. Int.* **2010,** *43*, 382–3869.

Donsi, F.; Annunziata, M.; Sessa, M.; Ferrari, G. Nanoencapsulation of Essential Oils to Enhance their Antimicrobial Activity in Foods. *LWT—Food Sci. Technol.* **2011,** *44*, 1908–1914.

EFSA, Scientific Opinion. Scientific Opinion on the Re-evaluation of Mixed Carotenes (E160a(i)) and Beta-Carotene (E160a(ii)) as a Food Additive. *EFSA J.* **2012,** *10,* 2593.

El-Said, M. M.; Haggag, H. F.; El-Din, H. M. F.; Gad, A. S.; Farahat, A. M. Antioxidant Activities and Physical Properties of Stirred Yoghurt Fortified with Pomegranate Peel Extracts. *Ann. Agric. Sci.* **2014,** *59*(2), 207–212.

Embuscado, M. W. Spices and Herbs: Natural Sources of Antioxidants: A Mini Review. *J. Funct. Foods* **2015,** *18,* 811–819.

FAO (Organización de las Naciones Unidas para la Alimentación y la Agricultura). Iniciativa mundial sobre la reducción de la pérdida y el desperdicio de alimentos. http://www.fao.org/3/a-i4068s.pdf (accessed Jun 12, 2014).

Fernández-López, J.; Viuda-Martos, M.; Sendra, E.; Sayas-Barberá, E.; Navarro, C.; Pérez-Alvarez, J. A. Orange Fiber as Potential Functional Ingredient for Dry-Cured Sausages. *Eur. Food Res. Technol.* **2007,** *226*(1–2), 1–6.

Friedman, M.; Henika, P. R.; Levin, C. E.; Mandrell, R. E. Antibacterial Activities of Plant Essential Oils and Their Components against *Escherichia coli* O157:H7 and *Salmonella entericain* apple juice. *J. Agric. Food Chem.* **2004,** *52,* 6042–6048.

Fung, D. Y. Microbial Hazards in Food: Food-Borne Infections and Intoxications. *Handbook of Meat Processing. Blackwell Publishing*: Ames, IA, 2010; pp 481–500.

Gad, A. S.; Kholif, A. M.; Sayed, A. F. Evaluation of the Nutritional Value of Functional Yoghurt Resulting from a Combination of Date Palm Syrup and Skim Milk. In *IV International Date Palm Conference 882*, 2010 March, 583–592.

Gill, A. O.; Holley, R. A. Disruption of *Escherichia coli, Listeria monocytogenes* and *Lactobacillus sakei* Cellular Membranes by Plant Oil Aromatics. *Int. J. Food Microbial.* **2006,** *108*(1), 1–9.

Gök, S. B.; Kalınkara, E. C.; Erdoğdu, Y. The Use of Plant Secondary Metabolites as Food Additives. *Curr. Opin. Biotechnol.* **2013,** *24,* S89.

Gök, V.; Akkaya, L.; Obuz, E.; Bulut, S. Effect of Ground Poppy Seed as a Fat Replacer on Meat Burgers. *Meat Sci.* 2011, *89*(4), 400–404.

Gómez, R.; Alvarez-Orti, M.; Pardo, J. E. Influence of the Paprika Type on Redness Loss in Red Line Meat Products. *Meat Sci.* 2008, *80*(3), 823–828.

González-Aguilar, G. A.; Ansorena, M. R.; Viacava, G. E.; Roura, S. I.; Ayala-Zavala, J. F. Plant Essential Oils as Antifungal Treatments on the Postharvest of Fruit and Vegetables. In *Antifungal Metabolites from Plants: Progress and Prospects*; Mehdi, R.-A., Mahendra, R., Eds.; Springer: Berlin-Heidelberg, Chapter 15, vol. XIV, 2013, *469*, pp 429–446.

Goulas, A. E.; Kontominas, M. G. Combined Effect of Light Salting, Modified Atmosphere Packaging and Oregano Essential Oil on the Shelf-Life of Sea Bream (*Sparus aurata*): Biochemical and sensory attributes. *Food Chem.* **2007,** *100*(1), 287–296.

Gramza, A.; Korczak, J. Tea Constituents (*Camellia sinensis* L.) as Antioxidants in Lipid Systems. *Trends Food Sci. Technol.* **2005,** *16,* 351–358.

Gutierrez, J.; Barry-Ryan, C.; Bourke, P. Antimicrobial Activity of Plant Essential Oils Using Food Model Media: Efficacy, Synergistic Potential and Interactions with Food Components. *Food Microbiol.* **2009,** *26*(2), 142–150.

Gyawali, R.; Ibrahim, S. A. Impact of Plant Derivatives on the Growth of Foodborne Pathogens and the Functionality of Probiotics. *Appl. Microbiol. Biotechnol.* **2012,** *95*(1), 29–45.

Gyawali, R.; Ibrahim, S. A. Natural Products as Antimicrobial Agents. *Food Control* **2014,** *46*, 412–429.

Hala, M. F.; Ebtisam, E. D.; Sanaa, I.; Badran, M. A.; Marwa, A. S.; Said, M. E. Manufacture of Low Fat UF-Soft Cheese Supplemented with Rosemary Extract (as Natural Antioxidant). *J. Am. Sci.* **2010,** *6*(10), 570–579.

Han, J.; Britten, M.; St-Gelais, D.; Champagne, C. P.; Fustier, P.; Salmieri, S.; Lacroix, M. Polyphenolic Compounds as Functional Ingredients in Cheese. *Food Chem.* **2011a**, *124*(4), 1589–1594.

Han, J.; Britten, M.; St-Gelais, D.; Champagne, C. P.; Fustier, P.; Salmieri, S.; Lacroix, M. Effect of Polyphenolic Ingredients on Physical Characteristics of Cheese. *Food Res. Int.* **2011b**, *44*(1), 494–497.

Hayek, S. A.; Gyawali, R.; Ibrahim, S. A. Antimicrobial Natural Products. In: *Microbial Pathogens and Strategies for Combating Them: Science, Technology and Education*; Méndez-Vilas, A. Ed.; Formatex Research Center; 2013, vol. 2, pp 910–921.

He, M.; Wu, T.; Pan, S.; Xu, X. Antimicrobial Mechanism of Flavonoids against *Escherichia coli* ATCC 25922 by Model Membrane Study. *Appl. Surf. Sci.* **2014,** *305*, 515–521.

Heipieper, H. J.; Keweloh, H.; Rehm, H. J. Influence of Phenols on Growth and Membrane Permeability of Free and Immobilized *Escherichia coli. Appl. Environ. Microb.* **1991,** *57*, 1213–1217.

Helal, A.; Tagliazucchi, D.; Verzelloni, E.; Conte, A. Gastro-Pancreatic Release of Phenolic Compounds Incorporated in a Polyphenols-Enriched Cheese-Curd. *LWT—Food Sci. Technol.* **2015,** *60*(2), 957–963.

Hygreeva, D.; Pandey, M. C.; Radhakrishna, K. Potential Applications of Plant Based Derivatives as Fat Replacers, Antioxidants and Antimicrobials in Fresh and Processed Meat Products. *Meat Sci.* **2014,** *98*(1), 47–57.

Hyldgaard, M.; Mygind, T.; Meyer, R. L. Essential Oils in Food Preservation: Mode of Action, Synergies, and Interactions with Food Matrix Components. *Front. Microbiol.* **2012,** *25*, 3–12.

Jang, S.; Shin, Y. J.; Song, K. B. Effect of Rapeseed Protein–Gelatin Film Containing Grapefruit Seed Extract on 'Maehyang' Strawberry Quality. *Int. J. Food Sci. Technol.* **2011,** *46*(3), 620–625.

Javidipour, I.; Vural, H.; Özbaş, Ö. Ö.; Tekin, A. Effects of Inter-esterified Vegetable Oils and Sugar Beet Fiber on the Quality of Turkish-Type Salami. *Int. J. Food Sci. Technol.* **2005,** *40*(2), 177–185.

Jiménez, J. B.; Orea, J. M.; Montero, C.; Urena, A. G.; Navas, E.; Slowing, K.; et al. Resveratrol Treatment Controls Microbialflora, Prolongs Shelf Life, and Preserves Nutritional Quality of Fruit. *J. Agric. Food Chem.* **2005,** *53*, 1526–1530.

Juneja, V. K.; Garcia-Davila, J.; Lopez-Romero, J. C.; Pena-Ramos, E. A.; Camou, J. P.; Valenzuela-Melendres, A. Modeling the Effects of Temperature, Sodium Chloride, and Green Tea and Their Interactions on the Thermal Inactivation of *Listeria monocytogenes* in Turkey. *J. Food Protect.* **2014,** *77*(10), 1696–1702.

Kaewprachu, P.; Osako, K.; Benjakul, S.; Rawdkuen, S. Quality Attributes of Minced Pork Wrapped with Catechin-Lysozyme Incorporated Gelatin Film. *Food Packag. Shelf-Life* **2015,** *3*, 88–94.

Kammerer, D.; Carle, R.; Schieber, A. Quantification of Anthocyanins in Black Carrot Extracts (*Daucus carrota* ssp. *sativus* var. *atrorubens* Alef.) and Evaluation of Their Colour Properties. *Eur. Food Res. Technol.* **2004,** *219*, 479–486.

Karaaslan, M.; Ozden, M.; Vardin, H.; Turkoglu, H. Phenolic Fortification of Yogurt Using Grape and Callus Extracts. *LWT—Food Sci. Technol.* **2011,** *44*(4), 1065–1072.

Karatzas, A. K.; Kets, E. P. W.; Smid, E. J.; Bennik, M. H. J. The Combined Action of Carvacrol and High Hydrostatic Pressure on *Listeria monocytogenes* Scott A. *J. Appl. Microbiol.* **2001,** *90*(3), 463–469.

Karre, L.; Lopez, K.; Getty, K. J. K. Natural Antioxidants in Meat and Poultry Products. *Meat Sci.* **2013,** *94*, 220–227.

Keweloh, H.; Weyrauch, G.; Rehm, H. N. Phenol-Induced Membrane Changes in Free and Immobilized *Escherichia coli. Appl. Microbiol. Biotechnol.* **1990,** *33*, 66–71.

Khare, A. K.; Biswas, A. K.; Sahoo, J. Comparison Study of Chitosan, EDTA, Eugenol and Peppermint Oil for Antioxidant and Antimicrobial Potentials in Chicken Noodles and their Effect on Colour and Oxidative Stability at Ambient Temperature Storage. *LWT—Food Sci. Technol.* **2014,** *55*(1), 286–293.

Kim, J. M.; Marshall, M. R.; Cornell, J. A.; Prestone III, J. F.; Wei, C. I. Antibacterial Activity of Carvacrol, Citral, and Geraniol against *Salmonella typhimurium* in Culture Medium and on Fish Cubes. *J. Food Sci.* **1995,** *60*(6), 1364–1368.

Kim, S. J.; Cho, A. R.; Han, J. Antioxidant and Antimicrobial Activities of Leafy Green Vegetable Extracts and Their Applications to Meat Product Preservation. *Food Control* **2013,** *29*(1), 112–120.

Kim, S. Y.; Kang, D. H.; Kim, J. K.; Ha, Y. G.; Hwang, J. Y.; Kim, T.; et al. Antimicrobial Activity of Plant Extracts against *Salmonella typhimurium, Escherichia coli* O157:H7, and *Listeria monocytogeneson* Fresh Lettuce. *J. Food Sci.* **2011,** *76*, 41–46.

Koketsu, M.; Satoh, Y. I. Antioxidative Activity of Green Tea Polyphenols in Edible Oils. *J. Food Lipids* **2007,** *1997*, 1–9.

Korczak, J.; Hes, M.; Gramza, A.; Jedrusek-Golin´ska, A. Influence of Fat Oxidation on the Stability of Lysine and Protein Digestibility in Frozen Meat Products. *EJPAU* **2004,** *7*, 1–13.

Kumar, N.; Pruthi, V. Potential Applications of Ferulic Acid from Natural Sources. *Biotechnol. Rep.* **2014,** *4*, 86–93.

Lai, P.; Roy, J. Antimicrobial and Chemopreventive Properties of Herbs and Spices. *Curr. Med. Chem.* **2004,** *11*(11), 1451–1460.

Lanciotti, R.; Corbo, M. R.; Gardini, F.; Sinigaglia, M.; Guerzoni, M. E. Effect of Hexanal on the Shelf Life of Fresh Apple Slices. *J. Agric. Food Chem.* **1999,** *47*, 4769–4776.

Lauro, G. J. Natural Colorants for Surimi Seafood. *Food Science and Technology*; Marcel Dekker: New York, USA, 2000; pp 417–444.

Li, Y.; Guo, C.; Yang, J.; Wei, J.; Xu, J.; Cheng, S. Evaluation of Antioxidant Properties of Pomegranate Peel Extract in Comparison with Pomegranate Pulp Extract. *Food Chem.* **2006,** *96*(2), 254–260.

Lin, S.; Pascall, M. A. Incorporation of Vitamin E into Chitosan and Its Effect on the Film Forming Solution (viscosity and drying rate) and the Solubility and Thermal Properties of the Dried Film. *Food Hydrocolloid* **2014,** *35*, 78–84.

Lopez-Romero, J. C.; Gonzalez-Rios, H.; Borges, A.; Simões, M. Antibacterial Effects and Mode of Action of Selected Essential Oils Components against *Escherichia coli* and *Staphylococcus aureus. Evid.-Based Compl. Altern.* **2015,** *2015*.

Luther, M.; Parry, J.; Moore, J.; Meng, J.; Zhang, Y.; Cheng, Z.; Yu, L. L. Inhibitory Effect of Chardonnay and Black Raspberry Seed Extracts on Lipid Oxidation in Fish Oil and Their Radical Scavenging and Antimicrobial Properties. *Food Chem.* **2007,** *104*(3), 1065–1073.

MacDougall, D. B. Colour in Food. Woodhead Publishing Limited: Cambridge, UK. 2002.

Mahmoud, B. S.; Yamazaki, K.; Miyashita, K.; Shin, I. I.; Suzuki, T. A New Technology for Fish Preservation by Combined Treatment with Electrolyzed NaCl Solutions and Essential Oil Compounds. *Food Chem.* **2006,** *99*(4), 656–662.

Maqsood, S.; Benjakul, S.; Balange, A. K. Effect of Tannic Acid and Kiam Wood Extract on Lipid Oxidation and Textural Properties of Fish Emulsion Sausages during Refrigerated Storage. *Food Chem.* **2012,** *130*(2), 408–416.

Mari, M.; Leoni, O.; Bernardi, R.; Neri, F.; Palmieri, S. Control of Brown Rot on Stonefruit by Synthetic and Glucosinolate-Derived Isothiocyanates. *Postharvest Biol. Technol.* **2008,** *47*, 61–67.

Mariem, C.; Sameh, M.; Nadhem, S.; Soumaya, Z.; Najiba, Z.; Raoudha, E. G. Antioxidant and Antimicrobial Properties of the Extracts from *Nitraria retusa* Fruits and Their Applications to Meat Product Preservation. *Ind. Crop. Prod.* **2014,** *55*, 295–303.

Martín-Diana, A. B.; Rico, D.; Barry-Ryan, C. Green Tea Extract as a Natural Antioxidant to Extend the Shelf-Life of Fresh-Cut Lettuce. *Innov. Food Sci. Emerg. Technol.* **2008,** *9*, 593–603.

McChesney, J. D.; Venkataraman, S. K.; Henri, J. T. Plant Natural Products: Back to the Future or into Extinction ?. *Phytochemistry* **2007,** *68*(14), 2015–2022.

Medina-Meza, I. G.; Barnaba, C.; Barbosa-Cánovas, G. V. Effects of High Pressure Processing on Lipid Oxidation: A Review. *Innov. Food Sci. Emerg.* **2014,** *22*, 1–10.

Mercadante, A. Z.; Capitani, C. D.; Decker, E. A.; Castro, I. A. Effect of Natural Pigments on the Oxidative Stability of Sausages Stored under Refrigeration. *Meat Sci.* **2010,** *84*(4), 718–726.

Mohamed, H. M.; Mansour, H. A. Incorporating Essential Oils of Marjoram and Rosemary in the Formulation of Beef Patties Manufactured with Mechanically Deboned Poultry Meat to Improve the Lipid Stability and Sensory Attributes. *LWT—Food Sci. Technol.* **2012,** *45*(1), 79–87.

Moreira, M. R.; Ponce, A. G.; Del Valle, C. E.; Roura, S. I. Effects of Clove and Tea Tree Oils on *Escherichia coli* O157:H7 in Blanched Spinach and Minced Cooked Beef. *J. Food Process Pres.* **2007,** *31*, 379–391.

Moreira, M. R.; Ponce, A. G.; del Valle, C. E.; Roura, S. I. Inhibitory Parameters of Essential Oils to Reduce a Foodborne Pathogen. *LWT—Food Sci. Technol.* **2005,** *38*, 565–570.

Moreno, D. A.; García-Viguera, C.; Gil, J. I.; Gil-Izquierdo, A. Betalains in the Era of Global Agri-Food Science, Technology and Nutritional Health. *Phytochem. Rev.* **2008,** *7*, 261–280.

Naveena, B. M.; Vaithiyanathan, S.; Muthukumar, M.; Sen, A. R.; Kumar, Y. P.; Kiran, M.; Shaju, V. A.; Chandran, K. R. Relationship between the Solubility, Dosage and Antioxidant Capacity of Carnosic Acid in raw and Cooked Ground Buffalo Meat Patties and Chicken Patties. *Meat Sci.* **2013,** *96*, 195–202.

Negi, P. S. Plant Extracts for the Control of Bacterial Growth: Efficacy, Stability and Safety Issues for Food Application. *Int. J. Food Microbiol.* **2012,** *156*(1), 7–17.

O'Brien-Nabors, L. *Alternative Sweeteners.* Marcel Dekker: New York, USA, 2001.

Ojagh, S. M.; Rezaei, M.; Razavi, S. H.; Hosseini, S. M. H. Effect of Chitosan Coatings Enriched with Cinnamon Oil on the Quality of Refrigerated Rainbow Trout. *Food Chem.* **2010,** *120*(1), 193–198.

Oliveira, D. M.; Melo, F. G.; Balogun, S. O.; Flach, A.; da Costa, L. A. M. A.; Soares, I. M.; de Oliveira Martins, D. T. Antibacterial Mode of Action of the Hydroethanolic Extract of *Leonotis nepetifolia* (L.) R. Br. Involves Bacterial Membrane Perturbations. *J. Ethnopharmacol.* **2015,** *172*, 356–363.

Ou, S.; Kwon, K. Ferulic Acid: Pharmaceutical Functions, Preparation and Applications in Foods. *J. Sci. Food Agric.* **2004,** *84*, 1261–1269.

Ouattara, B.; Sabato, S. F.; Lacroix, M. Combined Effect of Antimicrobial Coating and Gamma Irradiation on Shelf Life Extension of Pre-Cooked Shrimp (*Penaeus* spp.). *Int. J. Food. Microbiol.* **2001,** *68*(1), 1–9.

Owen, R. J.; Palombo, E. A. Anti-Listerial Activity of Ethanolic Extracts of Medicinal Plants, *Eremophila alternifolia* and *Eremophila duttonii*, in Food Homogenates and Milk. *Food Control* **2007,** *18*, 387–390.

Park, J. W. Coloring Technology for Surimi Seafood. In *ACS Symposium*. Oxford University Press: Oxford, UK, 2008, Vol. 983, pp 254–266.

Pazmiño-Durán, E. A.; Giusti, M. M.; Wrolstad, R. E.; Glória, M. B. A. Anthocyanins from *Oxalis triangularis* as Potential Food Colorants. *Food Chem.* **2001,** *75*(2), 211–216.

Perron, R. N.; Brumaghim, L. J. A. Review of the Antioxidant Mechanisms of Polyphenol Compounds Related to Iron Binding. *Cell Biochem. Biophys.* **2009,** *53*(2), 75–100.

Piqueras-Fiszman, B.; Spence, C. The Influence of the Feel of Product Packaging on the Perception of the Oral-Somatosensory Texture of food. *Food Qual. Prefer.* **2012,** *26*(1), 67–73.

Ponce, A. G.; Del Valle, C.; Roura, S. I. Shelf Life of Leafy Vegetables Treated with Natural Essential Oils. *J. Food Sci.* **2004,** *69*, 50–56.

Ponce, A.; Roura, S. I.; Moreira, M. D. R. Essential Oils as Biopreservatives: Different Methods for the Technological Application in Lettuce Leaves. *J. Food Sci.* **2011,** *76*, 34–40.

Rao, P. G. P.; Jyothirmayi, T.; Balaswamy, K.; Satyanarayana, Rao, D. G. Effect of Processing Conditions on the Stability of Annatto (*Bixa orellana* L.) Dye Incorporated into Some Foods. *LWT—Food Sci. Technol.* **2005,** *38*, 779–784.

Rashidinejad, A.; Birch, E. J.; Sun-Waterhouse, D.; Everett, D. W. Total Phenolic Content and Antioxidant Properties of Hard Low-Fat Cheese Fortified with Catechin as Affected by In Vitro Gastrointestinal Digestion. *LWT—Food Sci. Technol.* **2015,** *62*(1), 393–399.

Raybaudi-Massilia, R. M.; Mosqueda-Melgar, J.; Soliva-Fortuny, R.; Martin-Belloso, O. Control of Pathogenic and Spoilage Microorganisms in Fresh Cut Fruits and Fruit Juices by Traditional and Alternative Natural Antimicrobials. *Compr. Rev. Food Sci. F.* **2009,** *8*, 157–180.

Raybaudi-Massilia, R. M.; Rojas-Grau, M. A.; Mosqueda-Melgar, J.; Martin Belloso, O. Comparative Study on Essential Oils Incorporated into an Alginate-Based Edible Coating to Assure the Safety and Quality of Fresh-Cut Fuji apples. *J. Food Protect.* **2008,** *71*(6), 1150–1161.

Reddy, V. C.; Vidya Sagar, G. V.; Sreeramulu, D.; Venu, L.; Raghunath, M. Addition of Milk Does Not Alter the Antioxidant Activity of Black Tea. *Ann. Nutr. Metab.* **2005,** *49*(3), 189–195.

Rein, M. J.; Heinonen, M. Stability and Enhancement of Berry Juice Colour. *J. Agric. Food Chem.* **2004,** *52*, 3106–3114.

Ribas-Agustí, A.; Gratacós-Cubarsí, M.; Sárraga, C.; Guàrdia, M. D.; García-Regueiro, J. A.; Castellari, M. Stability of Phenolic Compounds in Dry Fermented Sausages Added with Cocoa and Grape Seed Extracts. *LWT—Food Sci. Technol.* **2014,** *57*(1), 329–336.

Rodríguez Vaquero, M. J.; Tomassini Serravalle, L. R.; Manca de Nadra, M. C.; Strasser de Saad, A. M. Antioxidant Capacity and Antibacterial Activity of Phenolic Compounds from Argentinean Herbs Infusions. *Food Control* **2010,** *21*, 779–785.

Schilter, B.; Andersson, C.; Anton, R.; Constable, A.; Kleiner, J.; O'Brien, J.; Renwick, A. G.; Korver, O.; Smit, F.; Walker, R. Guidance for the Safety Assessment of Botanicals and Botanical Preparations for Use in Food and Food Supplements. *Food Chem. Toxicol.* **2003,** *41*, 1625–1649.

Schrader, D.; Sabater-Luenzel, C.; Henze, T.; Batalia, M.; Heine, S. *U.S. Patent Application. 13/867, 361*, 2013.

Scotter, M. The Chemistry and Analysis of Annatto Food Colouring: A Review. *Food Addit. Contam.* **2009,** *26*, 1123–1145.

Serrano, A.; Librelotto, J.; Cofrades, S.; Sánchez-Muniz, F. J.; Jiménez-Colmenero, F. Composition and Physicochemical Characteristics of Restructured Beef Steaks Containing Walnuts as Affected by Cooking Method. *Meat Sci.* **2007,** *77*(3), 304–313.

Shahidi, F.; Zhong, Y. Lipid Oxidation and Improving the Oxidative Stability. *Chem. Soc. Rev.* **2010,** *39*(11), 4067–4079.

Shan, B.; Cai, Y. Z.; Brooks, J. D.; Corke, H. Antibacterial and Antioxidant Effects of Five Spice and Herb Extracts as Natural Preservatives of Raw Pork. *J. Sci. Food Agric.* **2009,** *89*(11), 1879–1885.

Shan, B.; Cai, Y. Z.; Brooks, J. D.; Corke, H. Potential Application of Spice and Herb Extracts as Natural Preservatives in Cheese. *J. Med. Food.* **2011,** *14*(3), 284–290.

Shan, B.; Cai, Y. Z.; Sun, M.; Corke, H. Antioxidant Capacity of 26 Spices Extracts and Characterization of Their Phenolic Constituents. *J. Agric. Food Chem.* **2005,** *53*, 7749–7759.

Sierra-Alvarez, R.; Lettinga, G. The Effect of Aromatic Structure on the Inhibition of Acetoclastic Methanogenesis in Granular Sludge. *Appl. Microbiol. Biotechnol.* **1991,** 34, 544–550.

Sirk, T. W.; Brown, E. F.; Sum, A. K.; Friedman, M. Molecular Dynamics Study on the Biophysical Interactions of Seven Green Tea Catechins with Lipid Bilayers of Cell Membranes. *J. Agric. Food Chem.* **2008,** *56*(17), 7750–7758.

Smith, J.; Hong-Shum, L. *Food Additives Data Book.* Blackwell Publishing: Oxford, UK. 2011.

Speijers, G.; Bottex, B.; Dusemund, B.; Lugasi, A.; Toth, J.; Amberg-Muller, J.; Galli, C. L.; Silano, V.; Rietjens, I. M. C. M. Safety Assessment of Botanicals and Botanical Preparations used as Ingredients in Food Supplements: Testing an European Food Safety Authority-tiered Approach. *Mol. Nutr. Food Res.* **2010,** *54*, 175–185.

Spillane, W. J. *Optimising Sweet Taste in Foods.* Woodhead Publishing Limited: Cambridge, UK. 2006.

Srinivasan, K. Antioxidant Potential of Spices and Their Active Constituents. *Cr. Rev. Food Sci.* **2014,** *54*, 352–372.

Stintzing, F. C.; Carle, R. Functional Properties of Anthocyanins and Betalains in Plants, Food, and in Human Nutrition. *Trends Food Sci. Technol.* **2004,** *15*(1), 19–38.

Stratford, M.; Eklund, T. Organic Acids and Esters. In: *Food Preservatives*; Russell, N. J., Gould, G. W. Eds.; Kluwer Academic/Plenum Publishers: London. 2003, pp 48–84.

Surh, Y. J. Molecular Mechanisms of Chemopreventive Effects of Selected Dietary and Medicinal Phenolic Substances. *Mutat. Res.* **1999,** *428*, 305–327.

Tajkarimi, M. M.; Ibrahim, S. A.; Cliver, D. O. Antimicrobial Herb and Spice Compounds in Food. *Food Control* **2010,** *21*, 1199–1218.

Tang, S. Z.; Kerry, J. P.; Sheehan, D.; Buckley, D. J. Antioxidative Mechanism of Tea Catechins in Chicken Meat Systems. *Food Chem.* **2002,** *76*, 45–51.

Tassou, C. C.; Drosinos, E. H.; Nychas, G. J. E. Effects of Essential Oil from Mint (*Mentha piperita*) on *Salmonella enteritidis* and *Listeria monocytogenes* in Model Food Systems at 4 and 10C. *J. Appl. Bacteriol.* **1995,** *78*, 593–593.

Tseng, A.; Zhao, Y. Wine Grape Pomace as Antioxidant Dietary Fiber for Enhancing Nutritional Value and Improving Storability of Yogurt and Salad Dressing. *Food Chem.* **2013,** *138*(1), 356–365.

Ultee, A.; Bennik, M. H. J.; Moezelaar, R. The Phenolic Hydroxyl Group of Carvacrol Is Essential for Action against the Food-Borne Pathogen *Bacillus cereus*. *Appl. Environ. Microb.* **2002,** *68*(4), 1561–1568.

Valencia, I.; O'Grady, M. N.; Ansorena, D.; Astiasaran, I.; Kerry, J. P. Enhancement of the Nutritional Status and Quality of Fresh Pork Sausages Following the Addition of Linseed Oil, Fish Oil and Natural Antioxidants. *Meat Sci.* **2008,** *80*(4), 1046–1054.

van den Berg, S. J. P. L.; Serra-Majem, L.; Coppens, P.; Rietjens, I. M. C. M. Safety Assessment of Plant Food Supplements (PFS). *Food Funct.* **2011,** *2*, 760–768.

Vasconsuelo, A.; Boland, R. Molecular Aspects of the Early Stages of Elicitation of Secondary Metabolites in Plants. *Plant Sci.* **2007,** *172*(5), 861–875.

Vega-Vega, V.; Silva-Espinoza, B. A.; Cruz-Valenzuela, M. R.; Bernal-Mercado, A. T.; González-Aguilar, G. A.; Vargas-Arispuro, I.; et al. Antioxidant Enrichment and Antimicrobial Protection of Fresh-Cut Mango Applying Bioactive Extracts from their Seeds Byproducts. *Food Nutr. Sci.* **2013,** *4*, 197–203.

Viacava, G. E.; Ansorena, M. R.; Roura, S. I.; González-Aguilar, G. A.; Ayala-Zavala, J. F. Fruit Processing Byproducts as a Source of Natural Antifungal Compounds. Chapter 16. In: *Antifungal Metabolites From Plants: Progress and Prospects;* Mehdi, R.-A., Mahendra, R. Eds.; Springer. XIV, **2013,** *469*, pp 447–462.

Vijayakumar, C.; Wolf-Hall, C. E. Evaluation of Household Sanitizers for Reducing Levels of *Escherichia coli* on Iceberg Lettuce. *J. Food Protect.* **2002,** *65*(10), 1646–1650.

Viuda-Martos, M.; Ruiz-Navajas, Y.; Fernández-López, J.; Pérez-Álvarez, J. A. Effect of Orange Dietary Fiber, Oregano Essential Oil and Packaging Conditions on Shelf-Life of Bologna Sausages. *Food Control* **2010,** *21*(4), 436–443.

Vural, H.; Javidipour, I. Replacement of Beef Fat in Frankfurters by Interesterified Palm, Cottonseed and Olive Oils. *Eur. Food Res. Technol.* **2002,** *214*(6), 465–468.

Walker, R. Criteria for Risk Assessment of Botanical Food Supplements. *Toxicol Lett.* **2004,** *149*, 187–195.

Wallace, T. C.; Giusti, M. M. Determination of Color, Pigment, and Phenolic Stability in Yogurt Systems Colored with Nonacylatedanthocyanins from *Berberis boliviana* L. as Compared to Other Natural/Synthetic Colorants. *J. Food Sci.* **2008,** *73*(4), C241–C248.

Wang, Y.; Li, F.; Zhuang, H.; Chen, X.; Li, L.; Qiao, W.; Zhang, J. Effects of Plant Polyphenols and α-tocopherol on Lipid Oxidation, Residual Nitrites, Biogenic Amines, and N-nitrosamines Formation during Ripening and Storage of Dry-Cured Bacon. *LWT—Food Sci. Technol.* **2015,** *60*, 199–206.

Weisburger, J. H.; Veliath, E.; Larios, E.; Pittman, B.; Zang, E.; Hara, Y. Tea Polyphenols Inhibit the Formation of Mutagens during the Cooking of Meat. *Mutat. Res.* **2002,** *516*, 19–22.

Wilson, A. E.; Bergaentzlé, M.; Bindler, F.; Marchioni, E.; Lintz, A.; Ennahar, S. In Vitro Efficacies of Various Isothiocyanates from Cruciferous Vegetables as Antimicrobial Agents against Foodborne Pathogens and Spoilage Bacteria. *Food Control* **2013,** *30*, 318–324.

Wu, X.; Beecher, G. R.; Holden, J. M.; Haytowitz, D. B.; Gebhardt, S. E.; Prior, R. L. Concentrations of Anthocyanins in Common Foods in the United States and Estimation of Normal Consumption. *J. Agric. Food Chem.* **2006,** *54*(11), 4069–4075.

Xia, E. Q.; Deng, G. F.; Guo, Y. J.; Li, H. B. Biological Activities of Polyphenols from Grapes. *Int. J. Mol. Sci.* **2010,** *11*(2), 622–646.

Yasin, N. M.; Abou-Taleb, M. Antioxidant and Antimicrobial Effects of Marjoram and Thyme in Coated Refrigerated Semi Fried Mullet Fish Fillets. *World J. Dairy Food Sci.* **2007,** *2*(1), 1–9.

Ye, C. L.; Dai, D. H.; Hu, W. L. Antimicrobial and Antioxidant Activities of the Essential Oil from Onion (*Allium cepa L.*). *Food Control* **2013,** *30*(1), 48–53.

Zarei, M.; Ramezani, Z.; Ein-Tavasoly, S.; Chadorbaf, M. Coating Effects of Orange and Pomegranate Peel Extracts Combined with Chitosan Nanoparticles on the Quality of Refrigerated Silver Carp Fillets. *J. Food. Process. Pres.* 2015.

Zeng, W. C.; He, Q.; Sun, Q.; Zhong, K.; Gao, H. Antibacterial Activity of Water Soluble Extract from Pine Needles of *Cedrus deodara*. *Int. J. Food Microbiol.* **2012,** *153*, 78–84.

Zhang, W.; Xiao, S.; Samaraweera, H.; Lee, E. J.; Ahn, D. U. Improving Functional Value of Meat Products. *Meat Sci.* **2010,** *86*, 15–31.

Zhou, C.; Zhang, L.; Wang, H.; Chen, C. Effect of *Amaranthus* Pigments on Quality Characteristics of Pork Sausages. *Asian Austral. Tral. J. Anim.* **2012,** *25*(10), 1493.

Zuo, Y.; Wang, C.; Zhan, J. Separation, Characterization, and Quantitation of Benzoic and Phenolic Antioxidants in American Cranberry Fruit by GC–MS. *J. Agric. Food Chem.* **2002,** *50*, 3789–3794.

CHAPTER 6

EFFECTS OF FOOD PROCESSING TECHNIQUES ON SECONDARY METABOLITES

VASUDHA BANSAL[1], MOHAMMED WASIM SIDDIQUI[2], MADAN LAL SINGLA[3], CHEERUVARI GHANSHYAM[3], and KAMLESH PRASAD[4]

[1]*Center of Innovative and Applied Bioprocessing (CIAB), Mohali, Punjab, India*

[2]*Department of Food Science and Post-Harvest Technology, Sabour, Bhagalpur 813210, Bihar, India*

[3]*Academy of Scientific and Innovative Research, CSIR—Central Scientific Instruments Organisation, Sector 30, Chandigarh 160030, India*

[4]*Department of Food Engineering and Technology, SLIET, Longowal 148106, Punjab, India*

CONTENTS

ABSTRACT

Secondary metabolites in the form of phenolic compounds are known to be potent functional compounds which are not consumed for the energy purposes rather consumed for medicinal properties. In today's world, where consumer needs are more inclined toward the natural products, there is a deep role of advanced processing techniques which can restore the maximum amount of functional compounds in the processed products.

6.1 INTRODUCTION

The wide consumption of botanical foods enriched with functional components has raised the concern for their safety. Functional components in the form of bioactive compounds like phenolics, flavonoids, hydroxycinnamic acids, tannins, and others, are referred as secondary metabolites. These compounds are not consumed for the energy or calorie drive purposes, but rather to strengthen the body, its metabolic activities and immunological functions (Bansal et al., 2014c). These bioactive compounds have also a long history in the therapeutic world in prevention of diseases due to health-rendering phytochemicals as phenolic acids, flavonoids, and vitamins.

Moreover, in the current scenario (as presented in Fig. 6.1), there has been an issue for public health as these secondary metabolites are vulnerable to some technological processing, packaging materials, and storage conditions which incite a direct impact on the nutritional quality and stability of the antioxidants. Also, loss of bioactive components results in deterioration in color, taste, and aroma of the food products. Juices, jams, jellies, squashes, purées, marmalades, and others, from the fresh fruits which are packed with strong antioxidants are in great demand due to their substantial advantages (Cortés et al., 2006; Goh et al., 2012). Therefore, in this aspect, it is utmost important to develop new processing methods, which safeguard their nutritional and sensory levels and also render protection to the bioactivity of the polyphenolic compounds present in their juices.

In recent times, the heat applications involving thermal processing of the foods are the most common methods for preservation due to their aptness to inactivate spoilage enzyme and microorganisms (Rawson et al., 2011). However, number of such processes involve high temperature treatment which affects the concentration of the secondary metabolites. The reason of favoring thermal processing techniques is the intangible operation and facile management of the equipment used (Soria & Villamiel, 2010). Nevertheless, thermal processing, specifically under the severe condition is likely to trigger physicochemical changes which further can minimize the bioavailability of secondary metabolites along with diminishing the organoleptic properties (Patras et al., 2009, 2010). Products of fruits which are contemplated as natural nutritional supplements over the world are commonly processed with conventional heat treatments and degradation of some of the secondary metabolites has been reported with the usage of high temperature which affects the contents of functional compounds and cause undesirable biochemical and nutritional changes (Buchner et al,. 2006; Suárez-Jacobo et al., 2011). Therefore, nonthermal techniques are becoming important processing methods (Khan et al., 2010; Ahmad & Langrish, 2012).

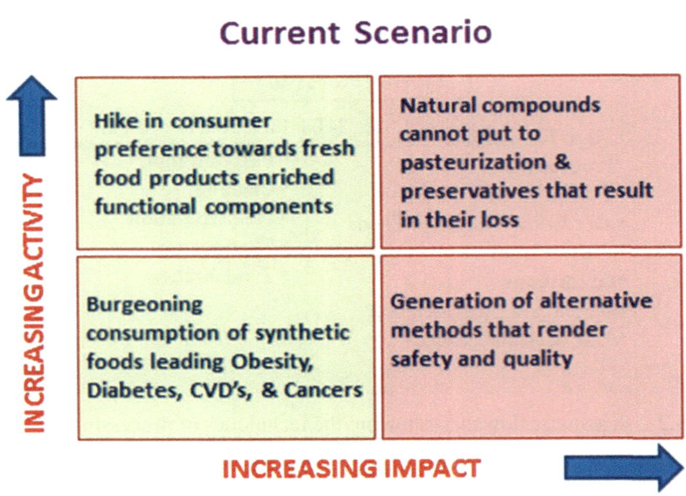

FIGURE 6.1 Schematic representation of current needs of consumer.

Juices, squashes, purées, and jellies of traditional fruits are commonly preserved using (1) thermal pasteurization methods, (2) addition of excessive sugars, and (3) use of chemical preservatives such as sorbates and benzoates. Phenolic constituents are heat labile and oxidized/degraded with harsh processing methods employed under thermal treatments in order to prolong the shelf life of fruit juices. These methods lead to significant degradation of phenolic constituents (Bansal et al., 2014a). The deleterious effects of abrasive processing methods and addition of artificial additives have drifted the interest of researchers toward the gentle yet reliable processing techniques (as shown in Fig. 6.2) which can preserve the freshness of natural food products possessing therapeutic role. Therefore, nonthermal treatments are alternative processing techniques which are emerging for attaining better shelf life with retained functional and organoleptic properties (Bansal et al., 2015).

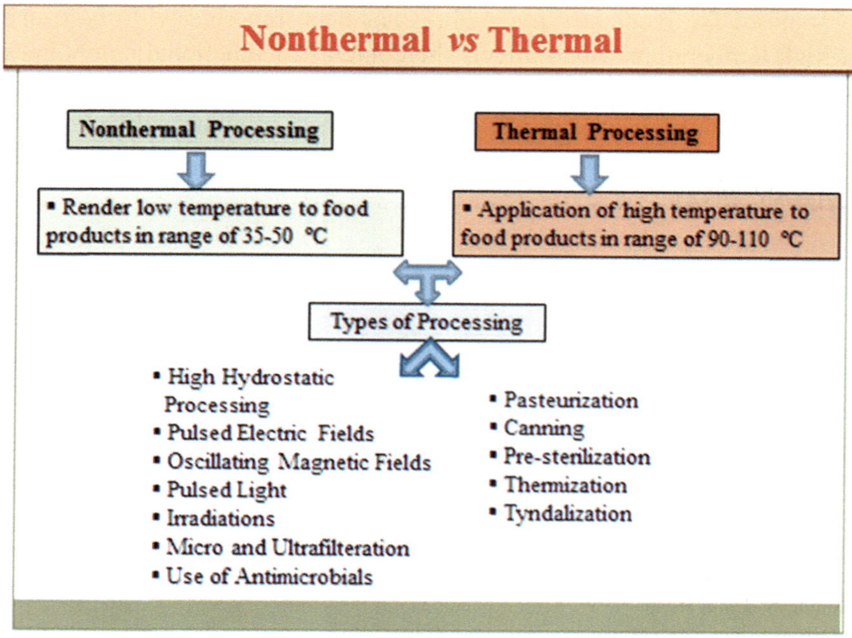

FIGURE 6.2 Schematic flowchart showing the techniques of processing.

6.2 THERMAL PROCESSING

Pasteurization is one of the widely used technologies which has been in usage since ages in the treatment of fruits and their diverse products. The key player of thermal operations for sustaining the quality of the products is the combination of time and temperature for reducing the significant number of microorganisms (Ramaswamy et al., 2005). Thermal processing is further categorized as low temperature long time at about 63°C for 30 min and high temperature short time at 75–90°C for 15–60 s (US FDA, 2000). Although higher temperature results in substantial killing of the microorganisms, it also leads to detrimental changes in food constituents, such as vitamins, proteins, and organoleptic characteristics such as color, flavor, and aroma (Rawson et al., 2011). In addition to nutritional losses, higher reduction of the functional components such as phenolic compounds is also the usual phenomenon (Chen et al., 2013) which can be regarded as one of the adverse effects of thermal pasteurization. These physical treatments involving the usage of elevated temperatures may cause the chemical and physical alterations affecting the concentration and transformation of phenolics (Rossi et al., 2010). These changes have led to investigation of the feasibility of nonthermal technologies for maximum inactivation of microbial load and minimum possible degradation of food quality (Lado & Yousef, 2002).

6.2.1 EFFECTS OF THERMAL PROCESSING ON SECONDARY METABOLITES

Thermal processing consists of heating methods that include the usage of electric heater, heat exchanger, conduction, and convention components (Pereira & Vicente, 2010). Since fruits are highly perishable in nature, their preservation is solely dependent on thermal processing. These methods are largely pertinent to the processing of juices. However, along with rendering the microbiological protection, heat treatments also result in deleterious reactions to food quality including nutrient loss, degradation of organoleptic properties such as color, texture, flavor, and also cause alterations in the bioactive content of the food products. Owing to their harshness, heat processes lead to the minimization of the phenolic compounds in the foodstuff (Elez-Martínez et al., 2006; Bansal et al., 2014b). Bioactive components are basically non-nutritive phytochemicals that act as scavengers in the human body against diseases (Oms-Oliu et al., 2013) and engulf the free radicals which are responsible for oxidative damage to proteins and degeneration of the cellular tissues. These phytochemicals are also called secondary metabolites that give color and flavor to the foodstuffs. Some of the important thermal studies on various traditional fruits which have been reported recently in literature are listed in Table 6.1.

TABLE 6.1 Effects of Thermal Processing on Secondary Metabolites.

S. No.	Fruit	Type	Conditions	Compounds affected	Reference
1.	Cashew apple	Juice	Heated at 60 and 90°C for 1, 2, and 4 h	↓Xanthophylls, ↓cis isomers of carotenoids, ↓total carotenoids	Zepka and Mercadante (2009)
2.	Durian	Juice	Drying at 88–130°C	↓Flavor volatile	Chin et al. (2010)
3.	Pitanga	Juice/pulp	Industrially pasteurized	↓Quercetin, ↓kaempferol, ↓myrecetin	Hoffmann-Ribani et al. (2009)
4.	Acerola	Juice/pulp	Industrially pasteurized	↓Quercetin, ↓kaempferol	Hoffmann-Ribani et al. (2009)
5.	Mulberry	Fruit	Stored at 70°C for 10 h	↓Ascorbic acid, ↓total anthocyanins	Aramwit et al. (2010)
6.	Noni	Juice/purée	Heating at 65 and 75°C for 24 h	↓Free radical scavenging activity	Yang et al. (2007)
	Noni	Juice/purée	Dehydrating at 50°C for 14 h	↓Free radical scavenging activity	Yang et al. (2007)
7.	Pink guava	Purée	Drying at 43.79, 50, 65, 80, and 86.21°C for 4–6 h	↓Lycopene, ↓antioxidant capacity	Kong et al. (2010)
8.	Jack fruit	Bulb slices	Hot air drying at 50, 60, and 70°C	↓Total carotenoids	Saxena et al. (2012)
9.	Cupuacu	Nectar	Thermally treated at 60–99°C for 0–240 min	↓Acsorbic acid, ↓dehydro-ascorbic acid	Vieira et al. (2000)
10.	Acai	Purée	Processed at 80°C for 1, 5, 10, 30, and 60 min	↓Polyphenolic content, ↓anthocyanins	Pacheco-Palencia et al. (2009)
11.	Pineapple	Juice	Thermally pasteurized at 80°C for 10 min	↓Ascorbic acid, ↓total phenolic content	Chia et al. (2012)
12.	Cranberry	Juice blend	Conventional pasteurization at 72°C for 26 s	↓Total polyphenols	Caminiti et al. (2011)

They are ubiquitously present in vegetables, fruits, seeds, wines, juices, and tea. Botanical supplements are the sources of phenolic constituents and majority of them are hydroxycinnamic acids and flavonoids. Beverages with plentiful of polyphenols especially flavonoids are predominantly protective against the risks of cancer and cardiac diseases (Klimczak et al., 2007). However, bioactive components are heat-sensitive and their concentration gets reduced at the temperature range of 70–110°C that is usually needed for pasteurization (Pereira & Vicente, 2010). This is the reason that processed fruits and vegetables are supposed to have lower antioxidant capacity in comparison to their fresh products (Odriozola-Serrano et al., 2009).

Different thermal methods such as drying, autoclaving, pasteurization, pressured-steam heating, water immersing, blanching showed the significant degradation of the bioactive compounds. Viña and Chaves (2008) reported the ruination of total flavonoids in boiled celery which is processed at the temperature of 50°C. Zhang et al. (2010) reported the deterioration of 25% flavonoids on steam heating the products of buckwheat for about 40 min. Similarly, the pasteurization treatment deteriorated naringin, rutin, quercetin, and naringenin content in juice of grape fruit (Igual et al., 2011). However, no impact on the content of quercetin and kaempferol was observed on thermal treatment of strawberry juices by Odriozola-Serrano et al. (2008) which may be accredited to the different composition of the fruit matrix for acting as heat barriers against degradation (Irina & Mohamed, 2012). The most intense degradation of one of the important flavonoids, quercetin, was observed on processing with thermal treatment (Buchner et al., 2006). Additionally, losses of bioactive compounds succeeding heat treatments are also dependent on the chemical structure of the molecules, nature of bonds which further can support in their sustenance.

In spite of improvement in thermal processing techniques, use of mild temperatures often causes degradation of nutritional/sensory characteristics, specifically to the heat-sensitive compounds. Thus, alternative nonthermal processing techniques are equivalent to thermal processing in their safety action and offer the quality characteristics as well.

6.3 EFFECTS OF NONTHERMAL PROCESSING TECHNIQUES ON SECONDARY METABOLITES

Nonthermal technology includes pulsed electric field (PEF), high hydrostatic pressure (HHP), oscillating magnetic fields, irradiation, use of ultrasound and ozone treatment, and so on. Owing to the effective role of HHP/PEF treatment on the bioactive compounds, these are established as techniques which render functionalities to the treated foods (Schilling et al., 2007). Such processing techniques are termed as nonthermal techniques, and have got good response worldwide due to their numerous benefits (as shown in Fig. 6.3). These methods employ the criterion of using lower temperature conditions than thermal techniques to keep organoleptic and nutritive aspects intact.

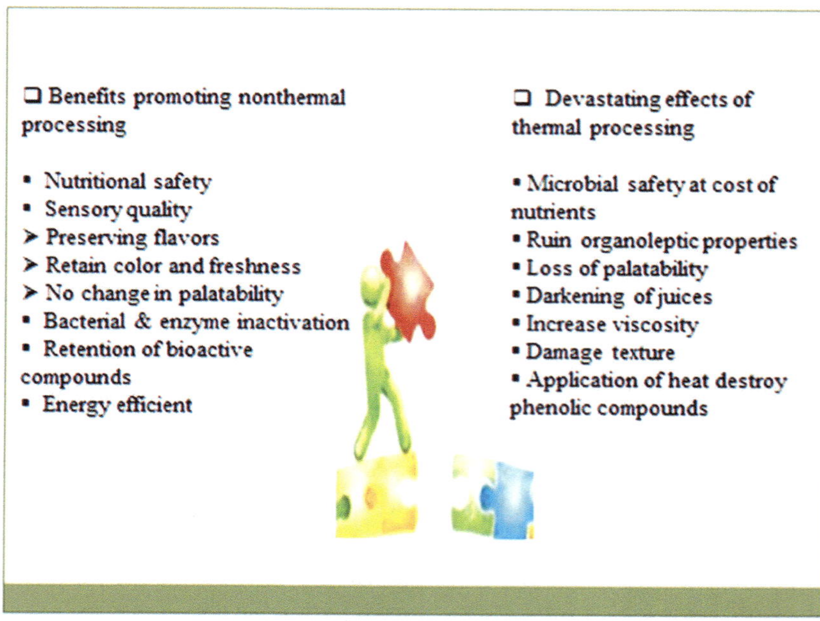

FIGURE 6.3 Schematic representation of gaps filled by nonthermal processing techniques.

6.3.1 PULSED ELECTRIC FIELD (PEF)

PEF induced recovery of plant oils and resulted in liberation of intracellular pigments and increased production of secondary metabolites in the plants of maize, olives, and soybeans (Guderjan et al., 2005). Production of secondary metabolites from wine grapes was exhibited where 13–28% of polyphenol content was found to be increased on application of 0.5–2.4 kV cm^{-1} with 50 pulses (Balasa et al., 2006). Further, release of intracellular anthocyanins from wine grapes was reported by Tedjo et al. (2002) where the application of electric field strength of 3 kV cm^{-1} at 50 pulses was studied which resulted in three fold increase in the amount of total polyphenolic content. The findings are in concordance with the theory that higher release of intracellular compounds is related to higher degree of permeabilization by PEF strength reported by Eshtiaghi and Knorr (2002). It was observed that high amount of amaranthin content (85%) from *Corallium rubrum* was released after PEF treatment at 1.6 kV cm^{-1} for 10 pulses. On the other hand, only 5.7% of anthraquinones from *Morinda citrifolia* was released on application of similar PEF treatment (Dörnenburg & Knorr, 1993). This signified the varied effect of PEF on the pigments present in the different compartment of the plant cells. On the contrary, the extractability of beetroot was investigated where the beetroot pigment, betalain, was found to be extracted on applying 270 pulses at electric field strength of 1 kV cm^{-1} and results exhibited no effects of compartments on different permeabilization (Fincan et al., 2004). Also, the color of the beetroot juice was observed to be enriched due to 90% of color release in isotonic solution. PEF treatment at 2.6 kV cm^{-1} for 5–100 pulses was applied to determine the content of β-carotene in carrot juice. However, no significant result was observed in the content (Geulen et al., 1994). The content of vitamin C and β-carotene was studied in bell peppers by applying PEF from 0.5 to 2.5 kV cm^{-1}. An increased amount of vitamin C was found at electric field strength of 2.5 kV cm^{-1} but lower reduction of β-carotene was exhibited owing to the usage of elevated temperature (Ade-Omowaye et al., 2002).

Thus, PEF with low to moderate intensity showed several advantages in the form of better efficiency of pressing, rapid extraction processes, and increased juice yield over other methods. However, due to numerous variations in PEF processing parameters such as electric field strength, treatment time, pulse width, frequency along with different designing of chambers makes the comparison and differentiation difficult and tangible. This demands regularity and standardization of equipment along with PEF processing parameters such as treatment chamber, treatment time, pulse duration, number of pulses, shape of pulses, energy input, and temperature variations.

Very little information is available related to the effect of high-intensity electric field on variegated and complex phenolic compounds which exist in the matrix of food. These compounds are further responsible for providing color to the fruits due to the higher concentration of anthocyanins and flavonoids. Inconsiderable number of studies have evaluated the effect of PEF on the qualitative and quantitative content of these compounds. The changes in color on the application of PEF to cranberry were reported by Jin and Zhang (1999) at 40 kV cm^{-1} for 150 μs and indicated nonsignificant differences in the color modifications of the juice. Similarly, flavonoid content of orange juice was observed at 35 kV cm^{-1} for 850 μs in the frequency range of 800 Hz, where variations were observed in the individual content of flavonoids (Sánchez-Moreno et al., 2005). However, nonsignificant decrease in the total carotenoids of the orange juice was observed by Cortés et al. (2006) applying the treatments at 25–50 kV cm^{-1} for 30–340 μs leading to prominent retention of antioxidant capacity of the juices. Their concentration was also found to be stable during the storage period for retaining the shelf life of orange juice. Table 6.2 provides a summary of studies on the effect of PEF treatment on the bioactive compounds.

In case of citrus juices, number of studies exhibited nonsignificant losses on the carotenoids and flavonoids (Yeom et al., 2000; Elez-Martinez et al., 2007; Cserhalmi et al., 2006). In contrast, some compounds were observed to be affected by PEF. The individual carotenoids like α-carotene, β-carotene, α-cryptoxanthin, and β-cryptoxanthin were decreased in the mixture of orange and carrot juice (Torregrosa et al., 2005). These variations were due to processing at long durations.

Lycopene is one of the major pigments, and tomatoes are its richest sources and it has been studied by different processing of PEF. Tomato juice was processed at 40 kV cm^{-1} for 57 μs and compared with thermal processing. Significant changes were observed in the lycopene concentrations and increased action of nonenzymatic browning was detected in thermal processing (Min et al., 2003). Watermelon juice was subjected to PEF (35 kV cm^{-1} for 50 μs) and exhibited maximal content of lycopene (Oms-Oliu et al., 2009). Cranberry is known for its higher content of anthocyanins and PEF (34 kV cm^{-1} for 93 μs) resulted in increased retention than thermal treatment.

Due to vulnerability of phenolic compounds to high temperature, vitamin content of foods also gets affected whereas high-intensity PEF have been observed to cause less change. An orange-based beverage fortified with protein processed with PEF at 40 kV cm^{-1} resulted in 4–13% reduction of vitamin C content (Sharma et al., 1998). The other authors reported negligible decrease in the content of vitamin C in the juices of grape (Wu et al., 2005), orange (Yeom et al., 2000; Hodgins et al., 2002; Plaza et al., 2006), gazpacho cold vegetable soup (Elez-Martinez et al., 2007), orange–carrot juice (Torregrosa et al., 2006), blueberry juice (Barba et al., 2012), and watermelon juice (Oms-Oliu et al., 2009).

TABLE 6.2 Effects of PEF Treatment on Bioactive Compounds.

Treated juice media	Compounds	Treatment conditions	Effects	Reference
Tomato	Carotenoids	40 kV cm^{-1} for 57 μs	No alterations in the concentration of lycopene	Min et al. (2003)
Orange	Carotenoids	25–40 kV cm^{-1} for 30–340 μs	Stability of compounds in comparison to thermal treatment	Cortés et al. (2006)
Orange-carrot blend	Carotenoids	25–40 kV cm^{-1} for 30–340 μs	Enhancement in the concentration of carotenoids compared to heat treatment	Torregrosa et al. (2005)
Orange	Flavonoids	35 kV cm^{-1} for 750 μs	No significant changes in the content of total and individual flavonone	Sánchez -Moreno et al. (2005)
Protein-fortified orange blend	Vitamin C	28 kV cm^{-1} for 100–300 μs	Limited the reduction to 4–13%	Sharma et al. (1998)
Grape	Vitamin C	65 kV cm^{-1} for 20 pulses	No reduction	Wu et al. (2005)
Orange	Vitamin C	80 kV cm^{-1} for 20 pulses	2.5% reduction	Hodgins et al. (2002)
Orange	Vitamin C	35 kV cm^{-1} for 750 μs	14–18% reduction	Plaza et al. (2006)
Orange	Vitamin C	35 kV cm^{-1} for 59 μs	No significant difference in the concentration as compared to thermal processing	Yeom et al. (2000)
Gazpacho vegetable soup	Vitamin C	15–35 kV cm^{-1} for 100–1000 μs	8.0–12.5% reduction	Elez-Martinez and Belloso (2007)
Orange-carrot blend	Vitamin C	25 kV cm^{-1} for 280–330 μs	13% reduction	Torregrosa et al. (2006)
Blueberry	Vitamin C	36 kV cm^{-1} for 100 μs	Losses of 31%	Barba et al. (2012)
Cranberry	Anthocyanins	34 kV cm^{-1} for 93 μs	Increased retention	Caminiti et al. (2011)
Watermelon	Lycopene	35 kV cm^{-1} for 50 μs	113% increased concentration	Oms Oliu et al. (2009)
Watermelon	Vitamin C	35 kV cm^{-1} for 50 μs	28% reduction	Oms Oliu et al. (2009)

PEF treatment resulted in retaining high level of phenolic compounds in fruit juices, thereby enhancing their stability during storage. Odriozola-Serrano et al. (2008) reported 49% less phenolic degradation by PEF than 55% caused in thermal pasteurization after 56 days of storage. Application on PEF was studied at 35 kV cm^{-1} using bipolar pulses at the frequency of 1200 pps (pulses per second) on apple juice and found only 14.49% degradation in total polyphenols as compared to 32.2% in thermal treatment (Aguilar-Rosas et al., 2007). Also, Puértolas et al. (2010) found that total phenolic content was 22% higher on application of PEF to *Cabernet sauvignon* red wines in comparison to untreated wines after 4 months of storage. The application of high-intensity PEF in food processing has been studied as a possible treatment to stimulate stress reactions in plant systems to enhance or stimulate the production of secondary plant metabolites like polyphenols (Toepfl, 2011) and offer consequent generation of reactive oxygen species (Vallverdú-Queralt, 2012). Reactive oxygen species are known to be the part of defense response of plants to stress and are required to trigger the synthesis of polyphenols (Shohael et al., 2006). Vallverdú-Queralt et al. (2012) reported the enhancement of polyphenolic compounds (phenolic acids and flavanones) in juices from moderate-intensity pulsed electric field-processed tomatoes and increased level of chlorogenic acid and naringenin-7-*O*-glucoside was observed when compared with untreated tomato juice. Lethal damage to cells induced by permeabilizing tissue structures improved the intracellular metabolite extraction. PEF treatment has added another dimension for liberating intracellular compounds by breaking the cell linings and resulting in their permeabilization (Luengo et al., 2013). Earlier studies reporting the PEF treatment-based better retention of phenolic compounds are listed Table 6.3.

TABLE 6.3 Exemplification of High Intensity PEF Processing on the Retention of Phenolic Contents of Fruit Juices.

S. No.	Fruit	Treatment conditions	Phenolic degradation (%)	Reference
1.	Orange juice	35 kV cm^{-1}, 1000 μs	15	Elez-Martínez et al. (2006)
2.	Longan juice	32 kV cm^{-1}, 90 s	18	Zhang et al. (2010)
3.	Apple juice	40 kV cm^{-1}, 100 μs	7	Noci et al. (2008)
4.	Apple juice	35 kV cm^{-1}, 4800 μs	14.5	Aguilar-Rosas et al. (2007)
5.	Apple juice	30 kV cm^{-1}, 6 ms	26	Ertugay et al. (2013)
6.	Watermelon	35 kV cm^{-1}, 50 μs	NSS	Oms-Oliu et al. (2009)

TABLE 6.3 *(Continued)*

S. No.	Fruit	Treatment conditions	Phenolic degradation (%)	Reference
7.	Tomato juice	35 kV cm^{-1}, 1500 μs	NSS	Odriozola-Serrano et al. (2009)
8.	Blueberry juice	36 kV cm^{-1}, 100 μs	NSS	Barba et al. (2012)
9.	Strawberry juice	35 kV cm^{-1}, 1700 μs	49	Odriozola-Serrano et al. (2008)
10.	Blend of apple and cranberry juice	34 kV cm^{-1}, 93 μs	NSS	Caminiti et al. (2011)
11.	Blend of soymilk and fruit juice	35 kV cm^{-1}, 800 μs and 1400 μs, 4 μs bipolar pulses	NSS	Morales-de La Peña et al. (2010)

NSS: Nonstatistically significant.

6.3.2 ULTRASONICATION PROCESSING

It involves the application of alternating electric currents in which ultrasonic waves are attained through ultrasonic transducer (Ortega-Rivas, 2012). Propagation of ultrasound to biological structure causes compressions and depressions and high energy is imparted to the treated medium. It is the high power of the ultrasound which damages the biological cell walls leading to destruction of living cells (Bermúdez-Aguirre et al., 2011). The other application of extracting bioactive compounds using ultrasonication has added to food technology. The energy generated from high powered ultrasound penetrates into the cellular material of plants which improves the mass transfer (Li et al., 2004; Vilkhu et al., 2008). The application of ultrasonication has been initiated in the food industry at the commercial front as well. The laboratory-based experimentation has lasted for 5 years in Europe and the USA (Patist & Bates, 2008). There are numerous applications where ultrasonication has shown realistic output such as emulsification, extraction, homogenization, crystallization, viscosity transformations, degassing, defoaming, extraction, pasteurization, and inactivation of enzymes and microbes.

The role of ultrasonication on the organic compounds of plants has captured the major attention owing to its nondevastating effect on the concentration or structure of secondary metabolites (Balachandran et al., 2006). It offers the advantages in terms of increased yield, productivity, enhanced quality, being environmental friendly, and decreased effect on physical and chemical properties of the secondary metabolites (Chemat & Khan, 2011). The usage of ultrasonic processing has been explored in herbal plants, oil, protein, and bioactives from the plants (such as polyphenols and flavanoids). Many of the reports have shown the increased yield of secondary metabolites from various plant-based products. Polyphenols, amino acid, and caffeine from green tea (Xia et al., 2006) increased soy protein and soy isoflavanones (Moulton & Wang, 1982) and also increased concentration of rutin from Chinese scholar tree (Paniwynk et al., 2001) and carnosic acid from rosemary (Albu et al., 2004).

Therefore, ultrasonication holds the potential for safeguarding the assemblage and concentration of secondary metabolites from all sources of plants by processing the heat-sensitive bioactive compounds at lower temperatures.

6.3.3 HIGH HYDROSTATIC PROCESSING

HHP processing is one of the nonthermal treatment techniques in which food product is treated with static pressure from 50 to 1000 MPa to solid/packaged liquid food product with varying treatment time of seconds to minutes (Williams, 1994). The paramount cause of the cell damage in this process is cell membrane permeabilization due to irreversible changes in the structural membranes of proteins (Suárez-Jacobo et al., 2011). The HHP processing is currently used for jams, sauces, soups, and others (Leistner & Gould, 2002). Reported HHP studies have showed minimal effect on the nutritional content and bioactive profile of food products (Oey et al., 2008). It was observed that permeabilization was affected by compression leading to reduction in the cross-sectional area of the bilayered membrane (Rendueles et al., 2011). Further, HHP has been observed to ameliorate the polyphenolic profile of the treated fruit products by releasing their secondary metabolites from the tissues under the membrane (Ancos et al., 2000). It has been reported that prolonged processing resulted in stimulating the anthocyanins from the food products.

HHP processing has been extensively noticed in the food processing with the greater advantage of its application of uniform pressure in all the directions (Oey et al., 2008). Further, HHP has been reported to have minimum effect on the content of secondary metabolites of various fruits and vegetables. Similarly the effects of high pressure (400–600 MPa) were investigated on the phytochemicals (such as anthocyanins and polyphenols) present in the pomegranate juice (Ferrari et al., 2010). More than the application of preservation, HHP has been used for the perseverance of the secondary metabolites of fruits and vegetables (Rawson et al., 2011). The diffusion of secondary metabolites takes place through the treatment of HHP which further increases the cell permeability of these compounds. The main cause behind the cell permeability is the breaking of hydrophobic bonds and salt bridges among the secondary metabolites which trigger the liberation of bioactive compounds from the peel, pulp, skin, or juice (Rawson et al., 2011). Therefore, HHP retains the antioxidant capacity in the treated plant-based supplements and is a prospective technique in the food and pharmaceutical industry.

6.3.4 IRRADIATION PROCESSING

The irradiation processing is of two types, ionizing and non-ionizing. The source of ionization can be X-rays, gamma rays, or high-energy electrons, whereas the non-ionizing radiations can be from ultraviolet rays, infrared rays, microwaves, and others. The minimal effect of irradiation over the organoleptic properties of food products and functional components of food has been reported (Wood & Bruhn, 2000; Siddiqui et al., 2014). This advantage makes irradiation to render least effect on nutrients, color, flavor, and content of secondary metabolites (Rawson et al., 2011).

Owing to that no major changes were observed in the concentration of carotenoids, flavonoids, and polyphenol content of mango (Reyes & Cisneros-Zevallos, 2007) with the range of application till 1–3.1 kGy. Likewise, ultraviolet treatment was performed on pineapple, banana, and guava (Alothman et al., 2009). The findings exhibited no noticeable change in reduction of total polyphenols. On the contrary, the level of flavonoids got increased. Similar study was reported by Lopez-Rubira et al. (2005), where nonsignificant alterations were observed in the concentration of anthocyanins and antioxidant capacity of pomegranate arils on the application of ultraviolet treatment (0.56–13.62 kJ/m^2). The irradiation processing of the fruits leads to biosynthesis of bioactive compounds due to the metabolic stress which stimulates the liberation of these metabolites. Therefore, it is discernible from the studies that irradiation has the upcoming major role in safeguarding the composition of functional constituents of heat-liable fruits and vegetables.

6.4 CONCLUSION

Replacing or complementing thermal processes to offer retained bioactive metabolites to consumers at appropriate cost is an interesting proposition. Therefore, nonthermal methods seem to be an alternative because of the low processing temperature employed. It can meet the suppositions for minimally processed foods of fresh quality, which maintains nutritional value in addition to color and flavor retention.

KEYWORDS

- **functional component**
- **phytochemicals**
- **bioavailability**
- **pasteurization**
- **secondary metabolites**

REFERENCES

Ade-Omowaye, B.; Rastogi, N.; Angersbach, A.; et al. Osmotic Dehydration of Bell Peppers: Influence of High Intensity Electric Field Pulses and Elevated Temperature Treatment. *J. Food Eng.* **2002,** *54*(1), 35–43.

Aguilar-Rosas, S.; Ballinas-Casarrubias, M.; Nevarez-Moorillon, G.; et al., Thermal and Pulsed Electric Fields Pasteurization of Apple Juice: Effects on Physicochemical Properties and Flavour Compounds. *J. Food Eng.* **2007,** *83*(1), 41–46.

Ahmad, J.; Langrish, T, Optimisation of Total Phenolic Acids Extraction from Mandarin Peels using Microwave Energy: The Importance of the Maillard Reaction, *J. Food Eng.* **2012,** *109*(1), 162–174.

Albu, S.; Joyce, E.; Paniwnyk, L.; Lorimer, P.; Mason, J. Potential for the Use of Ultrasound in the Extraction of Antioxidants from *Rosmarinus officinalis* for the Food and Pharmaceutical Industry. *Ultrason. Sonochem.* **2004,** *11*, 261–265.

Alothman, M.; Bhat, R.; Karim, A. A. UV Radiation-Induced Changes of Antioxidant Capacity of Fresh-cut Tropical Fruits. *Innovative Food Sci. Emerg. Technol.* **2009,** *10*, 512–516.

Ancos, B. de.; Gonzalez, E.; Cano, M. P. Effect of High-Pressure Treatment on the Carotenoid Composition and the Radical Scavenging Activity of Persimmon Fruit Purees, *J. Agric. Food Chem.* **2000,** *48*(8), 3542–3548.

Aramwit, P.; Bang, N.; Srichana, T. The Properties and Stability of Anthocyanins in Mulberry Fruits. *Food Res. Int.*, **2010,** *43*(4), 1093–1097.

Balachandran, S.; Kentish, S. E.; Mawson, R.; Ashokkumar, M. Ultrasonic Enhancement of the Supercritical Extraction from Ginger. *Ultrason. Sonochem.* **2006,** *13*, 471–479.

Balasa, A.; Toepfl, S.; Knorr, D. *Pulsed Electric Field Treatment of Grapes. Food Factory of the Future 3*. Elsevier: Gothenburg, Sweden, 2006.

Bansal, V.; Naiyyar, A. Md.; Siddiqui W. Md.; Ahmad, S. Md. *Res. J. Pharm., Biol. Chem. Sci.* **2014c,** *5*(3), 148–155.

Bansal, V.; Sharma, A.; Ghanshyam, C.; Singla, M. L. Coupling of Chromatographic Analyses with Pretreatment for the Determination of Bioactive Compounds in *Emblica officinalis* juice. *Anal. Methods*, **2014a,** *6*, 410–418.

Bansal, V.; Sharma, A.; Ghanshyam, C.; Singla, M. L. Optimization and Characterization of Pulsed Electric Field Parameters for Extraction of Quercetin and Ellagic Acid in *Emblica officinalis* Juice. *J. Food Meas. Charact.* **2014b,** *8*(3), 225–233.

Bansal, V.; Siddiqui, W. Md.; Rahman, S. Md. In *Minimally Processed Foods: Technologies for Safety, Quality, and Convenience*; Siddiqui, Md. W., Rahman, Md. S., Eds.; Springer: New York, 2015; 1st ed., Chapter: 01, pp 1–15.

Barba, F.; Jäger, H.; Meneses, N.; et al. Evaluation of Quality Changes of Blueberry Juice during Refrigerated Storage after High-Pressure and Pulsed Electric Fields Processing, *Innovative Food Sci. Emerg. Technol.* **2012,** *14*, 18–24.

Bermúdez-Aguirre, D.; Mobbs, T.; Barbosa-Cánovas, G. V. Ultrasound Applications in Food Processing. *Ultrasound Technologies for Food and Bioprocessing*; Springer: New York, 2011; pp 65–105.

Buchner, N.; Krumbein, A.; Rohn, S.; et al., Effect of Thermal Processing on the Flavonolsrutin and Quercetin, *Rapid Commun. Mass Spectrom.* **2006,** *20*(21), 3229–3235.

Caminiti, I. M.; Noci, F.; Muñoz, A.; et al. Impact of Selected Combinations of Non-Thermal Processing Technologies on the Quality of an Apple and Cranberry Juice Blend, *Food Chem.* **2011,** *124*(4), 1387–1392.

Chemat, F.; Khan, M. K. Applications of Ultrasound in Food Technology: Processing, Preservation and Extraction. *Ultrason. Sonochem.* **2011,** *18*(4), 813–835.

Chen, Y.; Yu, L. J.; Rupasinghe, H. Effect of Thermal and Non-Thermal Pasteurisation on the Microbial Inactivation and Phenolic Degradation in Fruit Juice: A Mini-Review, *J. Sci. Food Agric.* **2013,** *93*(5), 981–986.

Chia, S.; Rosnah, S.; Noranizan, M.; et al. The Effect of Storage on the Quality Attributes of Ultraviolet-Irradiated and Thermally Pasteurised Pineapple Juices. *Int. Food Res. J.* **2012,** *19*(3), 1001–1010.
Chin, S. T.; Hamid Nazimah, S. A.; Quek, S. Y.; et al. Effect of Thermal Processing and Storage Condition on the Flavour Stability of Spray-Dried Durian Powder. *LWT—Food Sci. Technol.* **2010,** *43*(6), 856–861.
Cortés, M.; Esteve, M.; Rodrigo, D.; et al. Changes of Colour and Carotenoids Contents during High Intensity Pulsed Electric Field Treatment in Orange Juices. *Food Chem. Toxicol.* **2006,** *44*(11), 1932–1939.
Cserhalmi, Z.; Sass-Kiss, A.; Tóth-Markus, M.; Lechner, N. Study of Pulsed Electric Field Treated Citrus Juices. *Innov. Food Sci. Emerg. Technol.* **2006,** *7*(1), 49–54.
Dörnenburg, H.; Knorr, D. Cellular Permeabilization of Cultured Plant Tissues by High Electric Field Pulses or Ultra High Pressure for the Recovery of Secondary Metabolites. *Food Biotechnol.* **1993,** *7*(1), 35–48.
Elez-Martínez, P.; Aguiló-Aguayo, I.; Martín Belloso, O. Inactivation of Orange Juice Peroxidase by High-Intensity Pulsed Electric Fields as Influenced by Process Parameters. *J. Sci. Food Agric.* **2006,** *86*(1), 71–81.
Elez-Martinez, P.; Martin-Belloso, O. Effects of High Intensity Pulsed Electric Field Processing Conditions on Vitamin C and Antioxidant Capacity of Orange Juice and Gazpacho, a Cold Vegetable Soup. *Food Chem.* **2007,** *102*(1), 201–209.
Elez-Martínez, P.; Soliva-Fortuny, R. C.; Martín-Belloso, O. Comparative Study on Shelf Life of Orange Juice Processed by High Intensity Pulsed Electric Fields or Heat Treatment. *Eur. Food Res. Technol.* **2006,** *222*(3–4), 321–329.
Ertugay, M. F.; Başlar, M.; Ortakci, F. Effect of Pulsed Electric Field Treatment on Polyphenol Oxidase, Total Phenolic Compounds, and Microbial Growth of Apple Juice. *Turk. J. Agric. For.* **2013,** *37*(6), 772–780.
Eshtiaghi, M.; Knorr, D. High Electric Field Pulse Pretreatment: Potential for Sugar Beet Processing. *J. Food Eng.* **2002,** *52*(3), 265–272.
Ferrari, G.; Maresca, P.; Ciccarone, R. The Application of High Hydrostatic Pressure for the Stabilization of Functional Foods: Pomegranate Juice. *J. Food Eng.* **2010,** *100*, 245–253.
Fincan, M.; DeVito, F.; Dejmek, P. Pulsed Electric Field Treatment for Solid–Liquid Extraction of Red Beetroot Pigment. *J. Food Eng.* **2004,** *64*(3), 381–388.
Geulen, M.; Teichgraeber, P.; Knorr, D. Zellaufscluss by Electrical High Voltage Pulses. *Food Technol.* ZFL, **1994,** *45*(7/8), 24–27.
Goh, S.; Noranizan, M.; Leong, C.; et al. Effect of Thermal and Ultraviolet Treatments on the Stability of Antioxidant Compounds in Single Strength Pineapple Juice throughout Refrigerated Storage. *Int. Food Res. J.* **2012,** *19*(3), 1131–1136.
Guderjan, M.; Töpfl, S.; Angersbach, A.; et al. Impact of Pulsed Electric Field Treatment on the Recovery and Quality of Plant Oils. *J. Food Eng.* **2005,** *67*(3), 281–287.
Hodgins, A.; Mittal, G.; Griffiths, M. Pasteurization of Fresh Orange Juice Using Low-Energy Pulsed Electrical Field, *J. Food Sci.* **2002,** *67*(6), 2294–2299.
Hoffmann-Ribani, R.; Huber, L. S.; Rodriguez-Amaya, D. B. Flavonols in Fresh and Processed Brazilian Fruits. *J. Food Compos. Anal.* **2009,** *22*(4), 263–268.
Igual, M.; García-Martínez, E.; Camacho, M. M.; Martínez-Navarrete, N. Changes in Flavonoid Content of Grapefruit Juice Caused by Thermal Treatment and Storage. *Innov. Food Sci. Emerg. Technol.* **2011,** *12*(2), 153–162.

Irina, I.; Mohamed, G. Biological Activities and Effects of Food Processing on Flavonoids as Phenolic Antioxidants. *Adv. Appl. Biotechnol.* **2012,** 101–124.

Jin, Z.; Zhang, Q. H, Pulsed Electric Field Inactivation of Microorganisms and Preservation of Quality of Cranberry Juice. *J. Food Process. Preser.* **1999,** *23*(6), 481–497.

Khan, M. K.; Abert-Vian, M.; Fabiano-Tixier, A. S.; et al. Ultrasound-Assisted Extraction of Polyphenols (flavanone glycosides) from Orange (*Citrus sinensis* L.) Peel. *Food Chem.* **2010,** *119*(2), 851–858.

Klimczak, I.; Małecka, M.; Szlachta, M.; et al. Effect of Storage on the Content of Polyphenols, Vitamin C and the Antioxidant Activity of Orange Juices, *J. Food Compos. Anal.* **2007,** *20*(3), 313–322.

Kong, K. W.; Ismail, A.; Tan, C. P.; et al. Optimization of Oven Drying Conditions for Lycopene Content and Lipophilic Antioxidant Capacity in a By-product of the Pink Guava Puree Industry using Response Surface Methodology. *LWT—Food Sci. Technol.* **2010,** *43*(5), 729–735.

Lado, B. H.; Yousef, A. E. Alternative Food-Preservation Technologies: Efficacy and Mechanisms. *Microbes Infect.* **2002,** *4*(4), 433–440.

Leistner, L.; Gould, G. W. *Hurdle Technologies: Combination Treatments for Food Stability, Safety and Quality: Combination Treatments for Food Stability, Safety, and Quality*; Springer, New York, 2002.

Li, H.; Pordesimo, L.; Weiss, J. High Intensity Ultrasound-Assisted Extraction of Oil from Soybeans. *Food Res. Int.* **2004,** *37*(7), 731–738.

Lopez-Rubira, V.; Conesa, A.; Allende, A.; Artés, F. Shelf Life and Overall Quality of Minimally Processed Pomegranate Arils Modified Atmosphere Packaged and Treated with UV-C. *Postharvest Biol. Technol.* **2005,** *37*, 174–185.

Luengo, E.; Álvarez, I.; Raso, J. Improving the Pressing Extraction of Polyphenols of Orange Peel by Pulsed Electric Fields. *Innovative Food Sci. Emerg. Technol.* **2013,** *17*, 79–84.

Min, S.; Jin, Z. T.; Zhang, Q. H. Commercial Scale Pulsed Electric Field Processing of Tomato Juice. *J. Agric. Food Chem.* **2003,** *51*(11), 3338–3344.

Morales-de, La Peña, M,; Salvia-Trujillo, L.; Rojas-Graü, M.; et al. Impact of High Intensity Pulsed Electric Field on Antioxidant Properties and Quality Parameters of a Fruit Juice–Soymilk Beverage in Chilled Storage. *LWT—Food Sci. Technol.* **2010,** *43*(6), 872–881.

Moulton, J.; Wang, C. A Pilot Plant Study of Continuous Ultrasonic Extraction of Soybean Protein. *J. Food Sci.* **1982,** *47*, 1127–1129.

Noci, F.; Riener, J.; Walkling-Ribeiro, M.; et al. Ultraviolet Irradiation and Pulsed Electric Fields (PEF) in a Hurdle Strategy for the Preservation of Fresh Apple Juice. *J. Food Eng.* **2008,** *85*(1), 141–146.

Odriozola-Serrano, I.; Soliva-Fortuny, R.; Hernández-Jover, T.; et al. Carotenoid and Phenolic Profile of Tomato Juices Processed by High Intensity Pulsed Electric Fields Compared with Conventional Thermal Treatments. *Food Chem.* **2009,** *112*(1), 258–266.

Odriozola-Serrano, I.; Soliva-Fortuny, R.; Martín-Belloso, O. Changes of Health-Related Compounds throughout Cold Storage of Tomato Juice Stabilized by Thermal or High Intensity Pulsed Electric Field Treatments. *Innovative Food Sci. Emerging Technol.* **2008,** 9(3), 272–279.

Odriozola-Serrano, I.; Soliva-Fortuny, R.; Martín-Belloso, O. Phenolic Acids, Flavonoids, Vitamin C and Antioxidant Capacity of Strawberry Juices Processed by High-Intensity

Pulsed Electric Fields or Heat Treatments. *Eur. Food Res. Technol.* **2008,** *228*(2), 239–248.
Oey, I.; Lille, M.; Van Loey, A.; et al. Effect of High-Pressure Processing on Colour, Texture and Flavour of Fruit-and Vegetable-Based Food Products: A Review. *Trends Food Sci. Technol.* **2008,** *19*(6), 320–328.
Oey, I.; Plancken, I. V.; Loey, A. V.; Hendrickx, M. Does High Pressure Processing Influence Nutritional Aspects of Plant Based Food Systems? *Trends Food Sci. Technol.* **2008,** *19*(6), 300–308.
Oms-Oliu, G.; Odriozola-Serrano, I.; Martín-Belloso, O. Metabolomics for Assessing Safety and Quality of Plant-Derived Food. *Food Res. Int.* **2013,** *54*(1), 1172–1183.
Oms-Oliu, G.; Odriozola-Serrano, I.; Soliva-Fortuny, R.; et al. Effects of High-Intensity Pulsed Electric Field Processing Conditions on Lycopene, Vitamin C and Antioxidant Capacity of Watermelon Juice. *Food Chem.* **2009,** *115*(4), 1312–1319.
Ortega-Rivas, E. Ultrasound in Food Preservation. *Non-thermal Food Engineering Operations*; Springer, 2012; pp 251–262.
Pacheco-Palencia, L. A.; Duncan, C. E.; Talcott, S. T. Phytochemical Composition and Thermal Stability of Two Commercial Açai Species, *Euterpe oleracea* and *Euterpe precatoria. Food Chem.* **2009,** *115*(4), 1199–1205.
Paniwynk, L.; Beaufoy, E.; Lorimer, P.; Mason, J. The Extraction of Rutin from Flower Buds of *Sophora japonica*. Ultrason. Sonochem. **2001,** *8*, 299–301.
Patist, A.; Bates, D. Ultrasonic Innovations in the Food Industry: From the Laboratory to Commercial Production. *Innovative Food Sci. Emerging Technol.* **2008,** *9*(2), 147–154.
Patras, A.; Brunton, N. P.; Donnell, C. O.; et al. Effect of Thermal Processing on Anthocyanin Stability in Foods: Mechanisms and Kinetics of Degradation. *Trends Food Sci. Technol.* **2010,** *21*(1), 3–11.
Patras, A.; Brunton, N.; Da Pieve, S.; et al. Effect of Thermal and High Pressure Processing on Antioxidant Activity and Instrumental Colour of Tomato and Carrot Purées. *Innovative Food Sci. Emerg. Technol.* **2009,** 10(1), 16–22.
Pereira, R.; Vicente, A. Environmental Impact of Novel Thermal and Non-Thermal Technologies in Food Processing. *Food Res. Int.* **2010,** 43(7), 1936–1943.
Plaza, L.; Sánchez-Moreno, C.; Elez-Martínez, P.; et al. Effect of Refrigerated Storage on Vitamin C and Antioxidant Activity of Orange Juice Processed by High-Pressure or Pulsed Electric Fields with Regard to Low Pasteurization. *Eur. Food Res. Technol.* **2006,** 223(4), 487–493.
Puértolas, E.; López, N.; Condón, S.; et al. Potential Applications of PEF to Improve Red Wine Quality. *Trends Food Sci. Technol.* **2010,** *21*(5), 247–255.
Ramaswamy, H. S.; Chen, C. C.; Marcotte, M. Novel Processing Technologies for Food Preservation. In: *Processing Fruits: Science and Technology*; Barrett, D. M., Ed.; CRC Press: Boca Raton, FL, 2005; pp 201–219.
Rawson, A.; Patras, A.; Tiwari, B.; et al. Effect of Thermal and Non Thermal Processing Technologies on the Bioactive Content of Exotic Fruits and their Products: Review of Recent Advances. *Food Res. Int.* **2011,** *44*(7), 1875–1887.
Rendueles, E.; Omer, M.; Alvseike, O.; et al. Microbiological Food Safety Assessment of High Hydrostatic Pressure Processing: A Review. *LWT—Food Sci. Technol.* **2011,** *44*(5), 1251–1260.

Reyes, L. F.; Cisneros-Zevallos, L. Electron-beam Ionizing Radiation Stress Effects on Mango Fruit (*Mangifera indica* L.) Antioxidant Constituents before and during Postharvest Storage. *J. Agric. Food Chem.* **2007,** *55*(15), 6132–6139.

Rossi, F.; Gaio, E.; Torriani, S. *Staphylococcus aureus* and *Zygosaccharomyces bailii* as Primary Microbial Contaminants of a Spoiled Herbal Food Supplement and Evaluation of their Survival During Shelf Life. *Food Microbiol.* **2010,** *27*(3), 356–362.

Sánchez-Moreno, C.; Plaza, L.; Elez-Martínez, P.; et al. Impact of High Pressure and Pulsed Electric Fields on Bioactive Compounds and Antioxidant Activity of Orange Juice in Comparison with Traditional Thermal Processing. *J. Agric. Food Chem.* **2005,** *53*(11), 4403–4409.

Saxena, A.; Maity, T.; Raju, P.; et al. Degradation Kinetics of Colour and Total Carotenoids in Jackfruit (*Artocarpus heterophyllus*) Bulb Slices during Hot Air Drying. *Food Bioprocess Technol.* **2012,** *5*(2), 672–679.

Schilling, S.; Alber, T.; Toepfl, S.; et al. Effects of Pulsed Electric Field Treatment of Apple Mash on Juice Yield and Quality Attributes of Apple Juices. *Innovative Food Sci. Emerging Technol.* **2007,** *8*(1), 127–134.

Sharma, S.; Zhang, Q.; Chism, G. Development of a Protein Fortified Fruit Beverage and Its Quality when Processed with Pulsed Electric Field Treatment. *J. Food Qual.* **1998,** *21*(6), 459–473.

Shohael, A.; Ali, M.; Yu. K.; et al. Effect of Light on Oxidative Stress, Secondary Metabolites and Induction of Antioxidant Enzymes in *Eleutherococcus senticosus* Somatic Embryos in Bioreactor. *Process Biochem.* **2006,** *41*(5), 1179–1185.

Siddiqui, W. Md.; Bansal, V.; Sharangi, A. B. Irradiation of Fruits, Vegetables and Spices for Better Preservation and Quality. In *Food Composition and Analysis: Methods and Strategies*; Haghi, A. K., Carvajal-Millan, E., Eds.; CRC Press, Taylor & Francis Group: USA, 2014; pp 227–250.

Soria, A. C.; Villamiel, M. Effect of Ultrasound on the Technological Properties and Bioactivity of Food: A Review. *Trends Food Sci. Technol.* **2010,** *21*(7), 323–331.

Suárez-Jacobo, Á.; Rüfer, C. E.; Gervilla, R. et al., Influence of Ultra-High Pressure Homogenisation on Antioxidant Capacity, Polyphenol and Vitamin Content of Clear Apple Juice, *Food Chem.* **2011,** *127*(2), 447–454.

Tedjo, W.; Eshtiaghi, M. N.; Knorr, D. Einsatznicht-thermischer Verfahrenzur Zellpermeabilisierung von Weintrauben und Gewinnung von Inhaltsstoffen. *FlüssigesObst.* **2002,** *9*, 578–583.

Toepfl, S. Pulsed Electric Field Food Treatment-Scale up from Lab to Industrial Scale. *Procedia Food Sci.* **2011,** *1*, 776–779.

Torregrosa, F.; Cortés, C.; Esteve, M. J.; et al. Effect of High-Intensity Pulsed Electric Fields Processing and Conventional Heat Treatment on Orange-Carrot Juice Carotenoids. *J. Agric. Food Chem.* **2005,** *53*(24), 9519–9525.

Torregrosa, F.; Cortés, C.; Esteve, M. J.; et al. Effect of High-Intensity Pulsed Electric Fields Processing and Conventional Heat Treatment on Orange-Carrot Juice Carotenoids. *J. Agric. Food Chem.* **2005,** *53*(24), 9519–9525.

Torregrosa, F.; Esteve, M.; Frígola, A.; et al. Ascorbic Acid Stability During Refrigerated Storage of Orange–Carrot Juice Treated by High Pulsed Electric Field and Comparison with Pasteurized Juice. *J. Food Eng.* **2006,** *73*(4), 339–345.

US FDA. 2000. *Center for Food Safety & Applied Nutrition. Bad Bug Book, Foodborne Pathogenic Microorganisms and Natural Toxins Handbook.* Available: http://www.fda.gov/Food/FoodSafety/FoodborneIllness/rneIllnessFoodbornePathogensNaturalToxins/BadBugBook/default.htm.

Vallverdú-Queralt A, Odriozola-Serrano I, Oms-Oliu G.; et al. Changes in the Polyphenol Profile of Tomato Juices Processed by Pulsed Electric Fields. *J. Agric. Food Chem.* **2012,** *60*(38), 9667–9672.

Vallverdú-Queralt, A.; Oms-Oliu, G.; Odriozola-Serrano, I.; et al. Effects of Pulsed Electric Fields on the Bioactive Compound Content and Antioxidant Capacity of Tomato Fruit. *J. Agric. Food Chem.* **2012,** *60*(12), 3126–3134.

Vieira, M. C.; Teixeira, A.; Silva, C. Mathematical Modeling of the Thermal Degradation Kinetics of Vitamin C in Cupuaçu (*Theobroma grandiflorum*) Nectar. *J. Food Eng.* **2000,** *43*(1), 1–7.

Vilkhu, K.; Mawson, R.; Simons, L.; et al. Applications and Opportunities for Ultrasound Assisted Extraction in the Food Industry—A Review. *Innovative Food Sci. Emerg. Technol.* **2008,** *9*(2), 161–169.

Viña, S. Z.; Chaves, A. R. Effect of Heat Treatment and Refrigerated Storage on Antioxidant Properties of Pre-Cut Celery (*Apium graveolens* L.). *Int. J. Food Sci. Technol.* **2008,** *43*(1), 44–51.

Williams A. New Technologies in Food Preservation and Processing: Part II. *Nutr. Food Sci.* **1994,** *94*(1), 20–23.

Wood, O. B.; Bruhn, C. M. Position of the American Dietetic Association: Food Irradiation. *J. Am. Dietetic Assoc.* **2000**, *100*, 246–253.

Wu, Y.; Mittal, G.; Griffiths, M. Effect of Pulsed Electric Field on the Inactivation of Microorganisms in Grape Juices with and without Antimicrobials. *Biosyst. Eng.* **2005,** *90*(1), 1–7.

Xia, T.; Shi, S.; Wan, X. Impact of Ultrasonic-Assisted Extraction on the Chemical and Sensory Quality of Tea Infusion. *J. Food Eng.* **2006,** *74*, 557–560.

Yang, J.; Paulino, R.; Janke-Stedronsky, S.; et al. Free-Radical-Scavenging Activity and Total Phenols of Noni (*Morinda citrifolia* L.) Juice and Powder in Processing and Storage. *Food Chem.* **2007,** *102*(1), 302–308.

Yeom, H.; Streaker. C.; Zhang, Q.; et al. Effects of Pulsed Electric Fields on the Activities of Microorganisms and Pectin Methyl Esterase in Orange Juice. *J. Food Sci.* **2000,** *65*(8), 1359–1363.

Zepka, L. Q.; Mercadante, A. Z. Degradation Compounds of Carotenoids Formed during Heating of a Simulated Cashew Apple Juice. *Food Chem.* **2009,** *117*(1), 28–34.

Zhang, M.; Chen, H.; Li, J.; et al. Antioxidant Properties of Tartary Buckwheat Extracts as Affected by Different Thermal Processing Methods. *LWT–Food Sci. Technol.* **2010,** *43*(1), 181–185.

Zhang, Y.; Gao, B.; Zhang, M,; et al. Pulsed Electric Field Processing Effects on Physicochemical Properties, Flavor Compounds and Microorganisms of Longan Juice. *J. Food Process. Preserv.* **2010,** *34*(6), 1121–1138.

CHAPTER 7

ULTRAVIOLET LIGHT STIMULATION OF BIOACTIVE COMPOUNDS WITH ANTIOXIDANT CAPACITY OF FRUITS AND VEGETABLES

ÁVILA-SOSA RAÚL[1], NAVARRO-CRUZ ADDI RHODE[1], VERA-LÓPEZ OBDULIA[1], HERNÁNDEZ-CARRANZA PAOLA[2], and OCHOA-VELASCO CARLOS ENRIQUE[1*]

[1]*Departamento de Bioquímica-Alimentos, Facultad de Ciencias Químicas, Benemérita Universidad Autónoma de Puebla, 14 Sur y Av. San Claudio, Ciudad Universitaria, Col. San Manuel, 72420 Puebla, Puebla, México*

[2]*Colegio de Ingeniería en Alimentos, Facultad de Ingeniería Química, Benemérita Universidad Autónoma de Puebla, 14 Sur y Av. San Claudio, Ciudad Universitaria, Col. San Manuel, 72420 Puebla, Puebla, México*

**Corresponding author, Tel: +52 222 295500;*
E-mail: carlosenriqueov@hotmail.com.

CONTENTS

ABSTRACT

This chapter discusses the effect of nonthermal ultraviolet light on the stimulation or emergence of bioactive compounds present in ruits and vegetables. Ultraviolet light is a nonthermal processing technique that puts its effect on the surface of the fruits and vegetables. Along with killing the microorganisms, it is also observed to liberate the functional compounds present on the surface (i.e., inside the peels) which, therefore, enhance the nutritional and functional value of the processed food.

7.1 INTRODUCTION

Through centuries, plants were used to obtain medicinal extracts. Approximately, 80% of the world population uses extracts for health care and 25% of medicines are derived from plants (Fowler, 2006). Fruits and vegetables as plant products show different secondary metabolites or bioactive compounds (Poiroux-Gonord et al., 2010; Rufino et al., 2010), some of them are provitamins, vitamins, pigments, flavonoids, and phenolic acids, among others (Jayaprakasha & Patil, 2007). According to the British Nutrition Foundation, secondary metabolites of plants are divided into four groups: terpenoids, alkaloids, phenolic compounds, and sulfur compounds (Goldberg, 2003). These compounds show different characteristics in human health; one of the most important is to scavenge free radicals, reducing oxidative stress level and the oxidation of biological molecules, avoiding the possibility of development of some chronic diseases like cardiovascular disease, cancer, and diabetes, among others (Almeida et al., 2011). Secondary metabolites of fruits and vegetables may be affected by several factors during the pre harvest (plant growth promoting and genetically modified organisms), harvest (conventional or organic agriculture), and post harvest (temperature, relative humidity, and light) practices (Bulley et al., 2009; Asami et al., 2003; Mittler, 2002); these factors stress the fruit and vegetable tissues and activate the synthesis of secondary metabolites via shikimic, malonic, and mevalonic acids (Poiroux-Gonord et al., 2010). Due to their biological activity, secondary metabolites are used in several applications such as medicinal, pharmaceutical, fragrance, pesticides, fungicides, and others (Cisneros-Zevallos, 2003).

Ultraviolet (UV) light is a non-ionizing irradiation and is part of the electromagnetic spectrum comprising between 200 and 400 nm (Gayán et al., 2012). UV light is divided in three ranges according to the wavelength, ultraviolet-C light (UV-C) between 200 and 280 nm, ultraviolet-B light (UV-B) between 280 and 320 nm, and ultraviolet-A light (UV-A) between 320 and 400 nm (Hollósy, 2002). UV-C light is widely used as a sterilized treatment of water and surface equipment (Koutchma, 2008); UV-B light can induce different damages in plant cells and stimulate photosynthesis and growth of plants (Kumari et al., 2009); and UV-A is the less studied, some reports indicated that UV-A can act as photosynthetically active radiation (Tsormpatsidis et al., 2008). The effect of UV light in fruits and vegetables is called hormetic effect that may stimulate the production of phenylalanine ammonia-lyase (PAL) and other enzymes that induce the production of secondary metabolites (Guerrero-Beltrán & Barbosa-Cánovas, 2004; Jagadeesh et al., 2011). The aim of this review is to explain the UV irradiation as a physical treatment to stimulate the production of secondary metabolites of fruits and vegetables and show the current application results.

7.2 APPLICATION OF ULTRAVIOLET LIGHT ON FRUITS AND VEGETABLES

7.2.1 PRINCIPLES OF UV LIGHT ON PLANT TISSUES

UV light has been applied as a physical treatment to stimulate the production of secondary metabolites (Alothman et al., 2009a). However, until the application of UV light as a microbial inactivation method of fruits and vegetables (Fig. 7.1), the interest of UV light as abiotic stress factor in order to increase the secondary metabolites of fruits and vegetables had been less studied (Jagadeesh et al., 2011). According to Stapleton (1992) and Mercier et al. (1993), UV light may affect plant cells in two ways: DNA damage (that may cause DNA mutation during replication) and physiological and biochemical processes (inducing biological stress and activate defense mechanisms of plants tissues).

The DNA damage by the UV-C light treatment has been extensively studied (Koutchma, 2008). DNA has a strong UV-C (260 nm) absorption; photons are highly energetic generating pyrimidine dimers and blocking the transcription and replication of the microbial cell, causing the death (Guerrero-Beltrán & Barbosa-Cánovas, 2004).

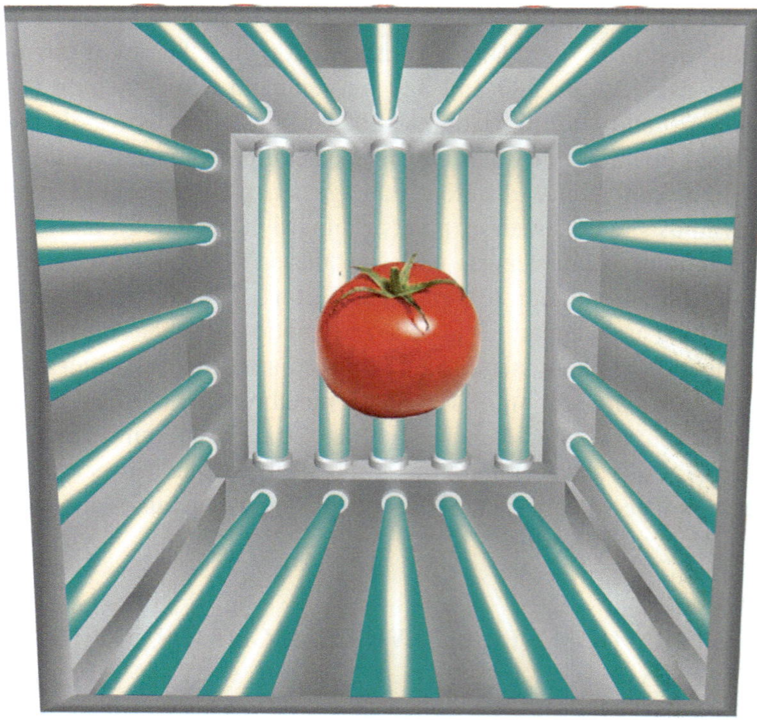

FIGURE 7.1 Ultraviolet-treatment chamber of fruits and vegetables.

As mentioned above, UV irradiation produces in fruits and vegetables a biological phenomenon known as hormesis, which affects morphological, physiological, and biochemical process and increases their phytochemical content (Charles & Arul, 2007). The hormesis effect in fruits and vegetables is affected by several factors such as UV light applied (UV-A, UV-B, and UV-C), irradiation dose (low irradiation doses promote the secondary metabolites production and high UV doses destroyed polypeptides of photosystem I and II), plant tissue (fruit, vegetable, herbs, flowers, and medicinal plants), physiological age (green or mature), pretreatment process (whole or fresh cut), among others (Charles & Arul, 2007; Alothman et al., 2009b; Carrasco-Ríos, 2009; Poiroux-Gonord et al., 2010; Jagadeesh et al., 2011; Bravo et al., 2012). As observed in Figure 7.2, the effect of UV light on plant tissues depends on different factors; however, the production of secondary metabolites due to UV light may be grouped in three different but not excluding pathways (Schreiner et al., 2014).

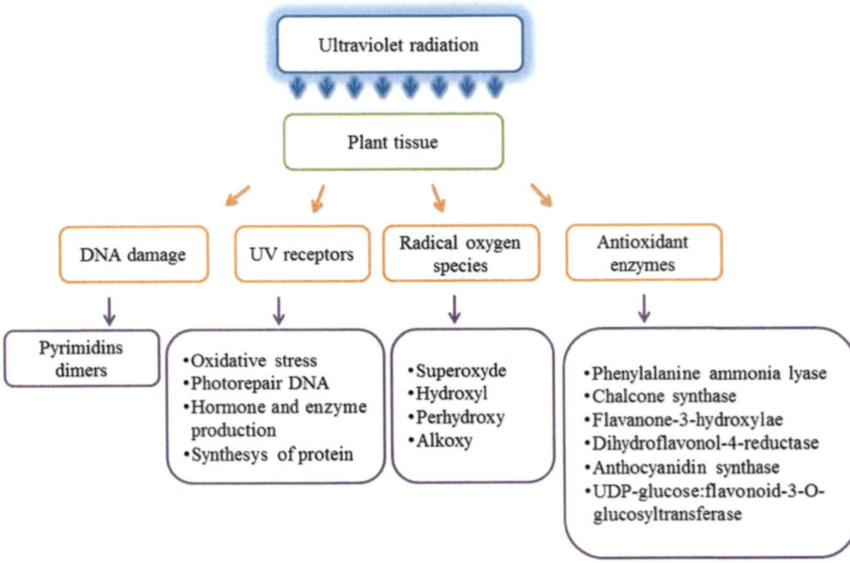

FIGURE 7.2 Effect of ultraviolet radiation on plant tissue.

7.2.1.1 ULTRAVIOLET PHOTORECEPTOR

UV-B light stresses plant tissues and promotes the expression of photomorphogenic response genes such as constitutive photomorphogenic 1 (COP1), elongated hypocotyl and ultraviolet resistances locus 8 (UVR8), which is recently identified as a UV-B receptor (Zhang & Björn, 2009). UVR8 is a seven-bladed β-propeller protein that makes use of tryptophan residues intrinsic to the protein as chromophore of UV-B absorption with a primary role as established for trp 285 (Jenkins, 2014). The sequence of UVR8 is similar to the human regulator of chromatin condensation (Kliebenstein et al., 2002). UVR8 commonly appears in plants as dimers, and the UV-B promotes the dimer breaking causing monomers (Singh et al., 2014). The UVR8 and COP1 are necessary for the UV-B photomorphogenic response and stimulating the gene production in order to defend the cell from UV-B exposure damage (Kataria et al., 2014).

7.2.1.2 ENZYME ACTIVATION

Secondary metabolites production pathways have not yet been completely elucidated; however, enzymes play an important role in the production of phytochemical compounds. Plant genomes (15–25%) are used to encode enzymes related to the secondary metabolites production in plants (Pichersky & Gang, 2000). Recent studies indicated that UV-B light induces PAL chalcone synthase (CHS), flavanone-3-hydroxylae (F3H), dihydroflavonol 4-reductase (DFR), anthocyanidin synthase (ANS), and UDP-glucose:flavonoid-3-*O*-glucosyltransferase, enzymes in flavonoid biosynthetic pathway (Ubi et al., 2006; Zhang & Björn, 2009). Moreover, some phenolic acids (coumaric, caffeic, vanillic, rosemarinic acids) are stimulated by UV-B light, and synthetized in phenyl propanoid and tyrosine pathways, using PAL, *trans*-cinnamate 4-hydroxylase, hydoxycinnamate, tyrosine transferase, hydroxyphenylpyruvate reductase, 4-hydroxyphenylpyruvic acid, rosmarinic acid synthase, 3-hydroxylases, and 3,4-dihydroxyphenyllactic acid (Alothman et al., 2009b; Zhang & Björn, 2009). Finally UV light promotes alkaloids, which are synthetized via mevalonate pathway, and enzymes such as tryptophan decarboxylase and strictosine synthase.

7.2.1.3 RADICAL OXYGEN SPECIES FORMATION AND ANTIOXIDANT ENZYMES

UV light on plants may generate reactive oxygen species (ROS), such as free radicals (superoxide, hydroxyl, perhydroxy, and alkoxy radicals) and non-radical (hydrogen peroxide and singlet oxygen) compounds (Gill & Tuteja, 2010). This may affect the inactivation of enzymes, lipid peroxidation, protein degradation, and DNA strand breaks in plants (Scandalios, 1993). UV irradiation might act as stress signals producing photooxidation products, to which plants react by stimulating defense mechanisms against oxidation, resulting in an increased antioxidant synthesis (Barka, 2001; Alothman et al., 2009a). Enzymatic ROS-scavenging system includes enzymes such as superoxide dismutase (SOD), catalase (CAT), ascorbate peroxidase (Asa-POD), glutathione reductase (GR), monodehydroascorbate reductase (MDAR), dehydroascorbate reductase (DHAR), glutathione peroxidase (GSH-POD), guaiacol peroxidase (G-POD), and glutathione-*S*-transferase (Gill & Tuteja, 2010). These enzymes may catalyze the dismutation of $^{+}O_2$ to hydrogen peroxide (H_2O_2), and CAT and peroxidases transform H_2O_2 to secondary metabolites (Foyer et al., 1994).

7.3 EFFECT OF ULTRAVIOLET LIGHT ON FRUITS

As mentioned above, several authors report the UV-light effects on fruits and vegetables; many of them are related to disinfestation or controlling foodborne microorganism's surface growth. However, the applications of UV light not only extend the postharvest shelf life of fruits and vegetables, but also act as abiotic stress (Table 7.1), increasing secondary metabolites and antioxidant capacity (González-Aguilar et al., 2001). The following sections present the effect of UV irradiation on secondary metabolites on some fruits and vegetables.

7.3.1 RED-BLUE FRUITS

Strawberry (*Fragaria* × *ananassa* D. cv. *Camarosa*) is a nonclimacteric fruit with fast development and ripening, which overlaps with senescence in the late stages (Beltrán et al., 2010). Allende et al. (2007) combined modified atmospheres, ozone, and UV-C light on strawberry preservation. The combination of O_3 and UV-C reduces phenols, procyanidins, and vitamin C concentration due to the generation of toxic molecules species that act as potent phytotoxic agents. However, the use of UV-C at different doses (0.43–4.30 kJ/m^2) can enhance antioxidant activity. In this sense, Erkan et al. (2008) evaluated the effect of UV-C light (0.43–4.3 kJ/m^2) on strawberry and demonstrated higher concentrations of total phenols, anthocyanin, and antioxidant enzymes including GSH-POD, GR, SOD, AsA-POD, G-POD, MDAR, and DHAR. The nonenzymatic components such as reduced glutathione (GSH) and oxidized glutathione (GSSG) also were increased by UV-C irradiation. Pombo et al. (2009) reported a high correlation between UV-C effect and the expression of a set of genes (FaExp2, FaExp4, FaExp5, FaCel1, FaPE1, and FaPG1) encoding for proteins and enzymes involved in cell wall degradation. They concluded that UV-C light affects gene transcription delaying strawberry softening.

TABLE 7.1 Effect of Ultraviolet Light on Secondary Metabolites with Antioxidant Capacity of Fruits and Vegetables.

Fruit or vegetable	UV light	Dose (kJ/m^2)	Secondary metabolites and antioxidant capacity	Source
Strawberry	UV-C	0.43–4.30	Promotes the increase of phenolic content by 1.1–1.4 mL/kg, anthocyanin content from 390.8 to 500.0 mg/kg, and antioxidant activity from 21.11 to 32.21 mmol/kg	Erkan et al. (2008)
Blueberry	UV-B	0.075 and 0.150 ($W h/m^2$)	Phenolic compounds increase from 25 to 28 mg GAE/100 g	Eichholz et al. (2011
	UV-A, B, and C	6	Total phenolic increase from 270 to 290 mg GAE/100g	Nguyen et al. (2014)
	UV-C	1 and 2	Total phenols and anthocyanin content increase by 10% and 10–20%, respectively	Perkins-Veazie et al. (2008)
	UV-C	2.15 and 4.30	Total phenols increase from 3.12 to 4.97 mg/g, anthocyanin increases from 2.02 to 3.11 mg/g, and antioxidant capacity increases from 30.5 to 43.8 μmol GAE/g	Wang et al. (2009)
Grape	UV-B	8.25 and 33 ($\mu W/cm^2$)	Stimulates secondary metabolites production with antioxidant compounds such as terpenoids, isoprenoids, xanthophylls and α-tocopherol	Gil et al. (2012)
	UV-C	2.4 and 4	The cold storage in combination with low UV-C induces stilbenes biosynthesis and increases the quercetin level	Crupi et al. (2013)
Papaya	UV-C	1.48	Flavonoid content increased from 15.6 to 23.2 mg quercetin/100 g and phenolic compounds increase from 85.8 to 100 mg GAE/100 g	Rivera-Pastrana et al. (2013)
Watermelon	UV-C	1.6–7.2	Increase the lycopene content and the antioxidant capacity in 25%	Artés-Hernández et al. (2010)
Pineapple	UV-C	2.158 J/m^2	Total flavonoids increase from 4.39 to 8.47 mg CEQ/100 g	Alothman et al. (2009a,b)

TABLE 7.1 *(Continued)*

Fruit or vegetable	UV light	Dose (kJ/m²)	Secondary metabolites and antioxidant capacity	Source
Banana	UV-C	2.158 J/m²	Polyphenol content increases from 72.21 to 134.16 mg/GAE/100 g, vitamin C increases from 0.19 to 0.21 mg AA/g, and total flavonoids increase from 23.72 to 37.22 mg CEQ/100 g	Alothman et al. (2009a,b)
Guava	UV-C	2.158 J/m²	Total phenols increase from 190.58 to 289.42 mg GAE/100 g, and total flavonoid content increases from 40.59 to 50.81 mg CEQ/100 g	Alothman et al. (2009a,b)
Mango	UV-C	0.112–1.12 (kJ/cm²)	Total phenols and flavonoid increase by 35% and 57%, respectively	González-Aguilar et al. (2007)
Carambola	UV-C	13	Phenolic compounds increase about 30%	Andrade-Cuvi et al. (2010)
Pear	UV-B/Vis		Improves the anthocyanin compounds	Zhang et al. (2012)
Apple	UV-B	0.20 W/m²	The anthocyanin, total flavonoids, ascorbic acid, and antioxidant capacity increase by 45, 33, 47, and 95%, respectively	Hagen et al. (2007)
Mandarin	UV-C	0.75–3.0	Total phenols increase by 2.4%	Shen et al. (2013)
Tomato	UV-B	6.08	Lycopene and β-carotene increase by 40% and 72%, respectively	Castagna et al. (2013)
	UV-B	10–80	Flavonoids, phenols, lycopene, and antioxidant capacity increase	Liu et al. (2011)
	UV-C	13.7	Lycopene content increases sixfold	Liu et al. (2009)
	UV-C	3.7	Ascorbic acid and total phenols increase 30% and 210%, respectively	Jagadeesh et al. (2011)
	UV-C	1–12.2	Lycopene content increase of 6.94 to 63.17 mg/kg, total phenolic compounds from 258.75 to 366.45 mg GAE/kg, and antioxidant activity (hydrophilic and lipophilic extracts) slightly increase by the UV-C treatment	Bravo et al. (2012)
	UV-C	2–16	Total phenolic and flavonoids increase by 21.2% and 227%, respectively	Liu et al. (2012)

TABLE 7.1 *(Continued)*

Fruit or vegetable	UV light	Dose (kJ/m^2)	Secondary metabolites and antioxidant capacity	Source
Pepper	UV-A, -B, and -C	6.1–5.7 (W/m^2)	Quercetin, rutin, and anthocyanin increase by 7–17%, 8–27%, and 11–20%, respectively	Mahdavian et al. (2008)
	UV-C	10	Antioxidant capacity increases by 8%	Andrade-Cuvi et al. (2011)
	UV-C	1–14	Antioxidant capacity (EC_{50}) increases from 0.135 to 0.151	Vicente et al. (2005)
Carrot	UV-B	141.4 mJ/cm^2	Carotenoids increase from 12.9 to 24.2 mg CAE/100 g and antioxidant capacity increases from 703.4 to 2335.7 μg Trolox/g	Du et al. (2012)
	UV-C	0.78 ± 0.36	Carotenoids increase threefold during storage	Alegria et al. (2012)
Broccoli	UV-B	2.2–16.4	Increases ascorbic acid from 0.33 to 0.71 g/100 g, total phenolic from 2.17 to 3.79 mg GAE/g, and flavonoid content from 0.235 to 0.266 mg CE/g	Topcu et al. (2015)
	UV-C	1.5–15	Total phenols increase by 25%	Martínez-Hernández et al. (2011)
Spinach	UV-C	0. 4–11.4	Chlorophyll *a* and *b* increase from 536 to 582 mg/kg and 185 to 219 mg/kg, respectively	Artés-Hernández et al. (2009)
Garlic	UV-C	0.1–2	Increases the total phenols from 537.82 to 559.08 mg GAE/kg, total flavonoids from 334.27 to 365.00 mg CAE/kg, and quercetin from 0.44 to 0.81 mg/g	Park and Kim (2015)
Onion	UV-C	2.5–40	Increases total phenol and anthocyanin content by 15% and 26%, respectively	Rodrígues et al. (2010)
	UV-C	2.4	Twofold increase in the flavonoids and antioxidant capacity	Rodov et al. (2010)

TABLE 7.1 *(Continued)*

Fruit or vegetable	UV light	Dose (kJ/m^2)	Secondary metabolites and antioxidant capacity	Source
Buckwheat sprout	UV-A and UV-B	250 and 890 W/m^2	Increases rutin concentration from 653 to 1034 mg/100 g, and the antioxidant capacity increases 1.6-fold	Tsurunaga et al. (2013)
Durum wheat	UV-C	40 μW/cm day	Increases the carotenoids, anthocyanin and flavonoids concentration by 2.5, 2.5, and 7.8-folds, respectively	Balouchi et al. (2009)

GAE: Gallic acid equivalent; CEQ: catechine quivalent; CAE: β-carotenoid equivalent.

Blueberry (*Vaccinium corymbosum*, cvs. Collins, Bluecrop) is a fruit with high concentration of phenols and anthocyanins (Sellappan et al., 2002). Perkins-Veazie et al. (2008) irradiated blueberry with UV-C light (1–4 kJ/m^2), and pointed out that low UV-C dose (1–2 kJ/m^2) increases phenolic compounds, while dose of 2–4 kJ/m^2 increases anthocyanin contents by 10%. Moreover, they informed that UV-treated blueberry delay decays by ripe rot. Similar results were obtained by Wang et al. (2009), total phenols, anthocyanin content, and antioxidant capacity increased after UV-C treatments (0.43–6.45 kJ/m^2); nevertheless during storage, these bioactive compounds were reduced. Eichholz et al. (2011) investigated the effect of UV-B (0.27 and 0.54 kJ/m^2) on volatile metabolites and phenolic compounds in blueberries. Phenolic compounds were increased at dose of 0.54 kJ/m^2; however, volatile metabolites were negatively affected by UV-B light. Nguyen et al. (2014) reported that blueberry treated with UV-B and UV-C show high concentration of phenolic compounds and anthocyanin than blueberries treated with UV-A light.

In grape (*Vitis vinifera* L.) leaf tissues, Gil et al. (2012) showed that UV-B light stimulated the formation of terpenes and other antioxidant compounds. They exposed grape leaf for 45 days (16 h/day) at two different UV-B doses [low UV-B (8.25 μW/cm^2) and high UV-B (33 μW/cm^2)] and concluded that low UV-B dose stimulated the formation of sitosterol, stigmasterol, and lupeol in young leaves. However, during the maturation of the leaves, di-terpenes, α- and γ-tocoferol, phytol, sesquiterpene E-nerolidol, carene, α-pinene, and terpinolene had a maximum accumulation under high-UV-B dose. The adaptive response induced by relatively low UV-B irradiations as suggested by terpene synthesis is related with membrane stability and the defense of abiotic and biotic stresses. Antioxidant activity in grape leaves stimulates the concentration in fruit. Crupi et al. (2013) evaluated the combined effect of time, temperature, and UV-C light as abiotic stress in redglobe table grape. They pointed out that storage at 4°C (48 h) and irradiation with 2.4 kJ/m^2, positively increased the content of cis (34 μg/g of skin) and trans (90 μg/g of skin) piceid, quercetin-3-*O*-galactoside, and quercetin-3-*O*-glucoside (15 and 140 μg/g of skin, respectively) up to threefold compared to control grape samples. Moreover, anthocyanin (cyanidin-3-*O*-glucoside and peonidin-3-O-glucoside) was positively affected after 24 h of storage at 4°C and irradiated for 5 min (4 kJ/m^2).

7.3.2 TROPICAL FRUITS

Papaya (*Carica papaya* L.) is a good source of iron, calcium, vitamins (Rivera-López et al., 2005) and carotenoids, such as β-carotene, lycopene, lutein, and zeaxanthine (Kaur & Kapoor, 2001). Different studies have been made in papaya to preserve its nutritional compounds and prevent microbial spoilage; however, few studies have been performed for maintaining their bioactive compounds. Rivera-Pastrana et al. (2013) analyzed the effect of UV-C irradiation and storage at 5 or 14°C, in peel and flesh of "Maradol" papaya, using dose of irradiation of 1.48 kJ/m^2. They reported that UV-C irradiation increased flavonoid content (15.6 to 23.2 mg quercetin/100 g compared to control), phenolic compounds (85.8–100 mg gallic acid/100 g) on papaya flesh, during the first 10 days of storage at 5°C. Antioxidant system of papaya was activated due to changes in enzymatic activities of SOD, CAT, and peroxidase.

Watermelon (*Citrullus lanatus* Thunb.) is an important horticultural crop highly cultivated around the world, and is a very significant source of ascorbic acid and lycopene. Artés-Hernández et al. (2010) analyzed the effect of UV-C light (2.8 kJ/m^2) and they observed a positive effect on lycopene content of fresh watermelon; although, the influence of UV-C treatment on fresh-cut watermelon has received little attention.

Alothman et al. (2009b) evaluated the effect of UV-C irradiation (2.158 J/m^2) on pineapple (*Ananas comosus* L.) and banana (*Musa paradisiaca* L.) at different times of exposure. They informed that pineapple showed an increase of total flavonoids of 92% after 30 min of UV-C irradiation; while in banana, the total phenol and flavonoids increased to 85% and 56%, respectively. However, a negative effect was observed in vitamin C. In fresh-cut pineapples, radiation tremendously decreased the content of vitamin C and induced browning throughout the storage period (Pan et al., 2012).

In fresh-cut "Tommy Atkins" mango (*Mangifera indica* L.), González-Aguilar et al. (2007) applied UV-C light at different doses (0.112–1.12 kJ/cm^2) and evaluated the residual effect during the storage. They concluded that increasing UV-C irradiation increased the phenol and flavonoid accumulation during the storage, due to a hypersensitive defense response of fresh-cut mango. Nevertheless, vitamin C and β-carotene were more affected by the UV-C irradiation and the concentrations decreased during storage.

Carambola (*Averrhoa carambola* L.) is an exotic fruit, highly valued in international markets, and popularly known as "star fruit." Little has been studied in carambola, related with UV light. Andrade-Cuvi et al. (2010) applied UV-C light in carambola fruits (13 kJ/m^2), irradiation treatment induced changes in antioxidant-enzymes activity and phenol accumulation (about 30%); however, flavonoids decreased during the storage. It might be related with antioxidant enzyme changes as a physiological response.

Guava (*Psidium guajava* L.) is a tropical fruit, native to Central America (Adsule & Kadam, 1995), and contains a lot of nutrients, such as phenolic compounds, vitamin A, and C with antioxidant properties (Temple, 2000). Alothman et al. (2009b) found an increase in phenols (52%) and flavonoids (25%) after 30 min of exposure to UV-C light (2.158 J/m^2) in contrast UV-C treatment decreased the vitamin C content (18%).

7.3.3 POMACEAS FRUITS

A combination of UV-B and visible irradiation was evaluated during the storage of Red Chinese pears (*Pyrus communis* L.) at two different temperatures (17 and 27°C) (Zhang et al., 2012). They concluded that higher temperature and UV-B/visible treatment enhanced the expression of PyMYB10 and five anthocyanin structural genes, PpPAL, PpCHI, PpCHS, PpF3H, and PpANS, that induced anthocyanin accumulation and the red color of pears after 48 h of irradiation.

Irradiation of Apples (*Malus domestica L.*) with UV-B and visible light was reported by Hagen et al. (2007). They evaluated the effect of UV-B (0.20 W/m^2) and visible light during 10 days (12 h per day) and concluded that the major effect of UV-B and visible irradiation was in peel than in flesh. In this sense, the epicatechin, procyanidins, anthocyanin, total flavonoids, chlorogenic acid, ascorbic acid, and antioxidant capacity increased by 33%, 30%, 45%, 33%, 250%, 47%, and 95%, respectively, compared to the apple maintained only with visible light.

7.3.4 CITRIC FRUITS

Minimally processed Satsu mandarin (*C. unshiu* Marc. cv. Owari) was treated with UV-C irradiation (0.75, 1.5, and 3.0kJ/m^2). Significant increases in flavonoids concentration (11.75–33.25% for hesperidin) and total phenols (5.7% to 8.1%) were found in mandarin treated with 1.5 and 3.0 kJ/m^2. The increase of flavonoids occurred during the first 3 days and diminished after 4 days of storage (Shen et al., 2013).

7.4 EFFECT OF ULTRAVIOLET LIGHT ON VEGETABLES

7.4.1 FRUIT VEGETABLES

Tomato (*Lycopersicon esculentum* L.) is a climacteric fruit, with high lycopene and carotenoids content. There are several reports of UV-C light increasing the content of lycopene and total phenolic compounds in tomatoes (Jagadeesh et al., 2011; Liu et al., 2009). Bravo et al. (2012) exposed mature green tomatoes (breaker stage) to different doses of UV-C irradiation (1.0, 3.0, and 12.2 kJ/m^2). Lycopene content increased almost twofold, total phenolic content and antioxidant capacity increased; moreover, other components (phenols and antioxidant capacity) can be enhanced with UV light. UV-C can induce some enzymes that are related to antioxidant protection as defense mechanisms like PAL and lipoxygenase (Barka, 2001). According to Liu et al. (2012), the optimal irradiation dose of UV-C light was between 4 and 8 kJ/m^2, because of increase in flavonoids, phenolic compounds, and antioxidant capacity. Moreover, the irradiation with UV-B light (10 to 80 kJ/m^2) applied as postharvest treatment increased the concentration of flavonoids, phenols, lycopene, and antioxidant capacity (Liu et al., 2011). Castagna et al. (2013) presented similar results, lycopene and β-carotene increased with UV-B light as a postharvest treatment.

Andrade-Cuvi et al. (2011) evaluated the effect of UV-C (10 kJ/m^2) on the ascorbic acid, dehydroascorbic acid, and antioxidant capacity (DPPH) in mature peppers (*Capsicum annuum* L.). They reported that UV-C irradiation did not affect the ascorbic acid and dehydroascorbic acid. The antioxidant capacity of red pepper showed a slight increase (8%) after 21 days of storage at 0°C, Vicente et al. (2005) obtained similar results. Mahdavian et al. (2008) treated pepper plants with UV-A, UV-B, and UV-C radiation (6.1, 5.8, and 5.7 W/m^2, respectively), increasing the concentration of quercetin (7–17%), rutin (8–27%), and anthocyanin (11–20%). Moreover, the effect of UV-C light was higher than UV-B and UV-A.

7.4.2 ROOT VEGETABLES

Carrot (*Daucus carota* L.) is one of the most important vegetable worldwide. Alegria et al. (2012) applied UV-C (0.78 ± 0.36 kJ/m^2) irradiation on carrots. They reported that carotenoid content significantly increased (threefold) after 7 days of storage at 5°C. Du et al. (2012) evaluated the effect of the UV-B light (141.4 ± 1.6 mJ/cm^2) on the antioxidant compounds of fresh-cut carrot products (baby carrots, carrot stix, shredded carrots, crinkle cut coins, and carrot chips). They reported that UV-B light significantly increased the chlorogenic acid, total soluble phenols, antioxidant capacity, and the PAL enzyme after 3 days of storage. Moreover, they informed that carrot chips were the presentation that showed the major bioactive compounds, antioxidant capacity, and PAL activity.

7.4.3 FLOWER VEGETABLES

Different reports indicate that UV-B treatment in some species of Brassicaceae show an increase in the glucosinolates, compounds with anticancer properties (Foggo et al., 2007; Kuhlmann & Müller, 2009). Martínez-Hernández et al. (2011) treated fresh-cut Chinese broccoli (*Brassica oleracea*) with UV-C at different irradiation doses (1.5, 4.5, 9.0, and 15 kJ/m^2). After UV-C irradiation total phenols and antioxidant capacity was significantly increased. During 19 days of storage (5 and 10°C), broccoli treated with low UV-C irradiation, (1.5 kJ/m^2) showed the higher concentration of total phenols. They also reported a reduction in chlorophyll degradation. In order to evaluate the effect of different abiotic stresses [UV-C (8 kJ/m^2) along with heat treatment (48°C during 3 h)] on some bioactive compounds and antioxidant capacity in broccoli, florets were observed during the storage. Lemoine et al. (2010) reported that phenols (10%) and antioxidant capacity (13%) increased after the combined treatment. However, after 7 days of storage (20°C), the broccoli florets treated showed an increase in total phenols, ascorbic acid, and antioxidant capacity compared to untreated florets. Moreover, the enzyme activity of PAL and SOD showed a significant increase in broccoli florets with abiotic stress. According to Topcu et al. (2015), UV treatment may be a good manner to increase the secondary metabolites with health benefits in broccoli florets.

7.4.4 LEAF VEGETABLES

Artés-Hernández et al. (2009) evaluated the effect of UV-C radiation (0. 4.5, 7.9, and 11.4 kJ/m^2) on minimally processed spinach leaves (*Spinacia oleracea* L.), they reported a negatively effect of irradiation on total phenols and antioxidant capacity; however, chlorophyll *a* and *b* were increased (from 536 to 582 mg/kg of fresh matter and from 185 to 219 mg/kg of fresh matter, respectively). Moreover, during 13 days of storage (5 and 8°C), spinach leaves treated with UV-C light did not show chlorophyll *a* and *b* decrease.

7.4.5 BULB VEGETABLES

Garlic (*Allium sativum* L.) is a popular plant use as food ingredient; they also possess antimicrobial, antioxidant, and anticarcinogenic properties and are used to prevent diseases such as hypercholesterolemia, diabetes, cardiovascular, and thrombosis (Bozin et al., 2008; Santhosha et al., 2013). In a study performed by Park and Kim (2015), garlic was treated with UV-C irradiation at different doses (0.1, 1, and 2 kJ/m^2); and reported that total phenols and flavonoids were not significantly affected. However, after 15 days of storage at 0°C, treatment of garlic with UV-C at a dose of 2 kJ/m^2 significantly increased the values of total phenols, total flavonoids, and quercetin (11%, 6%, and 506%, respectively).

Onions (*Allium cepa* L.) have been widely used as antioxidant, antiplatelet, antithrombotic, antiasthmatic, and antibiotic (Nile & Park, 2013). Onion is a good source of flavonols and anthocyanin (red onion). Rodrígues et al. (2010) treated white onion with UV-C light (2.5, 5, 10, 20, and 40 kJ/m^2), after one week of storage, onion treated with higher irradiation dose (20 and 40 kJ/m^2) showed significant increase in total phenol and anthocyanin content (15% and 26%, respectively). Moreover, Rodov et al. (2010) treated onion with UV-C light (2.4 kJ/m^2) and reported (after 5 days of storage) a twofold increase in flavonols and antioxidant capacity of hydrophilic extract. Individual flavonoids (quercetin and kaempferol) showed an increase of at least 293% and 287%, respectively. Higashio et al. (2005) pointed out that UV irradiation might enhance (50–70%) the onion quercetin content.

7.4.6 FUNGI VEGETABLES

Edible mushrooms have been treated with UV light to enhance some secondary metabolites with antioxidant capacity. Mau et al. (1998) increased levels of vitamin D_2 at low doses of UV-C (0.295 kgf/s^2). Koyyalamudi et al. (2011) reported that pulsed UV-C light may be a good manner to stimulate the secondary metabolite than continuous UV-C light. A study by Kalaras et al. (2012) evaluated the pulsed UV-B (0.791 J/cm^2/pulse) on sliced mushrooms. They reported that after three pulses of UV-B light, the vitamin D concentration increased by 32.2 μg/g (1300%) compared with untreated sliced mushroom. Moreover, 60 pulses of UV-B light increased the vitamin D_2 by 124 μg/g (5100%). Similar results were obtained with *Pleurotus* (Krings & Berger, 2014) and shiitake mushrooms (Ku et al., 2008) with UV-B (11.5 W/m^2, 60 min, 20°C; and 25 kJ/m^2, 25°C, respectively).

7.5 CEREALS

In buckwheat sprout (*Fagopyrum esculentum* Moench.), UV-A (250 W/m^2) and UV-B (>300 nm, 890 W/m^2) were applied; both irradiations increased anthocyanin and rutin concentration and enhanced the antioxidant capacity (Tsurunaga et al., 2013). Moreover, rutin content increased from 653 (grown in the dark) to 1034 (grown under UV-B > 300 nm) mg/100 g of dry weight. The antioxidant capacity was increased 1.6-fold in buckwheat sprout irradiated with UV-B light compared to control. However, they pointed out that wavelength below to 280 nm negatively affected the growth of the sprout promoting the death. Balouchi et al. (2009) informed that UV-C (40 μW/cm day) treatment in durum wheat (*Triticum durum* L.) showed positive effects on carotenoids, anthocyanin, and flavonoid concentration (2.5, 2.5, and 7.8-folds, respectively) than wheat irradiated with UV-A (18 μW/cm day) and UV-B light (25 μW/cm day). They reported a major increase in secondary metabolites in durum wheat treated with UV-C light compared to UV-A treatment.

7.6 CONCLUSION

As discussed throughout the chapter, UV light may be suitable for use as abiotic stress to stimulate the secondary metabolites of fruits and vegetables; however, the positive effect of the irradiation depends on several factors, such as the type of irradiation (UV-A, UV-B, and UV-C), dose (in general, low dose has the greatest effect on the secondary metabolites production), temperature, and time of storage, as the effects might be residual and, therefore, increase the bioactive compounds during the storage. It is also important to remark that although the increase of secondary metabolites of fruits and vegetables is very important, some metabolites are toxic or cause anti-feeding effects in humans. Therefore, the new trends in the field of secondary metabolites and the stimulation with UV light may be carried out considering the importance of these compounds in the human health.

KEYWORDS

- **secondary metabolites**
- **ultraviolet**
- **protein**
- **enzymes**
- **volatile metabolites**

REFERENCES

Adsule, R. N.; Kadam, S. S. Guava. In: *Handbook of Fruit Science and Technology: Production, Composition, Storage and Processing*; Salunkhe, D. K., Kadam S. S. Ed.; Marcel Dekker: New York, 1995; pp 419–433.

Alegria, C.; Pinheiro, J.; Duthoit, M.; Gonçalves, E. M.; Moldão-Martins, M.; Abreu, M. Fresh-Cut Carrot (cv. Nantes) Quality as Affected by Abiotic Stress (Heat Shock and UV-C Irradiation) Pre-Treatments. *LWT—Food Sci. Technol.* **2012,** *48*, 197–203.

Allende, A.; Marín, A.; Buendía, B.; Tomás-Barberán, F.; Gil, M. I. Impact of Combined Postharvest Treatments (UV-C light, gaseous O_3, superatmospheric O_2 and high CO_2) on

Health Promoting Compounds and Shelf-Life of Strawberries. *Postharvest Biol. Tecnol.* **2007,** *46*, 201–211.

Almeida, M. M. B.; Machado, P. H.; Campos, A.; Matias, G.; de Carvalho, C. E.; Arraes, G.; Gomes, T. Bioactive Compounds and Antioxidant Activity of Fresh Exotic Fruits from Northeastern Brazil. *Food Res. Int.* **2011,** *44*, 2155–2159.

Alothman, M.; Bhat, R.; Karim, A. A. UV Radiation-Induced Changes of Antioxidant Capacity of Fresh-Cut Tropical Fruits. *Innov. Food Sci. Emerg. Technol.* **2009a,** *10*, 512–516.

Alothman, M.; Bhat, R.; Karim, A. A. Effects of Radiation Processing o Phytochemicals and Antioxidants in Plant Produce. *Trends Food Sci. Tech.* **2009b,** *20*, 201–212.

Andrade-Cuvi, M. J.; Vicente, A. R.; Concellón, A.; Chaves, A. R. Changes in Red Pepper Antioxidants as Affected by UV-C Treatments and Storage at Chilling Temperatures. *LWT—Food Sci. Technol.* **2011,** *44*(7), 1666–1671.

Andrade-Cuvi, M. J.; Moreno-Guerrero, C.; Henríquez-Bucheli, A.; Gómez-Gordillo, A.; Concellón, A. Influencia de la radiación UV-C como tratamiento Postcosecha Sobre Carambola (*Averroha carambola* L.) mínimamente Procesada Almacenada en Refrigeración. *Rev. Iberoam. Tecnol. Postcos.* **2010,** *11*, 18–27.

Artés-Hernández, F.; Escalona, V. H.; Robles, P. A.; Martínez-Hernández, G. B.; Artés, F. Effect of UV-C Radiation on Quality of Minimally Processed Spinach Leaves. *J. Sci. Food Agric.* **2009,** *89*, 414–421.

Artés-Hernández, F.; Robles, P. A.; Gómez, P. A.; Tomás-Callejas, A.; Artés, F. Low UV-C Illumination for Keeping Overall Quality of Fresh-Cut Watermelon. *Postharvest Biol. Tecnol.* **2010,** *55*, 114–120.

Asami, B.; Hong, Y. J.; Barrett, D. M.; Mitchell, A. Comparison of the Total Phenolic and Ascorbic Acid Content of Freeze-Dried and Air-Dried Marionberry, Strawberry, and Corn Grown using Conventional, Organic, and Sustainable Agricultural Practices. *J. Agric. Food Chem.* **2003,** *51*, 1237–1241.

Balouchi, H. R.; Sanavy, S. A. M. M.; Emam, Y.; Dolatabadian, A. UV Radiation, Elevated CO_2 and Water Stress Effect on Growth and Photosynthetic Characteristics in Durum Wheat. *Plant Soil Environ.* **2009,** *55*(10), 443–453.

Barka, E. Protective Enzymes against Reactive Oxygen Species during Ripening of Tomato (*Lycopersicon esculentum*) Fruits in Response to Low Amounts of UV-C. *Aust. J. Plant Physiol.* **2001,** *28*(8), 785–791.

Beltrán, A.; Ramos, M.; Alvarez, M. Estudio de la vida útil de fresas (*Fragaria vesca*) Mediante Tratamiento Con Radiación Ultravioleta de Unda Uorta (UV-C) Resumen. *Rev. Tecnol. ESPOL-RTE.* **2010,** *23*(2), 17–24.

Bozin, B.; Mimica-Dukic, N.; Samojlik, I.; Goran, A.; Igic, I. Phenolics as Antioxidants in Garlic (*Allium sativum* L., Alliaceae). *Food Chem.* **2008,** *111*, 925–929.

Bravo, S.; García-Alonso, J.; Martín-Pozuelo, G.; Gómez, V.; Santaella, M.; Navarro-González, I.; Periago, M. J. The Influence of Post-Harvest UV-C Hormesis on Lycopene, β-Carotene, and Phenolic Content and Antioxidant Activity of Breaker Tomatoes. *Food Res. Int.* **2012,** *49*(1), 296–302.

Bulley, S. M.; Rassam, M.; Hoser, D.; Otto, W.; Schunemann, N.; Wright, M.; MacRae, E.; Gleave, A.; Laing, W. Gene Expression Studies in Kiwifruit and Gene Over-Expression in Arabidopsis Indicates that GDP-L-Galactose Guanyltransferase Is a Major Control Point of Vitamin C Biosynthesis. *J. Exp. Bot.* **2009,** *60*(3), 765–778.

Carrasco-Ríos, L. Efecto De La Radiación Ultravioleta-B En Plantas. *IDESIA*. **2009,** *27*(3), 59–76.

Castagna, A.; Chiavaro, E.; Dall'Asta, C.; Rinaldi, M.; Galaverna, G.; Ranieri, A. Effect of Postharvest UV-B Irradiation on Nutraceutical Quality and Physical Properties of Tomato Fruits. *Food Chem.* **2013,** *137*(1–4), 151–158.

Charles, M. T.; Arul, J. UV Treatment of Fresh Fruits and Vegetables for Improved Quality: A Status Report. *Stewart Postharest Rev.* **2007,** *3*, 6.

Cisneros-Zevallos, L. The Use of Controlled Postharvest Abiotic Stresses as a Tool for Enhancing the Nutraceutical Content and Adding-value of Fresh Fruits and Vegetables. *J. Food Sci.* **2003,** *68*, 1560–1565.

Crupi, P.; Pichierri, A.; Basile, T.; Antonacci, D. Postharvest Stilbenes and Flavonoids Enrichment of Table Grape cv Redglobe (*Vitis vinifera* L.) as Affected by Interactive UV-C Exposure and Storage Conditions. *Food Chem.* **2013,** *141*, 802–808.

Du, W. X.; Avena-Bustillos, R.; Breksa, A.; McHugh, T. Effect of UV-B Light and Different Cutting Styles on Antioxidant Enhancement of Commercial Fresh-Cut Carrot Products. *Food Chem.* **2012,** *134*, 1862–1869.

Eichholz, I. S.; Huyskens-Keil, S.; Keller, A.; Ulrich, D.; Kroh, L.; W.; Rohn, S. UV-B Induced Changes of Volatile Metabolites and Phenolic Compounds in Blueberries (*Vaccinium corymbosum* L.). *Food Chem.* **2011,** *126*, 60–64.

Erkan, M.; Wang, S. Y.; Wang, C. Y. Effect of UV Treatment on Antioxidant Capacity, Antioxidant Enzyme Activity and Decay in Strawberry Fruit. *Postharvest Biol. Technol.* **2008,** *48*, 163–171.

Foggo, A.; Higgins, S.; Wargent, J.; Coleman, R. Tri-Trophic Consequences of UVB Exposure: Plants, Herbivores and Parasitoids. *Oecologia* **2007,** *154*, 505–512.

Foyer, C.; Descouvrieres, P.; Kunert, K. J. Protection against Oxygen Radicals: An Important Defence Mechanism Studied in Transgenic Plants. *Plant, Cell Environ.* **1994,** *17*, 507–523.

Fowler, M. Plants, Medicine and Man. *J. Sci. Food Agric.* **2006,** *86*, 1797–1804.

Gayán, E.; Serrano, M. J.; Monfort, S.; Álvarez, I.; Condón, S. Combining Ultraviolet Light and Mild Temperatures for the Inactivation of *Escherichia coli* in Orange Juice. *J. Food Eng.* **2012,** *113*, 598–605.

Gill, S. S.; Tuteja, N. Reactive Oxygen Species and Antioxidant Machinery in Abiotic Stress Tolerance in Crop Plants. *Plant Physiol. Biochem.* **2010,** *48*, 909–930.

Gil, M.; Pontin, M.; Berli, F.; Bottini, R.; Piccoli, P. Metabolism of Terpenes in the Response of Grape (*Vitis vinifera* L.) Leaf Tissues to UV-B Radiation. *Phytochemistry*. **2012,** *77*, 89–98.

Goldberg, G. Plants: Diet and Health; Report of a British Nutrition Foundation Task Force; Blackwell Publishing: Oxford, U.K. 2003.

González-Aguilar, G. A.; Wang, C. Y.; Buta, J. G. Use of UV-C Irradiation to Prevent Decay and Maintain Postharvest Quality of Ripe "Tommy Atkins" Mangos. *Int. J. Food Sci. Technol.* **2001,** *36*, 775–782.

González-Aguilar, G. A.; Villegas-Ochoa, M. A.; Martínez-Téllez, M. A.; Gardea, A. A.; Ayala-Zavala, J. F. Improving Antioxidant Capacity of Fresh-Cut Mangoes Treated with UV-C. *J. Food Sci.* **2007,** *72*, 197–202.

Guerrero-Beltrán, J. A.; Barbosa-Cánovas. G. V. Review: Advantages and Limitations on Processing Foods by UV Light. *Food Sci. Technol. Int.* **2004,** *10*, 137–11.

Hagen, S. F.; Borge, G. I.; Bengtsson, G. B.; Bilger, W.; Berge, A.; Haffner, K.; Solhaug, K. A. Phenolic Contents and Other Health and Sensory Related Properties of Apple Fruit (Malus domestica Borkh., cv. Aroma): Effect of Postharvest UV-B Irradiation. *Postharvest Biol. Technol.* **2007,** *45*, 1–10.

Higashio, H.; Hirokane, H.; Sato, F.; Tokuda, S.; Uragami, A. Effect of UV Irradiation after the Harvest on the Content of Flavonoid in Vegetables. *Acta Hortic.* **2005,** *682*, 1007–1012.

Hollósy, F. Effects of Ultraviolet Radiation on Plant Cell. *Micron* **2002,** *33*, 179–197.

Jagadeesh, S. L.; Charles, M. T.; Gariepy, Y.; Goyette, B.; Raghavan, G. S. V.; Vigneault, C. Influence of Postharvest UV-C Hormesis on the Bioactive Components of Tomato during Post-treatment Handling. *Food Bioprocess Technol.* **2011,** *4*, 1463–1472.

Jayaprakasha, G. K.; Patil, B. S. In Vitro Evaluation of the Antioxidant Activities in Fruit Extracts from Citron and Blood Orange. *Food Chem.* **2007,** *101*, 410–418.

Jenkins, G . I. Structure and Function of the UV-B Photoreceptor UVR8. *Curr. Opin. Struct. Biol.* **2014,** *29*, 52–57.

Kalaras, M. D.; Beelman, R. B.; Elias, R. J. Effects of Postharvest Pulsed UV Light Treatment of White Button Mushrooms (*Agaricus bisporus*) on Vitamin D2 Content and Quality Attributes. *J. Agric. Food Chem.* **2012,** *60*, 220–225.

Kataria, S.; Jajoo, A.; Guruprasad, N. Impact of Increasing Ultraviolet-B (UV-B) Radiation on Photosynthetic Processes. *J. Photochem. Photobiol. B: Biology*. **2014,** *137*, 55–66.

Kaur, C.; Kapoor, H. C. Antioxidants in Fruits and Vegetables—The Millennium's Health (Review). *Int. J. Food Sci. Technol.* **2001,** *36*, 703–725.

Kliebenstein, D. J.; Lim, J. E.; Landry, L. G.; Last, R. L. Arabidopsis UVR8 Regulates Ultraviolet-B Signal Transduction and Tolerance and Contains Sequence Similarity to Human Regulator of Chromatin Condensation 1. *Plant Physiol.* **2002,** *130,* 234–243.

Koutchma, T. UV Light for Processing Foods. *IUVA News*, **2008,** *10*, 24–29.

Koyyalamudi, S. R.; Jeong, S. C.; Pang, G.; Teal, A.; Biggs, T. Concentration of Vitamin D2 in White Button Mushrooms (*Agaricus bisporus*) Exposed to Pulsed UV Light. *J. Food Compos. Anal.* **2011,** *24*, 976–979.

Krings, U.; Berger, R. G. Dynamics of Sterols and Fatty Acids during UV-B Treatment of Oyster Mushroom. *Food Chem.* **2014,** *149*, 10–14.

Ku, J. A.; Lee, B. H.; Lee, J. S.; Park, H. J. Effect of UV-B Exposure on the Concentration of Vitamin D2 in Sliced Shiitake Mushroom (*Lentinus edodes*) and White Button Mushroom (*Agaricus bisporus*). *J. Agric. Food Chem.* **2008,** *56*, 3671–3674.

Kuhlmann, F.; Müller, C. Development-Dependent Effects of UV Radiation Exposure on Broccoli Plants and Interactions with Herbivorous Insects. *Environ. Exp. Bot.* **2009,** 61–68.

Kumari, R.; Singh. S.; Agrawal, S. B. Effects of Supplemental UV-B Radiation on Growth and Physiology of *Acoruc calamus* L. (Sweet flag). *Acta Biol. Crac. Ser. Bot.* **2009,** *51*, 19–27.

Lemoine, M. L.; Chaves, A. R.; Martínez, G. A. Influence of Combined Hot Air and UV-C Treatment on the Antioxidant System of Minimally Processed Broccoli (*Brassica oleracea* L. var. Italica). *LWT—Food Sci. Technol.* **2010,** *43*, 1313–1319.

Liu, C.; Cai, L.; Lu, X.; Han, X.; Ying, T. Effect of Postharvest UV-C Irradiation on Phenolic Compound Content and Antioxidant Activity of Tomato Fruit During Storage. *J. Integr. Agric.* **2012,** *11*, 159–165.

Liu, C.; Han, X.; Cai, L.; Lu, X.; Ying, T.; Jiang, Z. Postharvest UV-B Irradiation Maintains Sensory Qualities and Enhances Antioxidant Capacity in Tomato Fruit During Storage. *Postharvest Biol. Technol.* **2011,** *59,* 232–237.

Liu, L. H.; Zabaras, D.; Bennett, L. E.; Aguas, P.; Woonton, B. W. Effects of UV-C, Light Red and Sun Light on the Carotenoid Content and Physical Qualities of Tomatoes During Post-Harvest Storage. *Food Chem.* **2009,** *115,* 495–500.

Mahdavian, K.; Ghorbanli, L.; Kalantari, K. M. The Effects of Ultraviolet Radiation on the Contents of Chlorophyll, Flavonoid, Anthocyanin and Proline in *Capsicum annuum* L. *Turk. J. Botany.* **2008,** *32,* 25–33.

Martínez-Hernández, G. B.; Gómez, P. A.; Pradas, I.; Artés, F.; Artés-Hernández, F. Moderate UV-C Pretreatment as a Quality Enhancement Tool in Fresh-Cut Bimi® Broccoli. *Postharvest Biol. Technol.* **2011,** *62,* 327–337.

Mau, J. L.; Chen, P. R.; Yang, J. H. Ultraviolet Irradiation Increased Vitamin D2 Content in Edible Mushrooms. *J. Agric. Food Chem.* **1998,** *46,* 5269–5272.

Mercier, J.; Arul, J.; Julien C. Effect of UV-C on Phytoalexin Accumulation and Resistance to *Botrytis cinerea* in Stored Carrots. *Phytopathology.* **1993,** *139,* 17–25.

Mittler, R. Oxidative Stress, Antioxidants and Stress Tolerance. *Trends Plant Sci.* **2002,** *7,* 405–410.

Nile, S. H.; Park, S. W. Total Phenolics, Antioxidant and Xanthine Oxidase Inhibitory Activity of Three Colored Onions (*Allium cepa* L.). *Front. Life Sci.* **2013,** *7,* 224–228.

Nguyen, C.; Kim, J.; Yoo, K.; Lim, S.; Lee, E. J. Effect of Prestorage UV-A, -B, and -C Radiation on Fruit Quality and Anthocyanin of 'Duke' Blueberries during Cold Storage. *J. Agric. Food Chem.* **2014,** *62,* 12144–12151.

Pan, Y. G.; Zu, H. Effect of UV-C Radiation on the Quality of Fresh-Cut Pineapples. *Procedia Eng.* **2012,** *37,* 113–119.

Perkins-Veazie, P.; Collins, J. K.; Howard, L. Blueberry Fruit Response to Postharvest Application of Ultraviolet Radiation. *Postharvest Biol. Technol.* **2008,** *47,* 280–285.

Pichersky, E; Gang, D. R. Genetics and Biochemistry of Secondary Metabolites in Plants: An Evolutionary Perspective. *Trends Plant Sci.* **2000,** *5,* 439–445.

Park, M. H.; Kim, J. G. Low-Dose UV-C Irradiation Reduces the Microbial Population and Preserves Antioxidant Levels in Peeled Garlic (*Allium sativum* L.) during Storage. *Postharvest Biol. Technol.* **2015,** *100,* 109–112.

Poiroux-Gonord, F.; Bidel, L.; Fanciullino, A. L.; Gautier, H.; Lauri-López, F.; Urban, L. Health Benefits of Vitamins and Secondary Metabolites of Fruits and Vegetables and Prospects to Increase Their Concentrations by Agronomic Approaches. *J. Agric. Food Chem.* **2010,** *58,* 12065–12082.

Pombo, M. A.; Dotto, M. C.; Martínez, G. A.; Civello, P. M. UV-C Irradiation Delays Strawberry Fruit Softening and Modifies the Expression of Genes Involved in Cell Wall Degradation. *Postharvest Biol. Technol.* **2009,** *51,* 141–148.

Rivera-López, J.; Vázquez-Ortiz, F. A.; Ayala-Zavala, J. F.; Sotelo-Mundo, R. R.; González-Aguilar, G. A. Cutting Shape and Storage Temperature Affect Overall Quality of Fresh-Cut Papaya cv. 'Maradol'. *J. Food Sci.* **2005,** *70,* S482–S489.

Rivera-Pastrana, D. M.; Gardea, A. A.; Yahia, E. M.; Martínez-Téllez, M. A.; González-Aguilar, G. A. Effect of UV-C Irradiation and Low Temperature Storage on Bioactive Compounds, Antioxidant Enzymes and Radical Scavenging Activity of Papaya Fruit. *J. Food Sci. Technol.* **2013,** *51,* 3821–3829.

Rodov, V.; Tietel, Z.; Vinokur, Y.; Horev, B.; Eshel, D. Ultraviolet Light Stimulates Flavonol Accumulation in Peeled Onions and Controls Microorganisms on Their Surface. *J. Agric. Food Chem.* **2010,** *58*, 9071–9076.

Rodrígues, A. S.; Pérez-Gregorio, M. R.; García-Falcón, M. S.; Simal-Gándara, J.; Almeida, D. P. F. Effect of Post-Harvest Practices on Flavonoid Content of Red and White Onion Cultivars. *Food Control.* **2010,** *21*, 878–884.

Rufino, M. S.; Alves, R.; Brito, E.; Pérez-Jiménez, J.; Saura-Calixto, F.; Mancini-Filho, J. Bioactive Compounds and Antioxidant Capacities of 18 Non-Traditional Tropical Fruits from Brazil. *Food chem.* **2010,** *121*, 996–1002.

Santhosha, S. G.; Jamuna, P.; Prabhavathi, S. N. Bioactive Components of Garlic and Their Physiological Role in Health Maintenance. *Food Biosci.* **2013,** *3*, 59–74.

Scandalios, J. G. Oxygen Stress and Superoxide Dismutase. *Plant Physiol.* **1993,** *101*, 7–12.

Schreiner, M.; Martínez-Abaigar, J.; Glaab, J.; Jansen, M. UV-B Induced Secondary Plant Metabolites. Potential Benefits for Plant and Human Health. *Optik & Photonik.* **2014,** 34–37.

Sellappan, S.; Akoh, C.; Krewer, G. Phenolic Compounds and Antioxidant Capacity of Georgia-Grown Blueberries and Blackberries. *J. Agric. Food Chem.* **2002,** *50*, 2432–2438.

Shen, Y.; Sun, Y.; Qiao, L.; Chen, J.; Liu, D.; Ye, X. Effect of UV-C Treatments on Phenolic Compounds and Antioxidant Capacity of Minimally Processed Satsuma Mandarin during Refrigerated Storage. *Postharvest Biol. Technol.* **2013,** *76*, 50–57.

Singh, S.; Agrawal, S. B.; Agrawal, M. UVR8 Mediated Plant Protective Responses under Low UV-B Radiation Leading to Photosynthetic Acclimation. *J. Photochem. Photobiol. B: Biology.* **2014,** *137*, 67–76.

Stapleton, A. Ultraviolet Radiation and Plants: Burning Questions. *Plant Cell* **1992,** *4*, 1353–1358.

Tsurunaga, Y.; Takahashi, T.; Katsube, T.; Kudo, A.; Kuramitsu, O.; Ishiwata, M.; Matsumoto, S. Effects of UV-B irradiation on the Levels of Anthocyanin, Rutin and Radical Scavenging Activity of Buckwheat Sprouts. *Food Chem.* **2013,** *141*, 552–556.

Ubi, B. E.; Honda, C.; Bessho, H.; Kondo, S.; Wada, M.; Kobayashi, S.; Moriguchi, T. Expression Analysis of Anthocyanin Biosynthetic Genes in Apple Skin: Effect of UV-B and Temperature. *Plant Sci.* **2006,** *170*, 571–578.

Vicente, A.; Pineda, C.; Lemoine, L.; Civello, P.; Martínez, G.; Chaves, A. UV-C Treatments Reduce Decay, Retain Quality and Alleviate Chilling Injury in Pepper. *Postharvest Biol. Technol.* **2005,** *35*, 69–79.

Temple, N. J. Antioxidants and Disease: More Questions than Answers. *Nutr. Res.* **2000,** *20*, 449–459.

Topcu, Y.; Dogan, A.; Kasimoglu, Z.; Sahin-Nadeem, H.; Polat, E.; Erkan, M. The Effects of UV Radiation during the Vegetative Period on Antioxidant Compounds and Postharvest Quality of Broccoli (*Brassica oleracea* L.). *Plant Physiol. Biochem.* **2015,** *93*, 56–65.

Tsormpatsidis, E.; Henbestb, R. G. C.; Davis, F. J.; Batteya, N. H.; Hadleya, P.; Wagstaffea, A. UV Irradiance as a Major Influence on Growth, Development and Secondary Products of Commercial Importance in Lollo Rosso Lettuce 'Revolution' Grown Under Polyethylene films. *Environ. Exp. Bot.* **2008,** *63*, 232–239.

Wang, Y. C.; Chen, C. T.; Wang, S. Y. Changes of Flavonoid Content and Antioxidant Capacity in Blueberries after Illumination with UV-C. *Food Chem.* **2009,** *117*, 426–431.
Zhang, W. J.; Björn, L. O. The Effect of Ultraviolet Radiation on the Accumulation of Medicinal Compounds in Plants. *Fitoterapia* **2009,** *80*, 207–218.
Zhang, D.; Yu, B.; Bai, J.; Qian, M.; Shu, Q.; Su, J.; Teng, Y. Effects of High Temperatures on UV-B/Visible Irradiation Induced Postharvest Anthocyanin Accumulation in "Yunhongli No. 1" (*Pyrus pyrifolia Nakai*) pears. *Sci. Hortic.* **2012,** *134*, 53–59.

INDEX

D

E

F

G

H

N

O

Q

R

S

For Product Safety Concerns and Information please contact our EU representative GPSR@taylorandfrancis.com Taylor & Francis Verlag GmbH, Kaufingerstraße 24, 80331 München, Germany

Batch number: 10726534

Printed by Printforce, the Netherlands